T0295166

Sarah Bowdich Lee (1791–1856) and Pioneering Perspectives on Natural History

Sarah Bowdich Lee (1791–1856) and Pioneering Perspectives on Natural History

Mary Orr

ANTHEM PRESS

Anthem Press
An imprint of Wimbledon Publishing Company
www.anthempress.com

This edition first published in UK and USA 2024
by ANTHEM PRESS
75–76 Blackfriars Road, London SE1 8HA, UK
or PO Box 9779, London SW19 7ZG, UK
and
244 Madison Ave #116, New York, NY 10016, USA

British Library Cataloguing-in-Publication Data
A catalogue record for this book is available from the British Library.

Library of Congress Control Number: 2023945051
A catalog record for this book has been requested.

ISBN-13: 978-1-83998-609-3 (Hbk)
ISBN-10: 1-83998-609-3 (Hbk)

Cover Credit: Courtesy of Hartley Library Special Collections, University of Southampton

This title is also available as an e-book.

CONTENTS

FIGURES AND TABLES

Figures

Tables

ACKNOWLEDGEMENTS

This book began life very unexpectedly in Salle V at the Bibliothèque Nationale de Paris in the summer of 2005, when my attention wandered from my work at the time on Flaubert's *Tentation de saint Antoine* (1874). As pivotal to unlocking the science behind its final tableau, I was researching the works of Georges Cuvier (1769–1832), especially his bitter 'Querelle des Analogues' (Quarrel of the Analogues) of 1830 with Étienne Geoffroy Saint-Hilaire. Looking for inspiration, I returned to the BNF catalogue and ran 'Cuvier' again as the author search term. His *Excursions dans les Isles de Madère et de Porto Santo* of 1826, which I had passed over many times before, suddenly commanded my fullest attention. Cuvier, I knew, had never travelled to Madeira. How could he be the author? Intrigued, I called up the book only to find the author name, 'Madame Bowdich', clearly on the inside title page, and both Cuvier and Humboldt as the writers of 'Notes' for it. If the rest is history regarding my mission to make Sarah Bowdich Lee better known in print since 2007, some remarkable people and organisations have supported the very much larger adventure of this book project and its research, and the even longer journey of its writing.

My especial thanks go first to Professors Jennifer Birkett, Máire Cross and Naomi Segal for encouraging my prioritisation of this project alongside my 'main' research in nineteenth-century French Studies. As I sought funding for it, thank you for your practical support and what you wrote as my referees for the successful Leverhulme Research Fellowship Application RF/1/RFG/2010/0435 in 2012–2014 that enabled me to undertake its foundational research.

The piecing together almost from scratch of the fascinating, illusive and complex story of this book in its many intersecting discipline domains required many hours in many libraries and archives around the world. A list would be too long of the knowledgeable and helpful librarians in special collections along the way, but to you all, thank you for your bibliographical expertise and real interest in my work, often translating into the additional documents that would speed to my desk well before closing times.

Three very different Visiting Research Fellowships throughout 2010 to 2012 brought vital perspectives to the international reach of the project. It was a particular pleasure to return to Queens' College, Cambridge to begin the research in Cambridge University Library. My warmest thanks go to its then President, the Rt. Hon. Lord Eatwell and Fellows, for welcoming me as a Distinguished Academic Fellow in Michaelmas 2010. My Visiting Fellowship at the Institute of Advanced Study, University of Western Australia, Perth, culminating in my first public lecture on Sarah, was a career pinnacle and lifetime memory. Professors Hélène Jaccomard (French/Modern Languages,

UWA) and Susan Takao (Deputy Director, IAS, UWA), you were exemplary hosts in August–September 2011. Hélène, thank you for inviting me to give a further plenary at the Australian Society of French Studies Annual Conference at UWA in December 2019! The third Visiting Fellowship at the Institute of Commonwealth Studies, School of Advanced Study University of London (January to end March 2012) exemplified the benefits for interdisciplinary projects of proactive engagement with the commonwealth of scholars in adjacent fields to one's own 'home' disciplines.

The exponentially larger and more challenging scope for the project than its Leverhulme Fellowship proposal and research then lay in the dissemination of its findings to wider specialist and informed general publics. My inaugural lecture in 2007 at the University of Southampton as its Professor of French Studies allowed me to don Sarah's 'hat' as I read out a key passage from her *Excursions* to conclude it. I also thank Camden History Society for their invitation to share my work on Sarah for its integral importance to the Borough. To find her local London feet and hear her voice behind her remarkable works was then to present her case for the first time in some 10 foundational articles and invited chapters from 2007 to 2022, to engage very different academic audiences in nineteenth-century history, literature and cultures of scientific endeavour in France and Britain. This book builds upon them, to address in its all-new chapter parts and their larger whole the lessons of Sarah's pioneering perspectives for her time, and their implications for today.

Drafts of chapters in the first two parts of this book were completed during two important semesters of institutional research leave, the first in 2015–2016 (semester two) at the University of Southampton, the second at the University of St Andrews in 2018–2019 (semester two). I thank both institutions and my colleagues in Modern Languages for supporting my work. But the project could only finally find its clearer voice in my further application to the British Academy/the Leverhulme Trust to complete it. My warmest thanks therefore go to two exceptional colleagues and long-standing friends of interdisciplinary research, who made all the difference to the proposal that I sent. In Derek Duncan, Professor of Italian at St Andrews, is the long-standing research friend par excellence. Thank you for that critical conversation in all senses, where you urged me to nail the project USPs! David Brown, Professor of British History at Southampton, is the 'big research project' supporter second to none, and beyond acting as the all-important named referee for my application in November 2019. Your insightful conversations and unstinting support throughout the final writing have immeasurably improved the result.

Nothing will match my shock when the British Academy's email arrived in March 2021 (during stringent Scottish lockdown), to announce the successful award of SRF21\210181, followed up by the Academy's announcement in May 2021 that I was the holder of the 2021–2022 'Donald Winch Fund Senior Research Fellowship in Intellectual History'. To the anonymous Reviewers and Research Awards Advisory Committee at the British Academy, thank you for your appraisal and support of my work for this Award, and for your enlarging wisdoms through it. The all-important classification for this project and its remits as 'intellectual history' provided the conceptual capstone for the book in its three interconnecting parts, and the externally validated equipping as I commenced the writing of chapters in the final part that could bring the project together.

Completion of this book at last in 2021–2022 could not have maintained its intensive momentum without the championing support of wonderful friends and family cheering me on to the finish line. To Neil, who has had to live with 'Mrs B' for far too long, thank you for making so much room for her in our journeying together. But it is the 'Mrs B' behind the page from whom I have learned the most. I have sought to put my feet into the paths of your own research, to see and hear with your eyes and ears and to understand how to recraft your voice. The pivotal importance of your hat in the crowd, and your most excellent pair of pioneering shoes set out the route. Your life–work blazes the trail for what can be possible despite the odds for intrepid women in 1823 and 2023 in world natural and intellectual history-making. This book is dedicated to the differences that you have made to major `environmental' concerns for your offspring, literal and intellectual.

Mary Orr, St Andrews, November 2023.

ABBREVIATIONS AND TERMS OF REFERENCE

This first study of the major contributions by Sarah née Wallis, Mrs T(homas) Edward Bowdich, then Mrs R(obert) Lee (1791–1856) to new knowledge of natural history focuses on her book-length publications differently concerning West Africa. Chapters below reference them using the following short forms, followed by the page number in the print edition below. All except *FWFGB* and *ENH2* are available in online facsimile free open access.

AnecA: *Anecdotes of the Habits and Instincts of Animals.* London: Grant and Griffith, 1852.

AnecBRF: *Anecdotes of the Habits and Instincts of Birds, Reptiles and Fishes.* London: Grant and Griffith, 1853.

AW: *The African Wanderers or the Adventures of Carlos and Antonio. Embracing Interesting Descriptions of the Manners and Customs of the Western Tribes and the Natural Productions of the Country.* London: Grant and Griffith, 1847.

EM: *Excursions in Madeira and Porto Santo, during The Autumn of 1823* … London: George B. Whittaker, 1825.

EMFr: *Excursions dans les Isles de Madère et de Porto Santo.* 2 vols. Paris: F. G. Levrault, 1826.

ENH: *Elements of Natural History. For the Use of Schools and Young Persons. Comprising the Principles of Classification, Interspersed with Amusing and Instructive Original Accounts of the most remarkable animals.* London: Longman, Brown, Green and Longmans, 1844.

ENH2: *Elements of Natural History. For the Use of Schools and Young Persons. Comprising the Principles of Classification, Interspersed with Amusing and Instructive Original Accounts of the most remarkable animals.* London: Longman, Brown, Green and Longmans, 1850.

FWFGB: *The Fresh-Water Fishes of Great Britain.* London: R. Ackermann, 1828–1838.

MBC: *Memoirs of Baron Cuvier.* London: Longmans Rees Orme Brown Green & Longman, 1833.

SSI: *Site of Special Scientific Interest.*

SSL: *Stories of Strange Lands and Fragments from the Notes of a Traveller.* London: Edward Moxon, 1835.

ST: *Sir Thomas, or the Adventures of a Cornish Baronet in North-Western Africa.* London: Grant and Griffith, 1856. Appendix 1 provides the fullest list to date of

Sarah's many publications in print during her lifetime. Posthumously published anthologies of her animal stories for the youngest readers are not included. Reprints are listed only for works published during her lifetime.

STEM(M): *Science, Technology, Engineering, Maths (and Medicine).*

Appendix 1 also exemplifies the often problematic question of authorship for women by their married name(s) that male writers never experience. Its recording of Sarah's double married and also foreign names on her French publications represents the larger remit of this book in its challenge to the monolingual bias in scholarship recovering women authors lost to natural and to literary history. Readers should use Appendix 1 to check the published surname(s) of Sarah's work(s) under discussion. I also follow the precedents of spelling for the plural of Bowdich as 'Bowdichs', first set by Donald deB. Beaver (see the 'References to Secondary Sources').

Naming and nomenclature also pose particular challenges in the domains of scientific endeavour, especially in the nineteenth century. This book will explore the implications and conventions for (new) species identification and naming in Latin and European vernaculars in Part 1. It therefore adopts usage of a capital for world species, such as Fishes. The larger fundamental issue, however, regards current discipline terminology and its related ideological freight: the natural and earth sciences (and their many subdisciplines) today were very differently configured in the nineteenth century, both in the remits of their subjects and by the persons contributing to them. These sciences were fully the domain of natural philosophers, of naturalists, and of non-institutional expert exponents exemplified in the term 'amateur'. This informed 'lover of the subject' was no dilettante or general public 'citizen scientist' in today's understanding. In underscoring the various intersecting nineteenth-century scientific contexts for Sarah's many contributions, this book and its title use the term, 'natural history', to cover the main and many fields of her scientific endeavours at the time of their undertaking. Natural history also neatly translates the equivalent French term *histoire naturelle*, contextualizing Sarah's French contributions and publications as well as her all-important training at the Paris Muséum National d'Histoire Naturelle.

'MRS SARAH', 1824: INTRODUCTION TO STANDOUT WOMEN IN COMPARATIVE NATURAL HISTORY

The Height of the Dry Season, 'Bathurst', The Gambia, 1824.

I see her now! She's on the quayside, waiting to board. Tugging her hand is her youngest of nearly two, Eugenia Keir, wanting up into her mother's arms. Running around her bags and skirts despite the shimmering heat is her elder daughter, Tedlie Hutchison – named after her father's closest friends on his expedition to Ashantee (Asante) – and her son, Edward Hope Smith (after his great-uncle). Or so Tedlie told me, as part of her vaguely-remembered long sea journey in the summer of 1822. She was not yet three, and her brother was only six months old when the family had travelled from Paris to Le Havre, to board a ship bound for Lisbon. Once there, her parents had set out immediately to examine the local rocks, plants and animals. Then another big ship had taken them to Madeira. How seasick little Edward and she had been! She chattered about bougainvillea, banana and other trees they played amongst in the garden of their home for more than a year in Funchal, near the English Church and the houses of Mr Veitch the British Consul General, and of Mr and Mrs Keir, who had taken such an interest in her parents' work. That was before the family set sail again last August, with Eugenia the new baby this time. When their ship reached Praia in the Cape Verde Islands, little Edward thought they had reached 'Africa' at last, only to learn that a further sea journey lay ahead to the Gambia, to the settlement the English called Bathurst (Banjul). They've been lodging here since November at Captain Findlay's residence at Government House, not far from the British garrison buildings, and the new hospital on St Mary's Island. Even Edward's little legs take him quickly to the long sandy Atlantic shore, dotted with its small fishing boats.

It's Captain Findlay talking to their mother, Mrs Sarah. She's the only English lady on the bustling quay, her dark dress and beribboned hat immediately standing out against the colourful melee of European and Gambian traders, including my many female compatriots in their finery, busy bartering, exchanging goods and instructing the porters in several languages. The British Customs House behind the little family is a hive of activity, like their ship for London. Aboard are the many crates and barrels carefully packed by Mrs Sarah. She had them registered and stamped some days ago, in preparation for their long sea journey. I've never quite understood her kind of trade in dead creatures sold for study in Europe, when they're better alive at the price for food! When I first agreed that she could sketch me – business won't get done if I sit! – she told me that she and Mr Edward were eagerly awaiting sea passage to Freetown (in Sierra Leone) since their scientific instruments had been sent on before them. Something to do with the plants and animals there, which Europeans haven't seen before. As here in the Gambia apparently from their excursions near and far since November, to collect and examine plants, shells, fish, birds and other creatures, including those the local children try to sell to them over again. Though I rarely saw Mr Edward. Always up and away long before dawn on any small vessel with crew that Captain Findlay could commandeer, whether British, Mandingo or a Joloff (Wolof) fishing boat heading up the River Gambia. Europeans need charts to navigate it! He was busy making them, and as far as possible upstream of the little island of the Fort St James British garrison (Kunta Kinte Island) opposite the French slave-market in Albreda, to the Portuguese settlements farthest upriver. Of course that Trade is still plied under cover of darkness. Small boats in full view of Fort St James transport their human cargo, children too, on the ebbing tide to the waiting slavers anchored offshore. So much for the British patrols in this region to uphold their Act (of Abolition, 1807) and its 'agreement'!

Mr Edward was often up into the night as well, to look at the stars from the veranda and grounds of Government House. After a long day in the heat, he should have known the real dangers of chills! From December he no longer rushed about, except in his fevered imagination. It was Sister Marcelline, one of the French Sisters of Charity from Goree nursing at the new British Hospital who told me, because General MacCarthy needed urgent care for the sick. Mr Edward tried so often to rise from his bed, while she and Mrs Sarah nursed him for days and nights on end. Though Sister Marceline was pleased to speak French with her. How determined he was particularly in these moments, to return to his work and his writing, although they were hers too. I have that on very good account from the servants at Government House! Nightfall was when Mr Edward and Mrs Sarah could pore together over their notes and sketches from his observations and collections upriver. And hers, frequenting the markets and beaches of St Mary's Island for her own collecting and drawing. A regular sight was Mrs Sarah for the local fishermen, women oyster pickers and wet market traders. Her intense interest and regular examinations of their fish catches was all they could talk about, especially when she wanted not the largest, but those she'd not seen before!

She's bidding her last goodbyes to Captain Findlay! My friends saw her in different places yesterday, up from first light with the children. They'd walked a last time from Government House along the dirt road to the walled enclosure among the scrub and wind-bent trees that face the Atlantic Ocean. It's the British cemetery, Mr Edward's resting place. She'd carefully brushed his plot of wind-blown sand these last weeks, to put blossoms and new shells on it that he knew so much about. Poor lady! Two graves she can never revisit! When drawing she told me about their first daughter, Florence, and how valiantly the little girl had fought the fever in 1817, when she and Mr Edward resided in Cape Coast Castle (Accra). Then came easier goodbyes in town. Friends saw her at the Wesleyan Mission with Mr Smith, and then Mr Martin at the new school in Bathurst. Now that the Quaker lady Hannah Kilham has departed, the Methodists have taken up her mission to found new schools in the settlement. And not only for the boys, but also for girls here, girls a little older than Tedlie. They're learning the rudiments of sewing, as well as counting, reading and writing. Then back at Government House, I hear that Eugenia had as usual charmed every well-wisher come to bid the family God-speed for their sea voyage.

Eugenia's almost asleep in Mrs Sarah's arms as she's stepping onto the gangplank, Tedlie in front tugging little Edward behind her by the hand. Oh! Mrs Sarah's looking back – has she seen my parasol? – her gaze now ahead again, to the hold; her crates and barrels, I suppose! What is to become of her, with her beloved husband gone, and their labours together in West Africa at an end? I can't see her trade in London in dead animals from these parts when she can't collect them! How will she support the children there, without the big extended family and long success in trade that I have? The British Government certainly won't offer to pay, if their Customs House here is anything to go by! Oh, they've lowered the gangplank! Goodbye, Mrs Sarah! Farewell! What will she do now?

***

There is no known portrait of 'Mrs Sarah', née Sarah Wallis, Mrs T. Edward Bowdich then Mrs R(obert) Lee, although one is readily accessible online of the 'late T. Edward Bowdich',[1] figuring in this imagined eye-witness cameo, and clearly on the inside title page below. The lack of an image for Sarah therefore epitomises and compounds the larger overlooking and occlusions of women in the history of science endeavour, especially in the early nineteenth century and overseas. By providing in place of a portrait the reported view of her instead – in profile in 1824 – I deliberately introduce Sarah through the eyes of her Gambian woman contemporary. She is none other than the unnamed historical 'woman with a parasol' (the 'Senhara, or Mulatto', *EM*, xi), whom Sarah drew for 'Costume of the Gambia' (*EM*, Plate IX) that illustrates our book cover. To see Sarah anew in 1824 and today in the eyes of a non-imperial woman eye-witness also introduces the 'pioneering perspectives' to be examined in this book. We will return in chapters below to Sarah's drawings and specimen collections, bespeaking her various engagements with West African natural history, including the knowledge of local women, and how she brought it to larger public attention in her many publications.

In classic mode for the major voyage of scientific exploration and the adventure story, Sarah's publication of the *Excursions in Madeira and Porto Santo* in London in 1825 proved her survival of the long homeward sea journey. This publication also provided an immediate answer to the Gambian woman's concluding question. For both classic travel genres, however, the new family twist put foremost in the imaginary portrait above is that the three Bowdich children did so too. The spring storms of 1824 made the Bay of Biscay particularly perilous, because the ship took in large amounts of sea-water. Cargo in the hold suffered extensive movement damage and waterlogging in consequence. When Sarah's crates and barrels finally reached the London quayside, her devastation would have known no bounds. The many new specimens that she had carefully selected, expertly preserved and packed after Edward's death – destined for the Natural History Museum collections of Paris and London – were unsalvageable and irreplaceable.[2] In their scientific importance lay the store of her children's future and her own in London through the further upholding and extending of her late husband's scientific reputation and publications.[3] All Sarah's best efforts to collect West African natural history new to Western science had come tragically and painfully to nothing.

In the history of (British) nineteenth-century geography, empire, science and scientific travel, this 'nothing' would immediately end Sarah's story and contributions to natural history if these had ever begun. As a cursory reading of the inside title page of the *Excursions* might confirm (Figure 0.1), its science was ultimately (only) by her late husband, Edward. Such automatic assumptions stand to reason in the unassailable contexts before 1860 for British history of scientific exploration. Leading exponents of the period, such as John Mackenzie, Felix Driver, David N. Livingstone and Charles Withers,[4] and historians of women's geography and travel such as Cheryl McEwan[5] all agree: it could not include women on several key grounds. Their exclusions from secondary and hence tertiary education in mathematics, anatomy and sciences compounded their lack of accessible independent means and of physiological suitability to pursue serious interests in the natural sciences at home or overseas. The

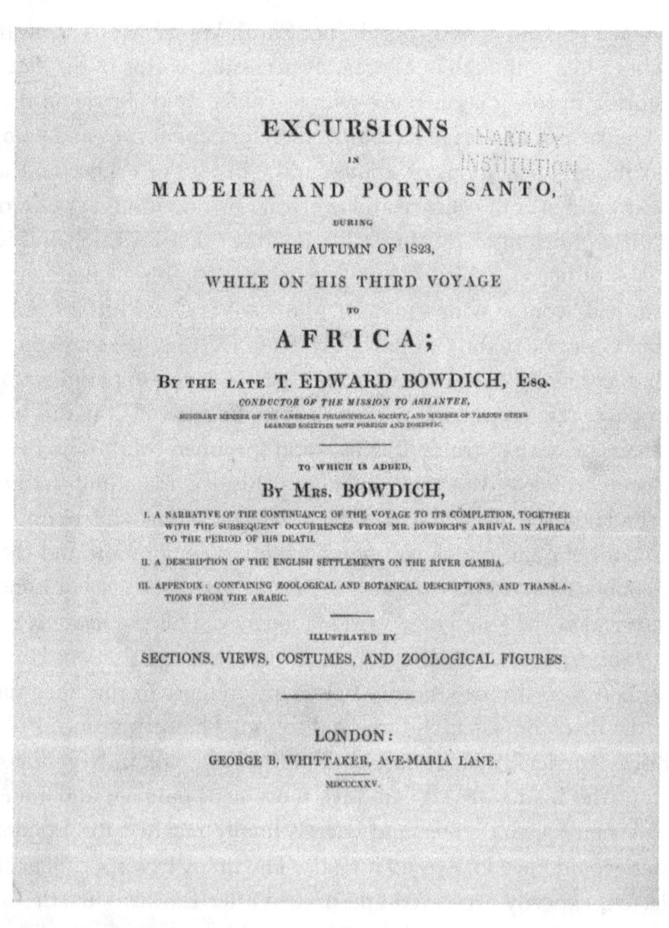
EXCURSIONS

IN

MADEIRA AND PORTO SANTO,

DURING

THE AUTUMN OF 1823,

WHILE ON HIS THIRD VOYAGE

TO

AFRICA;

BY THE LATE T. EDWARD BOWDICH, ESQ.

CONDUCTOR OF THE MISSION TO ASHANTEE,

HONORARY MEMBER OF THE CAMBRIDGE PHILOSOPHICAL SOCIETY, AND MEMBER OF VARIOUS OTHER LEARNED SOCIETIES BOTH FOREIGN AND DOMESTIC.

TO WHICH IS ADDED,

BY MRS. BOWDICH,

I. A NARRATIVE OF THE CONTINUANCE OF THE VOYAGE TO ITS COMPLETION, TOGETHER WITH THE SUBSEQUENT OCCURRENCES FROM MR. BOWDICH'S ARRIVAL IN AFRICA TO THE PERIOD OF HIS DEATH.

II. A DESCRIPTION OF THE ENGLISH SETTLEMENTS ON THE RIVER GAMBIA.

III. APPENDIX: CONTAINING ZOOLOGICAL AND BOTANICAL DESCRIPTIONS, AND TRANSLATIONS FROM THE ARABIC.

ILLUSTRATED BY

SECTIONS, VIEWS, COSTUMES, AND ZOOLOGICAL FIGURES.

LONDON:

GEORGE B. WHITTAKER, AVE-MARIA LANE.

MDCCCXXV.

Figure 0.1 Inside Title Page, *Excursions to Madeira and Porto Santo* … Courtesy of Archives and Special Collections, Hartley Library, University of Southampton.

measures of 'serious' – membership of a learned society and publication of scientific discoveries in its papers – could apply to women only most exceptionally, through influential and affluent male relatives and intermediaries.[6] Moreover, the rigours of genteel cultural travel as undertaken by intrepid women in the pre-Victorian period, whether unmarried or accompanying male relatives,[7] were no equipping for the many dangers and extreme challenges of scientific exploration overseas, especially in tropical climates. In consequence, experts of the period uphold that only in the later Victorian age could independent women with private means and education undertake extensive travels including scientific there, with Mary Kingsley (1862–1900) the unrivalled and acclaimed 'first' woman scientific traveller and explorer of West Africa.[8]

The contrary evidence to these accepted facts is the 'Mrs Bowdich' written as large on the title page of the *Excursions* as her late husband's name. But her very real existence in 1825 does more. Its larger challenge is to why leading (Anglophone) experts in nineteenth-century historical geography and scientific travel also alert to gender could not envisage and hence uncover her work. It had to be unusual, even standout and unmissable given the known bars above to women's engagements with science. As the best indicative experts above illustrate, however, the rosters of 'first' or 'great' scientists and explorers reveal the more fundamental blind spot that is nationality, whether 'English', 'French', 'Portuguese' etc. It is assumed to be a singular that is not only synonymous with linguistic identity but also indistinguishable from the national importance of the given imperial science and discovery. This unquestioned syllogism therefore also underpins the national histories of science of Portugal and France among other European nations implicated after the Napoleonic Wars in the geopolitical ambits of the *Excursions* in 1825. Each national science narrative then echoes and perpetuates variations of the unassailable reasons above that explain why 'French', 'Portuguese' etc. women could not exist in its endeavours, including in overseas scientific exploration.[9] In France, the Code Napoléon of 1804 apportioned to women, as to children (and the insane) second-class legal status (Articles 212–214). Primary education for girls was not made law in France until the Loi Falloux of 1850.[10] After the French Revolution, there could be no French 'ladies in the laboratory' in 1825.[11] In this allegedly blank period for women's scientific contributions, the title page above therefore hides Sarah's importance transnationally as well as nationally in plain sight in the region's longer interconnected Portuguese (Dutch, Danish), French and British imperial histories. If they and she are concealed – 'By the late ...' – and yet revealed – 'To which is added ...' – in the very page layout and differential font sizes of the larger Madeiran part of the Bowdichs' journey, Sarah's pioneering West African-Gambian (counter-)perspectives in the final part also illuminate her non-British, and multi-lingual, scientific preparation and acumen. As the opening report above sketched in, the Bowdichs' scientific training was several prior to their departure for West Africa in 1822 from Paris, not London. As I first exposed in 2007,[12] the key to Sarah's significance is therefore in her intercultural enlargement of natural history-making before 1860 in French and European as well as in British contexts. Her manifold achievements in snapshot on this title page in the field in (intercontinental) West Africa are principally endorsed through the further major expansions of the *Excursions* in its two-volume French edition in 1826. The extensive 'Notes' for it by both Georges Cuvier (1769–1832) and Alexander von Humboldt (1769–1859), the foremost figures in world science of the period in respectively the Paris Muséum National d'Histoire Naturelle (National Museum of Natural History, henceforth Muséum) and international scientific exploration, encapsulate the highest international reputational standing of Sarah's work and contributions to science in 1826, and these irrespective of the nationality or gender of the author.[13] To return to the major critical blind spot of nationality above not only for expert Anglophone, but also for Francophone, Lusophone etc. historians of nineteenth-century science, cultural geography, gender and science. Rosters of major protagonists in a national science automatically expand transnationally when the knowledge of languages, including the international language of science publication and dissemination, is overtly factored in. In the nineteenth century, the international vernacular for new scientific knowledge was not English, but French (and German), especially in the pre-1860 period. The transnational scientific world was therefore one's oyster for the (multi-lingual and French-speaking) Humboldt, Cuvier, T. Edward and Sarah Bowdich in 1825.

The comparative Modern Languages approach of this book therefore helps to explain why and how Sarah Bowdich Lee, her 1825/1826 *Excursions* and her larger corpus of expert scientific work thereafter could remain almost invisible until 2007 to the indicative monolingual (Anglophone and Francophone) experts in the different disciplines above. The more fundamental critical corollary is then the book's challenge to intercultural and interdisciplinary inquiry, especially when its subject was the prolific and established author of publications found in public, institutional and national libraries, and also available in electronic open access format in English and French. As the editors of *The Palgrave Handbook of Women in Science* state in 2022 in their Introduction, the larger recuperation of women at work in plain sight will come

> by challenging male-focused histories which have misunderstood past landscapes of science and, as a result, ignored women's contributions. Although we must be careful not to overestimate the numbers of women active in science, one startling realization [...] is that many of the women [...] were well known and respected in their own time yet have dropped out of sight, unlike many of their male peers.[14]

This book therefore takes as read Sarah's major presence in the first half of the nineteenth century through her important work(s) in international natural history-making on two counts. First, her readers numbered world expert and informed general public contemporaries on both sides of the Channel (and the Atlantic), as Chapters 3 and 8 will further spotlight. Her many sole-authored publications with major presses were also revised and reprinted in her lifetime due to their many merits. Second are my some ten publications since 2007 (in English and in French) listed in the references. They have made Sarah and some of her many works in different genres better known by galvanising overtly comparative Nineteenth-Century Studies approaches for the broader questions that these usefully raise concerning the language(s), cultures and literatures of science in the period: its 'past landscapes' above. As now the acknowledged leading specialist on Sarah Bowdich Lee's scientific contributions, my work is also taken as emblematic of research on women in nineteenth-century science who operated transnationally.[15] Readers of this book new to the importance of Sarah for her times and relevance for women in STEM(M) today may therefore take as read the French-international scientific contexts and parameters modelled in Cuvier at the Paris Muséum and in Humboldt for world scientific exploration that make her work(s) visible and important. These frames underpin this larger study and its three-part structure. Readers already familiar with Sarah through my work will also find no reprise in any chapter below. The arguments of this book are larger. What makes Sarah's contributions to both Francophone and Anglophone histories and cultures of science before 1860 ground-breaking individually – and when taken together revelatory of her pioneering perspectives on natural history-making of the period – can only emerge through examining *how* she developed her extensive corpus of indisputably sole-authored natural history from 1825 to 1856 despite seemingly impossible odds. The opening profiles of 'Mrs Sarah' in 1824, through both the imagined eyes of her known Gambian contemporary and the historical facts of her situation upon disembarkation in London, deliberately set her at double rock bottom. A widow with no salvageable scientific freight and three children under eight to support faced a dauntingly improbable immediate future in natural history-making, let alone its unbroken (British, European, international) pursuit spanning 30 years.

The key question put by the Gambian 'woman with the parasol', 'What will she do now?', therefore spearheads the intercultural and interdisciplinary inquiry and challenge of this book. Because informed by West African cultural perspectives and observed knowledge of 'Mrs Sarah' at work, her question anticipates the latter's larger story of success, not insurmountable loss in 1824 upon the death of her husband-scientist in versions of Western European 'male-focused histories' quoted above. These also include agenda-setting Anglophone studies of the 1980s and 1990s in women's history and history of science, of which *Uneasy Careers and Intimate Lives: Women in Science 1789–1979* is indicative.[16] It importantly located the many women at work in secondary roles to support the science of a father, brother or husband as his amanuensis, illustrator, translator, preparer. Sarah's extensive unheralded work as the 'illustrator' and 'translator' in T. Edward Bowdich's publications in Paris (listed in Appendix 1) before they departed for Sierra Leone should have made her more prominent than an honourable mention in surveys of women in science.[17] If the title page of the *Excursions* above then clearly exposes the primary authorship and not secondary editorial role of 'Mrs Bowdich', as the widow no longer cloaked behind her more famous husband, widowhood cannot then explain her expert, unusually broad-ranging production of natural history works from 1825 until her death in 1856. When she 'married Robert Lee, an assize clerk on 29 July 1826', according to Donald deB. Beaver her main biographer,[18] hers alone was the scientific work and its publications thereafter.

Sarah's case therefore neatly highlights the doubly problematic status of the woman author's name, especially of science publications as Appendix 1 exemplifies. Her anonymous works such as *Taxidermy* (in several editions),[19] her occlusion and omission by the titular man, T. Edward Bowdich, her married name or, indeed, change to a further married name as 'Mrs R. Lee' individually and collectively render such a woman's contributions invisible or much less traceable. Double (married) nomenclatures also complicate identification of such women in library catalogues, electronic searches and national biographies, let alone through their further titles and names in translation. The 'Mrs Sarah' in the Gambian woman's framing view allows me throughout this book to encapsulate in her forename her many other published names, so as to maintain the integrity and independence of Sarah's authorial identity from her double married surnames across her life's work in English and in French. That the Gambian woman's appellation acknowledges her white contemporary's equal status as a married woman with children also qualifies any potential diminishment or over-familiarisation that comes with the use of a forename only for women authors. 'Mr Edward' has the same treatment in the Gambian woman's portrait.

But Sarah's case then also highlights the problematic status of a woman's contributions to science authored independently or conjointly, as examined by Helena Pycior and others in *Creative Couples in the Sciences* in 1996.[20] The quality and extent of the woman's scientific work in such couples are too rarely considered equal to the man's, and only most exceptionally deemed superior. The slippage between the person doing the science and the science itself returns the default for both: the undertakings by primary men of privileged, that is gentlemanly then professional scientific class and rank set the bar.[21] Its exclusion of women such as Sarah is their perpetual loss to 'male-focused' history above, taken also to be a universal behind its 'national' figurations.

In consequence, the further key blind spot for intercultural and interdisciplinary inquiry on women in science concerns rank, as encapsulated in the word 'genteel' in the contexts of (Anglophone and European) 'genteel travel cultures' above, and framed in Sarah's profile as the English 'lady' in the Gambia in 1824. Adopted from the French, the OED defines it as '[b]elonging to the gentry', 'suited to the station of a gentleman or gentlewoman'. Only recently have historians of British nineteenth-century science begun to re-evaluate how the allegedly synonymous 'gentleman science' (and related 'clergyman science') for (all) science endeavour of the period fails to account for the contributions of Britons who were not part of the leisured, Anglican, British Establishment. Many were excluded principally by their creed, rather than necessarily by their class, occupational status or need to earn an income.[22] Their ranks importantly included those serving British colonial interests abroad in military, mercantile, administrative, missionary, medical and educational capacities.[23] Their extensive roster then additionally reveals and establishes 'ladies' in their midst, for example the Quaker missionary Hannah Kilham, also glimpsed in the Gambia in the opening portrait.[24] Overseas postings therefore provided the necessary financial and geographical platform that enabled informed 'leisure' pursuit of natural history interests overseas by (British) men, and a select sub-group of intrepid women.[25] They numbered many of non-Anglican creed, often correlating with their non-English place of birth in Scotland, Wales and Ireland. As Michael Watts has stated unequivocally, 'The history of late Georgian and Victorian England and Wales cannot be understood without a knowledge of Nonconformity.'[26]

Creed is therefore the critical parameter too often overlooked in nineteenth-century Anglophone, but also Francophone and European history and geography of (imperial) science, and histories of women in these fields. Two 'French' examples are salient here. When Cuvier first arrived in 1795 at the newly established Paris Muséum of 1793,[27] it was by invitation of its first Chair of Zoology Étienne Geoffroy Saint-Hilaire (1772–1844). Although the state was now separated from the Catholic Church in France, Cuvier could not have been appointed as the *aide naturaliste* (research assistant) to Jean-Claude Mertrud (1728–1802, Chair of Comparative Anatomy), had freedom of religious expression not been reinstated in 1795. A Lutheran of modest background, Cuvier had spent the Revolution in exile in Normandy, using this time to investigate molluscs and other invertebrates. When, as Mertrud's successor from 1802, Cuvier then set about transforming the Muséum's dedicated galleries and collections in comparative anatomy, he was its only Lutheran chair. His training of research assistants (in the specialist laboratories for dissection, preservation, classification and description in word and image) and publications developed his extensive review and reclassification of the work by eminent (Deist) Muséum predecessors, such as Buffon and Lacépède on world comparative anatomy. Cuvier's (four-volume) *Recherches sur les ossemens fossiles des quadrupèdes* (*Fossil Bones*) of 1812 and 1817 (four-volume) *Règne animal* (Animal Kingdom) – reclassifying the Fishes as vertebrates rather than as a separate order – became definitive reference points for future national museum-based work on world natural history and vertebrate palaeontology (including human by the 1840s). Cuvier's curation of the Muséum's galleries of Comparative Anatomy incorporated its ever-increasing specimen collections (also including humans).[28] Paris was the destination of preference over other major Western collections for new *fauna* sent by explorer-collectors overseas.

Their foremost representative, and paradigm of the 'French' scientific explorer undertaking major comparative work inter-continentally, was the German Alexander von Humboldt. He was also a baptised Lutheran, and was inspired by the unity and value of all things in nature, a stance that included his abhorrence of slavery.[29] In emulating and developing the late eighteenth-century international model for scientific exploration set by Cook's three Voyages (1768–1771, 1772–1775, 1776–1779) that included the scientific expertise of important nonconformists,[30] Humboldt had returned to Paris in 1819–1823 when the Bowdichs were at work at its Muséum. He was bringing to press in French the extensive findings from his scientific travels of 1799–1804 to 'the Equinoctial Regions of the New Continent' with Aimé Bonpland (1773–1858).[31] The mentorships by Cuvier and Humboldt of T. Edward, and more unusually Sarah Bowdich, were not only important scientifically. The factors of creed and socio-economic nonconformity potentially count more largely to explain why these two world experts in natural history invested personal interest in the intensive scientific preparation and training of this unusual British couple. Shared scientific belief systems can encompass and be propelled by private, and deeply-held religious and humanitarian worldviews. For example, both Cuvier and Humboldt considered that science had no sex.

To return to the frameworks of the landmark *Uneasy Careers and Intimate Lives: Women in Science 1789–1979* above. The double undertakings of women in their intimate life and in secondary roles with a scientist relative turn upon his 'gentleman' science. If there is no equivalent independent 'gentlewoman' science, the important corollary is that the 'secondary' work in science before 1860 by women such as Mary Lyell (1808–1873) and Elizabeth Gould (1804–1841) stands to be additionally, and negatively, recalibrated by the status of the 'intimate' man. The 'gentleman' science of Charles (1797–1849) for the first but not the second in husband John (1804–1881) adversely screens Elizabeth's extensive contributions to scientific illustration overseas. Sarah's invisibility by 'genteel' rank in agenda-setting Anglophone studies of the 1980s and 1990s in women's history and history of science, historical geography and overseas travel operates similarly, despite her various entries under 'Lee', 'Wallis' and 'Bowdich Lee' in major biographical dictionaries.[32]

In revealing the blind spots in the formulation and understanding of 'genteel travel cultures' for both sexes, creed then further illuminates gaps in the otherwise wide critical interest since the 1980s in Anglophone 'Romantic' and 'Victorian' travel writing for cultural, literary and more recently transnational studies.[33] The title page of the *Excursions* above, and Sarah's later travel fictions, *The African Wanderers* (1847) and *Adventures in Australia* (1851), made her clearly an important exponent of nineteenth-century (women's) travel writing forms. It is then striking that major handbooks on (Anglophone) travel published since the 1990s have ignored her work, especially when they treat both factual travel writing, which the *Excursions* clearly is, and semi-autobiographical adventure genres, including for 'Juveniles', which would include her travel fictions. Of course, unusual women travellers not fitting, or actively breaking, accepted social moulds and modes of travel writing may evade mainstream critical notice if vanity, deluxe or other small press formats with limited print-runs were the main outlets for their works. These will be less accessible, and lesser-cited in consequence.[34] Indeed 'small print' fate could account for the minimal critical attention since 1982 to 'Sarah Lee' specifically as a travel writer of West Africa, including in Silke Strickrodt's only book-length study in 1998 to date,[35] and hence lack of larger interdisciplinary uptake thereafter. However, Mary Louise Pratt's much acclaimed and still widely influential *Imperial Eyes: Travel Writing and Transculturation* of 1992 drew important attention to 'Sarah Lee' – specifically the *Stories of Strange Lands* (1835) examined in Chapter 5 – to investigate its 'rather predictable division of labor between female and male writers' in early nineteenth-century travel accounts of (West) Africa.[36] Subsequent Anglophone travel criticism has picked up from Pratt the 'genteel' Maria Graham (*née* Dundas, later Maria, Lady Callcott, 1785–1842) and the well-known 'first' woman explorer of West Africa, Mary Kingsley (1862–1900), but not 'Sarah Lee'. Moreover, Pratt's own reading of *Stories of Strange Lands* strangely fails to connect it to the pivotal theoretical insight of her book designating as 'contact zones' the 'social spaces where disparate cultures meet, clash, and grapple with each other, often in highly asymmetrical relations of domination and subordination—like colonialism, slavery, or their aftermaths as they are lived out across the globe today.'[37] Sarah clearly inhabited the several 'contact zones' of both the Europe(s) and the Africa(s) of Pratt's book. For her and other recent critics of 'lady' travellers in 'skirts', 'crinolines' and 'petticoats', the default remains those of privileged, gentlewomanly ranks.[38]

If significant non-Anglican, and nonconformist 'ladies' in scientific endeavour at home and overseas have failed to make the cut in recent travel and science handbooks, their many productions in creative genres and non-creative formats such as translations further defy their bars.[39] Book and literary historians then agree: cultural and material constraints have perennially determined a woman's entry into writing in any genre. But the socially-validated modes of propriety that govern women's authorship in given periods also determine how the 'lady' author adroitly negotiates them.[40] For example, her form of self-inscription on the title page (including use of a pseudonym), her deployment of a dedication or a self-effacing preface (illustrating the modesty *topos*) could uphold and protect her reputation and authority (to write). In ways without equivalent for similar works of literature or science by men, the person of the woman's pen as much as her work remains the object of particular public scrutiny, censure and self-censorship as Chapter 3 explores. To return to the term 'genteel' as the tacit standard for female authorial proprieties. It indicates not only the 'lady' writer's reputational position but also her rhetorical armoury. She more consciously gauges and engages the appropriateness of her topic and treatment of it in her acute awareness of the (publishable) acceptability of her work for her target readerships. Sarah's self-inscription above on the title page of her *Excursions* is indicative of how she negotiated her allegedly second and hence secondary authorship as cover for her prominent (and foremost) female author's place and authority.

Knowledge precisely of the long history of adept self-inscription practices by women writers therefore spearheaded Anglophone literary criticism of the nineteenth century, principally by Barbara Gates and Ann Shteir in the 1990s, seriously to investigate women's scientific as well as creative writing. Shteir's seminal *Cultivating Women, Cultivating Science: Flora's Daughters and Botany in England, 1760–1860* of 1996 spearheaded their pivotal co-edited collection of 1997, *Natural Eloquence: Women Reinscribe Science.* It immediately uncovered a plethora of women writing in various spheres of science, with Gates's ground-breaking *Kindred Nature: Victorian and Edwardian Women Embrace the Living World* further underscoring the importance of 'creative' outlets for women's work and writing of science.[41] Indeed, creative forms often circumvented and surmounted the otherwise prohibitive barriers of the period to female authority in scientific matters (because deemed 'unladylike'). Gates and Shteir were therefore among the first to identify and collect the work(s) of 'Sarah Bowdich Lee' among important unheralded women in nineteenth-century botany, natural history and public understanding of science.[42] These women, like their sisters in the creative arts, often played down and displaced the authorship, authority, original contribution or importance of their science material through their strategic choice of 'creative' genres for its expression. The Robinsonade and the improving nature or adventure story are indicative forms of creative science popularisation. Its more serious face, Victorian popular science writing, was also penned by women. Where Bernard Lightman could then identify Sarah's work within the important cohort he terms 'maternal popularizers',[43] final chapters of this book will investigate Sarah's much larger distinctiveness. Her deployment of 'popular' forms disseminated expert natural history knowledge first, not second hand.

This book now rises to the recent rallying charge of the Palgrave *Handbook* quoted above in its direct response to the major question and expectations of the Gambian 'woman with a parasol'. What Sarah can 'do' in her major corpus of book-length publications differently concerning West Africa from 1825 until her death is the focus of the following nine main chapters, also for the sum of their parts. But the 'how?' of these prolific achievements cannot be explained by the facts for (genteel) women in the history of science, geography and travel of the period, even by concerted applications of 'history from below'[44]: Sarah was not 'below' enough. Her 'intimate lives' also cannot explain her much more 'un-easy career', to return to the title keywords above of the landmark work on women in science by Abir-Am and Outram. Sarah remained the family breadwinner and principal initiator of new natural history, whether as the dependent 'Mrs Bowdich' or the independent 'Mrs Lee', for more than three decades, through creatively maintaining her pioneering perspectives and productions in her fields. Her primary authority and roles therefore have to be accounted for otherwise, including in earlier-formed expertise and in innate 'scientific' intelligence, temperament and predisposition. These attributes are glimpsed in the Gambian woman's report of Sarah's intensive 'occupations' in natural history-making. If these reframe and support her situation and onwards prolific output, they also disclose the alternative models, means and possibilities for women's work before 1860 in natural history-making in the field. This book therefore additionally tests Sarah's case as 'history between the bars' (my term) in science, not least because its intercultural and interdisciplinary focus can attend to Sarah's 'Nonconformity' to return to Watts's watchword as much more fruitfully inclusive of others like her than are narratives of 'exceptional' women.[45] Indeed, nonconformity provides the necessary *passe-partout* – and I use the French term for 'master key' advisedly – that opens understanding of Sarah's multi-stranded engagements in expert natural history-making in France and Britain in 1820–1856. As an outsider-insider contributor to them by default in terms of nationality, class, gender and creed, her diversity and freedoms of informed response in her work(s) in natural history from 1825 – her 'history between the bars' in other words – will unfold in its array in the chapters that follow.

Indeed, the 'nonconformity' framing Sarah's early life as 'Miss Wallis', glimpsed but not elaborated in the unsurpassed biography by deB. Beaver, prefaces the innate and acquired qualifications that will underpin her works:

> [...] born on 10 September 1791 in Colchester, Essex, the third of four children and the only daughter of John Eglonton Wallis (*c*.1766–1833), grocer and linen draper, and Sarah (*c*.1770–1839), daughter of Edward Snell (*c*.1744–1786) and his wife, Ann Wood (*c*.1750–1772). Her parents were nonconformists and owned considerable property. She learned to ride and fish, and enjoyed exploring the countryside.
>
> Sarah Wallis's father went bankrupt in 1802, and later (probably in 1806 or 1807) moved the family to London. There Sarah met Thomas Edward Bowdich (1791–1824).[46]

The significant gap literally between the above paragraphs denotes the many unknowns of Sarah's life story, and the necessary gaps in consequence that any biographical approach to Sarah's 'twenty-one books' (unlisted) by deB. Beaver's tally would therefore encounter, but also founder upon. Yet the same gap, when filled briefly below by the striking implications of 'nonconformity' for Sarah's lifework in natural history-making, makes the different critical focus on her works of natural history from 1825 until her death so fruitful for this book. Its account of Sarah's contributions to science as unmissable for 'history between the bars' lies in the distinctions for (comparative) natural history of her different works, but not as deB. Beaver classified them: her 'informed' natural history on the one hand; her 'undistinguished' fiction on the other. Rather, the profit that has been my textual attention to works as apparently different in their subject matter and form as *Taxidermy* (1820), 'The Booroom Slave', *Adventures in Australia* (1851), or *Playing at Settlers* (1855),[47] means that I have excluded renewed engagement with them for reasons of space in order to concentrate on those, especially concerning West African subjects, awaiting a first major critical review of their pioneering contributions to natural history.

If understanding of 'nonconformists' in fact fills the gap in the separated statements in deB. Beaver's entry above and provides all-important missing links between his last three sentences, his label requires a small theological qualification that pinpoints the striking independence and cultivation of Sarah's mind and spirit as one with the family's change of fortunes. Important 'French' implications are also in its nexus. The Wallis family were wealthy Unitarians, also centrally involved in Colchester's flourishing mercantile community in the early nineteenth century.[48] An avowedly anti-doctrinal church (like the Quakers), Unitarian central tenets in modern parlance include a questioning search for truth (faith in God coexists with reason, science, philosophy and the wisdom of other faith traditions); tolerance and belief in the inherent worth and salvation of every person (meaning opposition to slavery, to sexual and other inequality/discrimination); respect and reverence for the interdependence of all life in the universe (of which humans are a part), and social responsibility.[49] These tenets would have framed Sarah's upbringing and education little differently from her brothers as the only daughter. She would have learned mathematics, geography, history, Latin and French alongside drawing, literature and music. Her 'nature' pursuits of riding and fishing are also significant.[50] At 16 (in 1807) and of marriageable age, Sarah's unusually broad education was largely complete when her father moved the family to London.

The Wallis family's straightened fortune resulting in bankruptcy was as much social as economic because of prevailing political prejudice and even ostracism. Britain's state of war with France until 1815[51] meant that Unitarians and other nonconformists were highly suspect for their 'Republican' values in the eyes of the British Establishment.[52] Poor harvests in 1802, and food and cloth shortages thereafter due to the Napoleonic Wars – including blockades on inward transportation of goods from Holland through Harwich and other ports – made Unitarian merchants in Essex particularly vulnerable to the loss of stock, custom and hence ruin. These material losses were compounded by legal and religious discrimination. Indeed, toleration of Dissenters including Unitarians only became an Act of Parliament in 1813, when the Whig Abolitionist (and Unitarian) campaigner, William Smith (1756–1835) the grandfather of Florence Nightingale (1820–1910), lobbied for 'The Doctrine of the Trinity Act', known also as the Unitarian Relief Act and The Unitarian Toleration Bill.

When John Wallis moved his family from Colchester to London, however, he could rely on connections through London's founding Unitarian Church in Essex Street (The Strand) and the Newington Green Unitarian Church.[53] Its minister from 1808, Rochemont Barbauld, was the husband of Anna Laetitia Barbauld (1743–1825); earlier, Joseph Priestley had been the minister, and Mary Wollstonecraft had attended sermons there that would influence her work on behalf of the education and rights of women. London Unitarians were thus integrally involved in educational reform and in leading scientific endeavour, as well as in anti-slavery campaigning. Within London's Unitarian community, Miss Sarah Wallis would have had ample opportunity to develop her fine mind, many talents and interests in both natural history and the betterment of society, and to meet others of similar Abolitionist views such as T. Edward Bowdich, son and partner in a Bristol hat business.[54]

Sarah and Edward married in the (Anglican) Church of St Mary Newington on 9 January 1813, six months before Edward's family position in Bowdich, Son and Luce was definitively wound up: 'The Partnership subsisting between us the undersigned (Thomas Bowditch (*sic*), Thomas Edward Bowdich, T. Luce, jun.) as Hat-Manufacturers, N° 3 Clare Street, [Bristol] is this day dissolved by mutual consent [...]'.[55] In their unusual marriage of equal heart and mind, the Bowdich 'power couple' also had shared experience of transcending material adversity as part of the courage of their convictions and defiance of social conventions.[56] Their (joint) commitment to travel and work together for the greater causes of science, emancipation and education would not be through the auspices of missionary or Church societies, however. Unitarians did not operate them. Rather, the seasoning of straightened family fortune fitted them for the patterns of their unusual married life and work together in independently mounted natural history endeavour (1813–1824). 'To boldly go' therefore epitomises Edward's, but also Sarah's, first overseas travels to British West Africa. In 1814 (before the end of the Napoleonic Wars), Edward had jumped at a writership opportunity in the service of the Royal African Company at Cape Coast Castle in the Gold Coast (now Ghana), where his maternal uncle Mr Hope Smith was the governor-in-chief of its settlements. Unusually for the time, and especially so soon after giving birth in London to Florence their first child, Sarah set sail with her baby from Liverpool to join Edward at Cape Coast Castle. Upon arrival, she discovered that he had made a similar family decision to sail to London. As glimpsed in the opening portrait, little Florence succumbed to and died of tropical fever during the remainder of Bowdichs' Asante sojourn.[57] It resulted in the publication of T. Edward Bowdich's *Mission to Ashantee* (1819), still well-regarded by historians today.[58]

When no reward then came for it, Edward's pique and attacks on the African Company catalysed the Bowdichs' departure for the Paris Muséum in what was also a double decision not to be deterred in London from pursuit of science. If its scientific societies, collections and training were as closed to Edward because of his mercantile rank as to Sarah by her sex alone, Paris doors by contrast were open. Paris also offered what London could not: world specimen collections, world-class expertise in all branches of natural history and the latest scientific know-how for collecting in the field.[59] The fruits of the Bowdichs' intensive four-year Paris sojourn and training under Cuvier, Humboldt and other leading experts were concomitantly scientific and financial, to prepare their independent journey of exploration to Sierra Leone. In their English publications (see Appendix 1A) disseminating the Muséum's latest classifications, methods and descriptions of new collections in its many 'departments' of natural history/*histoire naturelle* was their provisioning to encounter species new to known world coverage.[60] Only the important scientific nonconformism of the Bowdich partnership could make of themselves human 'Muséums' in a delimited time for encountering the many fields of natural history in West Africa.

Notable in Sarah's pre-1825 experiences in Paris was then her rising (again) to new challenges that included overcoming personal family tragedy. The Gambian woman's account above not only highlights it in the earlier death of Florence at Cape Coast Castle, but also reports from its outset the Bowdichs' very unusual undertaking *en famille* of their scientific journeys to West Africa. Their Paris sojourn saw Sarah's fullest participation in Muséum study, illustration, translation and publication of works by 'T. Edward Bowdich', as well as acquisition of Arabic as the *lingua franca* of the region.[61] All these endeavours, however, were also managed around her (annual) pregnancies, confinements and child-rearing. Tedlie Hutchison and Edward Hope in the opening portrait were born in Paris, as were two other children who died in early infancy. The timing of the Bowdichs' longer stay in Madeira than was its intended stopover for Sierra Leone revolved around Sarah's last-known pregnancy and the birth of Eugenia Kerr.[62] Her arrival then necessitated the Bowdichs' stay in the Gambia to secure onwards passage to Sierra Leone. Only through building an independent family life in natural history-making outside 'gentlemanly' models for institutional and explorational science could the Bowdichs respond to such new opportunities overseas as their next at home.

The sketching in here of the Bowdichs' determined application of their nonconformist convictions for independent scientific endeavour informing their joint ventures from 1813 to 1824 also immediately enlightens how we review Sarah as Edward's widow in 1824 on that London quayside with their three children. If she was bereft of her beloved partner-in-science and of her crates, she was eminently equipped to overcome these calamities since she had lived *her* science work *en famille* since 1813 in Asante, Paris, Madeira and the Gambia. It would now have a London residence. Through a subscription fund, friends and family assisted Sarah to set up a new home for her young children in Camden (Figure 0.2), where she lived at various addresses in close proximity in the Borough – No. 33, No. 12 and No. 27 Burton Street; 27 Regent Square – until her final illness and death in 1856, nursed in her last days in Erith (Kent) by Eugenia. The latter then emigrated with her husband, Dr Swayne, to New South Wales Australia, the rich natural history and 1851 gold-rush devastations of which her mother had so recently captured in both the 1851 and revised *Adventures in Australia* of 1853.[63]

Sarah's courage not to depart from her life in (French) natural history in 1824, and to contribute at the forefront of its nascent disciplines thereafter, therefore made her a highly unusual British woman for her time. She could not be otherwise, because British and London science remained more closed to her than is the cul de sac of Burton Street in this figure. The further bitter irony of its Camden location lay in being only blocks away on foot from the British Museum, and from the scientific collections of private

Figure 0.2 'Impasse, or Rebuilding a Life in Natural History at Burton Street, Camden, London'. Copyright Neil Mackenzie.

individuals in gentleman science in which 'ladies' could not participate. Sarah would have had to wait until 1828 for London Zoo to open, and for Kew Park to become Kew National Botanical Gardens in 1840. The British Association for the Advancement of Science had yet to be born (in 1831), and also did not admit women to its sections until after 1836.[64] When the pension Sarah had hoped for in respect of Edward's death in the Gambia failed to materialise (her obituary reveals that she received a very small pension of £50 only two years before her death), this serious financial disappointment came as no ultimately defeating blow: she had weathered similar, or worse, before. Sarah did not then write science from 1825 'for survival' as Donald deB. Beaver contends.[65] Rather her 'genteel' creative writing – the listed Gift Book story titles produced from the 1820s to the 1840s in Appendix 1 – met her family's financial needs and, more importantly, resourced her time and independent pursuit of expert (French) natural history work and publication from 1825 no longer hidden by authorial covers of 'T. Edward Bowdich' (and partner).

To the question in expectation put by the 'woman with the parasol', 'What will [Sarah] do now?' is the story in this book. Her astonishing response in 1824 was not only to prepare the *Excursions* for different double publication in London and Paris in 1825/1826, but also through it to launch her unbroken, creative production over thirty years of book-length works in expert natural history for multiple audiences. These productions were the ongoing realities of her immediate and longer 'narrative of its continuance' from the title page of the *Excursions*. Because the French edition of the *Excursions* remains a closed book in Francophone as well as Anglophone nineteenth-century history of science, geology, exploration and scientific travel and hence women's history in these domains,[66] it is the subject variously of the first, fourth and seventh chapters because this momentous text identifies Sarah's many French-inspired pioneering perspectives in expert natural history. The original 1825 London-English imprint then becomes newly readable for its author's larger international significance through the lenses of the expanded two-volume French edition of 1826.

These comparative optics and structure of this book therefore take their cue from Sarah's authorship and double pen, lying in plain sight in the title page information of the *Excursions* (in both versions). Here the tensions and priorities that constitute the quintessential components of the scientific travel account for its informed reader lie in its choreography and counterpoint of the familiar, unfamiliar and hitherto unknown. In Sarah's case, they come in the unmistakable wrappers of the serious (world-class) scientific exploration-travelogue format *à la Humboldt*. Indeed, the Humboldian content coverage, clear geographies and authority findings that include expert illustrations and appendices guarantee their validity for the cultural norms of the genre and for new reports of challenging materials that whet contemporary reader interest, curiosity and (re-)appraisal. That Sarah's 'added' travel account of the Gambia appears to be of minor interest – the title page layout might infer it is supplementary, separated, accidental, niche, unimportant – rather than of larger intercontinental import, deftly operates the sleights of Sarah's foremost expert (female) scientific hands.

The three-part structure of this book therefore concertedly investigates the multiple importance of Sarah's 'West African' texts within her ensuing corpus in its (comparative) French and British scientific and cultural contexts. The book follows her 'excursions' – and excursus expositions and supplements within each part – to determine the pioneering perspectives that constitute her larger significance in world natural history-making. Opening the first part, 'Canvassing Cuvier', Chapter 1 addresses the French *Excursions* in the light of its 'Notes' by Cuvier for the first time. Cuvier only published the 'Prospectus' for his multi-volume new natural history of world fish (the *Histoire naturelle des poissons*, 1828–1848) in 1826. Our first chapter therefore reconsiders Sarah's published contributions to West African ichthyology in the 1825 London imprint *ahead of* his discipline-defining work, and why her forename matters in modern scientific (re-)classification. The necessary attention in the first chapter to scientific technicality that trades as objective fact and international standard will appear dry and the hardest read compared with other chapters. Yet for Sarah's work to stand for two centuries by modern international standards, despite their inbuilt automatic bars to pre-1870 scientific history, only makes her place in 1825/1826 and today the more towering. Reader, bear with nomenclature and scientific detail as signposts for 'natural history between the bars' that this book unpacks! Chapter 2 then turns to Sarah's *Fresh-Water Fishes of Great Britain* (1828–1838) to ascertain how it was similarly ahead of its time and of Cuvier's multi-volume *Histoire naturelle des poissons* in the nascent fields of (European) freshwater ichthyology. By these lights, Chapter 3 returns more concertedly to Sarah's international reputation in natural history, by examining the English and American editions of her *Memoirs of Baron Cuvier* (1833) as the different bookends upon the death of Cuvier in 1832 to its concomitantly published French edition. The larger purviews than are my seminal studies of Sarah's texts in Part One allow its chapters also to investigate the further nadirs in Sarah's life-long commitment to natural history-making: 1833 is its significant turning point.

The second part of the book, 'Harnessing Humboldt', returns to Humboldt's 'Notes' constituting his epilogue essay to the French *Excursions* less for their important geological debates concerning Madeira as I have previously investigated,[67] and more for their modelling of expert connective world natural history-making through comparative field observation. The seeming afterthought that is the Gambia in the *Excursions* is addressed in Chapter 4, which examines Sarah's first country profile of it as a 'plant geography' à la Humboldt. Her innovations to his model also highlight the latter's incompleteness of view. Sarah's geographical inquiry therefore returns Chapter 4 differently to the ingrained 'scientific' prejudices of her times and today treated in Chapter 1. But the fourth chapter also allows the book to integrate the importance of her contributions to botany as part of plant geographical knowledge also treated in Chapters 5 and 6, when the book's strategic coverage could devote no chapter-length space to her important *Trees, Plants and Flowers: their Beauties, Uses and Influences* of 1854. Informed scholarly concentration on women's contributions to botany as the supposedly more 'feminine' and 'genteel' pursuit for women in the long nineteenth century – whether in leisure botanising, flower drawing or plant collecting – has also counterproductively displaced more major interest in the all-important place of women in other less 'feminine' fields of science. Too often the realities of women's works in 'rugged' geography and geology and 'hard' natural history are precluded and occluded, especially when undertaken 'overseas'. To emphasise the latter in this book, Chapter 4 also reintegrates as one in its Humboldtian purview the vital intercultural and interdisciplinary scientific lenses that understand intercontinental species and habitats only when 'botany' and 'geology' inter-reflect the other. Chapter 5 similarly opens space in the book's focus on Sarah's sole-authored 'story' publications after 1825 to reconsider their larger contributions also to the Bowdichs' first Voyage to West Africa and earlier publications by 'T. Edward Bowdich'. But Sarah's multi-perspectival collection, *Stories of Strange Lands* of 1835, is additionally informed by the Act of the Abolition of Slavery of 1834, and by the Bowdichs' second and aborted journey to 'West Africa' only as far as the Gambia. Chapter 6 then develops the journey format of natural science exploration, to remould Sarah's lived report in doubled non-fictional and fictional forms in this chapter's first major examination of *The African Wanderers* (1847). In arguing that it forges an important new literary-scientific genre for 'juvenile' target readerships, Chapter 6 prepares the final part of the book: 'Opening Access to Expert Natural History'.

Where the first two parts of the study draw out Sarah's important reapplications in her works of the foremost science of her two differently eminent French mentors, the third part attends to the lesser-known commitments that Cuvier and Humboldt shared in their wider imparting and 'mentoring' of new and next-generational science. Their uses of scientific image, and non-technical yet scientifically informed language, count centrally in their works for informed general public audiences, and in Sarah's. Chapter 7 not only investigates her further pivotal contributions to scientific illustration in the *Excursions*, but it also uncovers and recovers for the first time her many unacknowledged scientific drawings in print elsewhere in her corpus. Chapter 8 then reconnects word and image as quintessential to the clear instruction with first-hand authority of 'text-book' natural history. This chapter's analysis of *Elements of Natural History* of 1844, because penned by a woman, focusses on Sarah's interests in scientific method as integral to form and content. That the important stimuli of (informed) interest, curiosity and new observation in the field pertain to and are encouraged in the budding, as well as expert amateur naturalist in Sarah's 'textbook' works is the subject of Chapter 9: the powers of authoritative scientific observation and its trained development come through model natural history anecdotes, understood as keenly drawn expert reports from the field. In the double-volume *Anecdotes*, Sarah's promotion of respect for participants of all ages and ranks in foremost natural history endeavour at home and abroad is then her call not for a 'citizen science' in its equivalent nineteenth-century guise of 'popularisation', but rather for what in today's parlance are 'equality, diversity and inclusion' (and decolonisation) agendas for scientific work. Our final chapter therefore concludes with their surprising disclosure by antithesis in Sarah's cautionary tale for imperial science that is *Sir Thomas the Cornish Baronet* (1856). Here in her last word before her death was a warning about the disastrous consequences of high colonial, professionalising natural science and 'bioprospecting' that ignored its local and global 'biospheres'. These modern terms in quotation marks signal the early timeliness of Sarah's works for experts and 'juveniles' alike, and why their pioneering perspectives matter today.

In the nine chapter titles and their findings in this book, Sarah's contributions to both Francophone and Anglophone histories and cultures of science prove groundbreaking individually and as a corpus, because of how she could undertake them before 1860 *outside* the conformist rules and bars for their endeavour in two (rival) national contexts. Only when Sarah's many firsts are added up, however, is the array and consistency of their quality contribution the more remarkable. This sum then presses out the 'how' that sustained her work more concertedly before as well as after 1825, and as revelatory of the pioneering perspectives that inspired and renewed it. They clearly exceed Sarah's circumstances and sex, and transcend the seemingly impossible odds against her life in natural history-making. The larger intercultural challenge of the book is then to draw conclusions and lessons from her case as no exception (that proves the rule), but rather as indicative of others with committed independence of informed scientific view. To reconnect the variously intertwining West African and European 'contact zones' (in Pratt's terms) that informed Sarah's travel(-led) writing of new natural history over three decades, and at the forefronts of its various new nineteenth-century sub-disciplines – ichthyology in Part One; plant geography and early 'anthropology' in Part Two and in different genres in Part Three – is to return to the *passe-partout* of Sarah's committed religious, scientific and creative nonconformity, and how she expressed it. In also making Sarah's case in the contexts of the opening line of her profile in the Gambian woman observer's portrait, her corpus cannot be disentangled from her children buried in and surviving West Africa. Sarah's striking maternal nonconformism therefore also inspires further research on nineteenth-century women in science and draws lessons for women in STEM(M) today. Conclusions also take up the necessity that is the intercultural broadening of (women's) intellectual history. To magnify the lenses of Sarah's nonconformity is everywhere to see her make 'history between the bars' that reveals a larger diversity of respected figures with much more to tell twenty-first century 'interdisciplinarity'. Our text-focused study therefore holds up the malleable 'unitarian' forms of Sarah's natural history 'productions' as a corpus that connects nature and narrative for larger understanding of natural worlds. Perhaps its creative-scientific integrity is its largest critical challenge today to binary divides, especially what constitutes acceptable forms for 'proper' scientific and literary-artistic endeavours and representation.

The larger trial of Sarah's situation in London in 1824 was therefore not as the Gambian woman had supposed the lack of extended family or 'business' networks (in Paris). Rather, Sarah's rank and status now constrained the many freedoms that she had enjoyed outside Britain, because they circumscribed proprieties for a young widow with children to support. Once more Paris will prove her scientific and financial life-line and outlet, however, and the place of larger recognition for Sarah's unprecedented natural history work in West Africa. To follow Sarah's 'continuance of her narrative' in the publication in 1826 of the French edition of her *Excursions* is now also to discover her London independence in natural history from 1825 to 1856 in the 'Mrs Sarah' who can no longer be overlooked in the histories of French and British science nearly two centuries later.

Notes

1 'Engraving of Thomas Edward Bowdich, after a painting by William Derby'. Source http://www.britannica.com/ebc/art-9018. A portrait by T. A. Woolnoth illustrating 'Not Quite Attentive' in Sarah's *The Juvenile Album, or Tales from Far and Near.* London: Ackermann and Co, 1841 (26) has been erroneously attributed as being a likeness of Sarah in 'Evocadora imagen biográfica de Sarah Bowdich Lee' in a blog by Salvador Pérez: https://www.taxidermidades.com/2016/07/taxidermy-obra-de-la-naturalista-e-ilustradora-sarah-bowdich-lee.html.

2 The Bowdichs had sent on some new bird specimens to the Paris collections. See Justin J. E. F. Jansen, 'The Bird Collection of the Muséum National d'Histoire Naturelle, Paris, France: The First Years (1793–1825)', *Journal of the National Museum (Prague)* 184, no. 5 (Nov. 2015): 81–111.

3 See Appendix 1, Section A.

4 See John M. Mackenzie, *The Empire of Nature: Hunting, Conservation and British Imperialism.* Manchester: Manchester University Press, 1988, and his *Museums and Empire: Natural History, Human Cultures and Colonial Identities.* Manchester: Manchester University Press, 2009; Felix Driver, *Geography Militant: Cultures of Exploration and Empire.* Oxford: Blackwell, 2001 and David N. Livingstone, 'The History of Science and the History of Geography: Interactions and Implications', *History of Science* 22, no. 3 (Sept. 1984): 271–302 and *Putting Science in its Place: Geographies of Scientific Knowledge.* Chicago: Chicago University Press, 2003. See also the more recent work, David N. Livingstone and Charles W. J. Withers, eds., *Geographies of Nineteenth-Century Science.* Chicago: University of Chicago Press, 2011, and Carl Thompson, 'Earthquakes and Petticoats: Maria Graham, Geology, and Early Nineteenth-Century "Polite" Science', *Journal of Victorian Culture* 17, no. 3 (Sept. 2012): 329–46.

5 See Cheryl McEwan, 'Gender, Science and Physical Geography in Nineteenth-Century Britain', *Area* 30, no. 3 (1998): 215–23, and *Gender, Geography and Empire: Victorian Women Travellers in West Africa.* Aldershot and Burlington: Ashgate, 2000.

6 See Carl Thompson, 'Maria Graham and the Chilean Earthquake of 1822: Contextualizing the First Female-Authored Article in *Transactions of the Geological Society*', in C. V. Burek and B. M. Higgs, eds., *Celebrating 100 Years of Female Fellowship of the Geological Society: Discovering Forgotten Histories.* Geological Society Special Publications 506. London: Geological Society, 2021.

7 See Patricia Fara, *Pandora's Breeches: Women, Science and Power in the Enlightenment.* London: Pimlico, 2004; Carl Thompson, *The Suffering Traveller and the Romantic Imagination.* Oxford: Clarendon Press, 2007, and Nigel Leask, *Curiosity and the Aesthetics of Travel Writing, 1770–1840.* Oxford: Oxford University Press, 2002. On women's travel and art connoisseurship see Caroline Palmer, '"I will tell nothing that I did not see": British Women's Travel Writing,

Art and the Science of Connoisseurship, 1776–1860', *Forum for Modern Language Studies* 51, no. 3 (July 2015): 248–68.

8 See Alison Blunt, *Travel, Gender and Imperialism: Mary Kingsley and West Africa*. New York: The Guilford Press, 1989, and Dea Birkett, *Spinsters Abroad: Victorian Lady Explorers*. Oxford: Basil Blackwell, 1994.

9 See, for example Margaret Alic, *Hypatia's Heritage: A History of Women in Science from Antiquity to the Late Nineteenth Century*. London: The Women's Press, 1986; Jean-Pierre Poirier, *Histoire des Femmes de Science en France: du Moyen Age à la Révolution*. Paris: Pygmalion Gérard Watelet, 2002; Eric Sartori, *Histoire des Femmes Scientifiques de l'Antiquité au XXe siècle: les Filles d'Hypatie*. Paris: Plon, 2006. The exception will always prove the rule. See *A Woman of Courage: The Journal of Rose de Freycinet on her Voyage around the World, 1817-1820*, ed. and trans. by Marc Serge Rivière. Canberra: National Library of Australia, 1996.

10 Girls' education was still under the control of the Catholic Church in 1850, because there was no training of women teachers otherwise. The 1833 Loi Guizot established secular primary education for boys, with a further decree of 1836 extending it to girls by discretion, not law. There was no secondary education for girls, except privately for those in bourgeois families.

11 This phrasing borrows the title of Mary R. S. Creese, with Thomas M. Creese, *Ladies in the Laboratory? American and British Women in Science, 1800–1900: A Survey of their Contributions to Research*. Lanham and London: The Scarecrow Press, 1998 (revised 2004). Illustrating the point in France literally is Marie-Anne Paulze Lavoisier (1758–1836) who, having collaborated in her chemist husband's laboratory experiments, saw his execution at the hands of the Revolutionaries and their scientific instruments and property confiscated.

12 See Mary Orr, 'Pursuing Proper Protocol: Sarah Bowdich's Purview of the Sciences of Exploration', *Victorian Studies* 49, no. 2 (Winter 2007): 277–85. All subsequent citations in this book of my many publications on Sarah, listed under 'Orr, Mary' in the secondary references, will take as their short form 'Orr', followed by the identificatory first title keywords.

13 As the acknowledgements clarify, I stumbled on Sarah when researching a very different monograph on religion and science in Flaubert's *Tentation de Saint Antoine* (Temptation of Saint Anthony), see Mary Orr, *Flaubert's Tentation: Remapping Nineteenth-Century French Histories of Religion and Science*. Oxford: Oxford University Press, 2008. It was my knowledge of Cuvier and Humboldt as major scientific reference points for Flaubert that allowed me immediately triangulate the importance of 'Mme Bowdich' with their respective authorship of 'Notes' to the 1826 *Excursions*.

14 Claire G. Jones, Alison E. Martin and Alexis Wolf, 'Women in the History of Science: Frameworks, Themes and Contested Perspectives', in Claire G. Jones, Alison E. Martin and Alexis Wolf, eds., *The Palgrave Handbook of Women in Science since 1660*. Springer Nature: Switzerland AG, 2022, 3–24 (5).

15 See Carl Thompson, 'Women Travellers, Romantic-era Science and the Banksian Empire', *Notes and Records* 73 (2019): 431–55, the final part constituting an acknowledged résumé of my ground-breaking (French Studies) framing of Sarah's 'science' of his title. Further evaluative exposés of my work are the last section of Angela Byrne's 'The Scientific Traveller' opening chapters of *The Routledge Research Companion to Travel Writing*, ed. Alasdair Pettinger and Tim Youngs. Abingdon and New York: Routledge, 2019, and Katharine Turner's edited curation of 'Women Writing Travel: A Virtual Special Issue', *Forum for Modern Language Studies* 59, no. 2 (April 2023): 315–19 (reprising Mary Orr, 'Amplifying Women's Intelligence through Travel: Inna's Tale in "The Booroom Slave" by Sarah Bowdich', *Forum for Modern Language Studies* 51, no. 3 (July 2015): 269–86). See also Orr, 'Catalysts, Compilers and Expositors', in *The Palgrave Handbook of Women and Science* (505–28) cited in note 14.

16 Pnina G. Abir-Am and Dorinda Outram, eds., *Uneasy Careers and Intimate Lives: Women in Science 1789–1979*. New Brunswick: Rutgers University Press, 1987.

17 See, for example the work of Mary R. S. Creese, cited in note 11.

18 Donald deB. Beaver, 'Lee [*née* Wallis; *other married name* Bowdich], Sarah'. Oxford: The Oxford Dictionary of National Biography (most recently updated, 8 July 2021). This second marriage is also dated elsewhere in 1827 and 1829.

19 See Orr, 'The Stuff of Translation'.

20 Helena M. Pycior, Nancy G. Slack and Pnina Abir-Am, eds., *Creative Couples in the Sciences*. New Brunswick: Rutgers University Press, 1996.
21 For indicative examination of the gender of science, see Ruth Barton, '"Men of Science": Language, Identity and Professionalization in the Mid-Victorian Scientific Community', *History of Science* xli (2003): 73–119; Evelyn Fox Keller, *Reflections on Gender and Science*. New Haven and London: Yale University Press, 1985; Sally Gregory Kohlstedt, 'In from the Periphery: American Women in Science, 1830-1880', *Signs* 4, no. 1: 81–96; and Londa Schiebinger, 'The History and Philosophy of Women in Science: A Review Essay', *Signs* 12, no. 2 (1987): 305–32, 'Gender and Natural History', in Nicholas Jardine, James A. Secord and Emma Spary, eds., *Cultures of Natural History*, 163–77. Cambridge: Cambridge University Press, 1996, and 'Has Feminism Changed Science?' *Signs* 25, no. 4 (Summer 2000): 1171–75.
22 See David Lowther, 'Un-gentlemanly Science', in Bernard Lightman and Bennett Zon, eds., *Victorian Culture and the Origin of Disciplines*, 111–34. New York: Routledge, 2020, building differently on the work of Anne Secord, 'Science in the Pub: Artisan Botanists in Early Nineteenth-Century Lancashire', *History of Science* 32, no. 3 (Sept. 1994): 269–315 and her 'Corresponding Interests: Artisans and Gentlemen in Nineteenth-Century Natural History', *The British Journal for the History of Science* 27, no. 4 (Dec. 1994): 383–408.
23 See Andrew Porter, '"Cultural Imperialism" and Protestant Missionary Enterprise, 1780-1914', *Journal of Imperial and Commonwealth History* 25, no. 3 (1997): 367–91.
24 See Elizabeth Provost, 'Assessing Women, Gender, and Empire in Britain's Nineteenth-Century Protestant Missionary Movement', *History Compass* 7, no. 3 (2009): 765–99 and Alison Twells, '"So distant and wild a scene": Language, Domesticity and Difference in Hannah Kilham's Writing from West Africa, 1822–1832', *Women's History Review* 4, no. 3 (Dec. 1995): 301–18.
25 See, for example the renowned Scottish botanist, Robert Brown (1773–1858), an ardent Episcopalian and Jacobite, training and serving abroad as an Army surgeon, or Louisa Anne Meredith (1812–1895). For a study of the scientific 'lady' see Patricia Phillips, *The Scientific Lady: A Social History of Women's Scientific Interests, 1520–1918*. London: Weidenfeld and Nicolson, 1990. Unusually, 'Sarah Wallis' is included (116).
26 Michael Watts, *The Dissenters*. Vol. II, *The Expansion of Evangelical Nonconformity*. New York: Oxford University Press, 1995, 4.
27 See Stéphanie Deligeorges, Alexandre Gady and Françoise Labelette, *Le Jardin des Plantes et le Muséum National d'Histoire Naturelle*. Paris: Éditions du Patrimoine, 2004.
28 See Claude Blankaert, Claudine Cohen, Pietro Corsi and Jean-Louis Fischer, eds., *Le Muséum au premier siècle de son histoire*. Paris: Éditions du Muséum National d'Histoire Naturelle, 1997. Among the spoils of Napoleon's European Campaigns sent to the Paris Muséum were rival national and private scientific collections and menageries (for example, from Holland and Italy).
29 I am the first to draw attention to Humboldt's 'Notes' as in fact an extensive epilogue essay, and to explore their significance. See Orr, 'New Observations on a Geological Hotspot Track: *Excursions in Madeira and Porto Santo* (1825) by *Mrs* T. Edward Bowdich', *Centaurus* 56, no. 3 (August 2014): 135–66.
30 Central roles were played in Cook's Voyages by 'gentlemen naturalists' like Joseph Banks (1743–1820), alongside nonconformists such as the Lutheran Swede, Daniel Solander (1733–1782), Linnaeus's disciple and then Banks's archivist), and the Scottish Quaker, Sydney Parkinson (1745–1771).
31 For one of the few Anglophone studies of Bonpland, see Stephen Bell, *A Life in Shadow: Aimé Bonpland in Southern South America 1817–1858*. Stanford: Stanford University Press, 2010. Bonpland significantly arrived to study at the Paris Muséum in 1795.
32 See deB. Beaver, 'Lee [*née* Wallis]', cited in note 18 and the entry, 'BOWDICH LEE, Sarah Eglonton (née Wallis: 1791–1856)', in Mary R. S. Creese, *The Dictionary of Nineteenth-Century British Scientists*, Vol. 1 (A-C), ed. Bernard Lightman. Bristol: Thoemmes Continuum, 2004, 243–44, extending her entry in Creese and Creese, cited in note 11 (1998, 128–30, 383, 396, 419; revised 2004, 225–27).
33 For the former, see for example Tim Fulford, Debbie Lee and Peter J. Kitson, *Literature, Science and Exploration in the Romantic Era*. Cambridge: Cambridge University Press, 2004, and for the latter, Kathryn N. Jones, Carol Tully and Heather Williams, *Hidden Texts, Hidden Nation:*

(Re)Discoveries of Wales in Travel Writing in French and German (1780–2018). Liverpool: Liverpool University Press, 2020.

34 Sarah's *Fresh-Water Fishes of Great Britain* exemplifies the latter, its original deluxe small print run determining its slight contemporary and later critical reappraisal.

35 See the first short piece by Susan Greenstein, 'Sarah Lee: The Woman Traveller and the Literature of Empire', in David F. Dorsey, Phanuel A. Egejuru and Stephen H. Arnold, eds., *Design and Intent in African Literature*, 133–37. Washington, DC: African Literature Association & Three Continents Press, 1982; and the short monograph by Silke Strickrodt, *'Those wild Scenes': Africa in the Travel Writings of Sarah Lee (1791–1856)*. Gliennicke/Berlin and Cambridge, MA: Galda-Witch Verlag, 1998.

36 See Mary Louise Pratt, *Imperial Eyes: Travel Writing and Transculturation*. London and New York: Routledge, 1992, 5 (and 106–07 for her comments on Lee's work).

37 Pratt, *Imperial Eyes*, 4.

38 See Birkett, *Spinsters Abroad*; Christel Mouchard, *Adventurières en crinolines*. Paris: Editions du Seuil, 1987, and Thompson, 'Earthquakes and Petticoats', 329–46.

39 Women's work in science in the guise of translation of works by men is a further important genre. See, as an exemplary study, Alison E. Martin, *Nature Translated: Alexander von Humboldt's Works in Nineteenth-Century Britain*. Edinburgh: Edinburgh University Press, 2018.

40 See Mary Poovey, *The Proper Lady and the Woman Writer: Ideology as Style in the Works of Mary Wollstonecraft, Mary Shelley, and Jane Austen*. Chicago: Chicago University Press, 1984 and Elizabeth Eger, Charlotte Grant, Clíona Ó Gallchoir and Penny Warburton, eds., *Women, Writing and the Public Sphere, 1700–1830*. Cambridge: Cambridge University Press, 2001.

41 See Ann B. Shteir, *Cultivating Women, Cultivating Science: Flora's Daughters and Botany in England, 1760-1860*. Baltimore: Johns Hopkins University Press, 1996; Barbara T. Gates and Ann B. Shteir, eds., *Natural Eloquence: Women Reinscribe Science*. Madison: University of Wisconsin Press, 1997; and Barbara T. Gates, *Kindred Nature: Victorian and Edwardian Women Embrace the Living World*. Chicago: The University of Chicago Press, 1998.

42 See Shteir, *Cultivating Women*, for her reference to Sarah's *Trees, Plants and Flowers* and *Elements of Natural History*, and 'Elegant Recreations? Reconfiguring Science Writing for Women', in Bernard Lightman, ed., *Victorian Science in Context*. Chicago and London: University of Chicago Press, 1997 referencing Sarah's *Elements of Natural History* (236–55). Gates's *Kindred Nature*, 77–79 promotes Sarah's *Excursions* of 1825 and *Fresh-Water Fishes of Great Britain*.

43 Bernard Lightman, *Victorian Popularizers of Science: Designing Nature for New Audiences*. Chicago and London: University of Chicago Press, 2007, Chapter 3.

44 See the description of 'history from below' among other resources in the online, 'Making History' project: https://archives.history.ac.uk/makinghistory/themes/history_from_below.html (last accessed 14 June 2022). The retrieval of Mary Anning (1799–1847) in the history of palaeontology is a case in point.

45 Important though rosters of famous, greatest or exceptional women (in science) are, they uphold the model in science of great men, first discoveries and contributions that 'change the world', at the cost of ignoring other vital protagonists and ways of undertaking science.

46 See note 18, and Donald deB. Beaver, 'Writing Natural History for Survival – 1820–1856: The Case of Sarah Bowdich, later Sarah Lee', *Archives of Natural History* 26, no. 1 (1999): 19–31.

47 See respectively Orr, 'The Stuff of Translation', 'Amplifying Women's Intelligence', '*Adventures in Australia* (1851) by Mrs R Lee' and 'Rethinking the Pioneering Text'.

48 See Orr, '"Women Peers in the Scientific Realm: Sarah Bowdich (Lee)'s Expert Collaborations with Cuvier, 1825–1833". In "Women and Science", edited by Claire Jones and Susan Hawkins', *Notes and Records* 69, no. 1 (March 2015): 37–52. and http://www.ukunitarians.org.uk/colchester/colhistory.html.

49 See https://www.unitarian.org.uk/pages/faith.

50 The heroine of *Playing at Settlers* (1855) reprises much of Sarah's own nonconformist upbringing. See Orr, 'Rethinking the Pioneering Text'.

51 For Ronald Hyam, 1815 therefore marks the beginning of Britain's 'Imperial Century'. See Ronald Hyam, *Britain's Imperial Century, 1815–1914*. BT Batsford: University of Michigan Press, 1976. See, in similar vein, Simon Smith, *British Imperialism, 1750–1970*. Cambridge: Cambridge University Press, 1998.

52 Many supported the values of the French Revolution and of Reason and rejected Trinitarian, and hence Anglican, articles of faith (including loyalty to the Crown) in consequence.
53 The 'Unitarian Society for promoting Christian Knowledge' (1791) provided such networks. For their investigation, see Chapter 10 of Stuart Andrews's *Unitarian Radicalism; Political Rhetoric, 1770-1814*. Basingstoke: Palgrave Macmillan, 2003, 107–15.
54 See John Westby Gibson and Felix Driver. 2021. 'Bowdich, Thomas Edward (1791–1824)'. Oxford Dictionary of National Biography. Oxford: Oxford University Press.
55 *The London Gazette*, 27 July 1813, p. 1496. Thomas Luce's earlier unsuccessful business dealings, concerning mining interests in Bideford, may also have contributed. *The London Gazette*, 1812, Part 1, 349.
56 I draw on the title key word, 'courage', from Rose de Freycinet's diary-travelogue, the story of her life as an 'adventuress', disguised as her sea-captain husband's ship boy, to accompany him on his Australian voyages of discovery. See note 9.
57 See Sarah's later account, 'A Scene in Negroland', in Appendix 1.
58 See most recently the magisterial study of West Africa by Toby Green, *A Fistful of Shells: West Africa from the Rise of the Slave Trade to the Age of Revolution*. London: Allen Lane, 2019 referencing 'Thomas Edward', but nowhere 'Sarah' Bowdich.
59 See Adrian Desmond, 'The Making of Institutional Zoology in London, 1822–1836: Part 1', *History of Science* 23, no. 2 (June 1985): 153–85 and 'The Making of Institutional Zoology in London, 1822–1836: Part 2', *History of Science*, 23 no. 3 (Sept. 1985): 223–50.
60 Orr, 'Women Peers', 41. See also Appendix 1, Section A.
61 An Appendix to the *Excursions* demonstrates the profit gained from Sarah's learning Arabic in Paris.
62 See Orr, 'New Observations'.
63 See Orr, '*Adventures in Australia* (1851) by Mrs R. Lee'.
64 See Rebekah Higgitt and Charles W. J. Withers, 'Science and Sociability: Women as Audience at the British Association for the Advancement of Science, 1831–1901', *Isis* 99 (2008): 1–27.
65 See deB. Beaver, 'Writing Natural History for Survival', note 46.
66 One possible reason is that it remains catalogued under Cuvier (and Humboldt) at the BNF, yet it has also escaped the notice of specialist historians and biographers of both men.
67 See Orr, 'New Observations'.

Part One
CANVASSING CUVIER

Chapter One

A FIRST NATURAL HISTORY OF THE FISHES OF WEST AFRICA IN THE *EXCURSIONS DANS LES ISLES DE MADÈRE ET DE PORTO SANTO* (1826)

When the Bowdichs set out from Madeira for Sierra Leone via the Gambia in 1822, European scientific knowledge of its *flora* and *fauna* was sparse and hence highly sought after. Anna Maria Falconbridge's journal account of 1794, *Two Voyages to Sierra Leone during the Years 1791-2-3 [...]*,[1] offers rare insights into what travelogues term the 'habits and customs' of the colony and its hinterlands. Posterity places this widow of a surgeon to slaving ships firmly on the side of anti-abolition.[2] Her descriptions provide incidental glimpses, however, of generic species of *flora* and *fauna* that also testify to the necessarily piecemeal work of European science in Sierra Leone and the wider West African region:

> Our Botanist and Mineralist (*sic*) have, as yet, made little proficiency in those branches of natural philosophy; the confusion of the colony has retarded them as well as others; they are both Swedes, and considered very eminent in their professions. The Mineralist is about to make an excursion into the interior country, and is very sanguine in his expectations. He has but slightly explored the country hereabouts, and been as slightly rewarded the only fruits of his researches are a few pieces of iron oar (*sic*), richly impregnated with magnetism, with which the mountains abound.
>
> The Botanist, is preparing a garden for experiments, and promised himself much amusement and satisfaction, when he can strictly attend to his business. His garden is now very forward, but is attended with considerable expence (*sic*). Letter IX (Aug. 25 1792)

By contrast, the Bowdichs envisaged their independently-funded scientific exploration of Sierra Leone from Freetown as a primary endeavour, markedly different from economic, commercial, educational or religious civilising mission, and from colonial 'garden' and acclimatisation projects above. Their four-year training under Cuvier, Humboldt and others at the Paris Muséum therefore included their published translation of the most recent travel accounts to West Africa, such as Gaspard Mollien's *Travels in the Interior of Africa* (see Appendix 1A). It augmented their knowledge of the region accrued from their own earlier sojourn and account of the 'habits and customs' of Ghana and the Gabon in T. Edward Bowdich's *Mission to Ashantee* (1819), to confirm where they could fill the missing gaps in scientific knowledge ahead of departure for Sierra Leone. Two new fields were not among the usual coverage in scientific travels, or in museum instructions for scientific exploration: marine and freshwater invertebrates (particularly molluscs) and vertebrates. Under Lamarck and Cuvier, the Bowdichs were prepared directly to serve these 'departments' of the Muséum's collections. New Macaronesian and West African malacology specimens would immediately supplement them and Lamarck's definitive seven-volume *Histoire naturelle des animaux sans vertèbres, présentant les caractères généraux et particuliers de ces animaux* (1815–1822). Indeed T. Edward Bowdich's two-volume *Elements of Conchology* of 1821 (see Appendix 1A) condensed its most recent developments and contents. The many plates of conchology drawings for it all signed 'S. Bowdich lithog' were no mere copies.[3] These plates attest to Sarah's trained scientific eye concerning the anatomical and technical details that are vital for field identification and future classification purposes. For reasons of space, I draw malacology to the attention of other specialists as a further field of natural history to which Sarah contributed in comparative scientific contexts that are those of this book.[4] The largest gap of all in European knowledge of West African *fauna*, however, concerned the region's marine and freshwater vertebrates.

In 1822 Cuvier's intended global coverage for his *Histoire naturelle des poissons* (*Natural History of Fishes*) was only in early preparation stages. The Bowdichs therefore had no equivalent Muséum publication for Fishes to the work above by Lamarck for invertebrates, or by Cuvier for animals and birds to remake into a portable Muséum manual 'for students and travellers'. Any additions to knowledge of (West) African fish species that their Sierra Leone expedition could provide would immediately make Paris foremost among European scientific collections, as well as enhance knowledge of Fishes already known in the 'Atlantic'. Before the Bowdichs' departure from Madeira to Sierra Leone, Sarah had published the updated third edition of her anonymous *Taxidermy*,[5] specifically to include the expert preparation of Fishes for transportation to national museum collections:

> In long voyages we must furnish ourselves with small casks, holding from four to eight gallons, and bound with iron. [...] We fill about two-thirds of one of these little barrels with spirits. *We take notes of the fish to be preserved: where it was caught; whether male or female; if good or bad to eat; if salted in the country, &c.* This done, we wrap the fish in a piece of linen, and sew it; we then attach a little plate of wood, on which we have engraved *the number corresponding to our note* with a sharp knife; we then put the fish in the cask, which we close hermetically, that the spirit may not evaporate. [...] As we deposit a bed of fish, we put a bed of cotton or new flax, to prevent the rubbing and tossing about in the conveyance. (*Taxidermy,* 1823, 86–90, emphasis added)

Out of water, fish swiftly lose their colours and spoil after death, but especially so in hot climates. In Cape Verde and the Gambia, Sarah would have applied the above instructions to the letter and sketched her chosen specimen as its further identifier. The *Excursions* therefore preserve Sarah's account of new West African Fishes from her corresponding 'note' of each; her sketches furnished her signed and published 'Figures' in the *Excursions* (*EM* and *EMFr*). In 1825/1826, it underscored not only the exigencies above of scientific collection of Fishes overseas but also the practical necessities of Sarah's expert selection of the most important among them, as our opening portrait imparted. The fifth chapter of the *Excursions* therefore opens with Sarah's direct challenge to museum (and Paris Muséum) science, to qualify the severally expert material arts and sciences of gathering new fish knowledge in the field, including in her telling footnote:

> COULD I have afforded to have invited the fishermen and peasantry to bring me specimens of all the fishes, birds &c., they knew, or might meet with promising a fair price, *I might have done much more for zoology in general.* A traveller who has only his (*sic*) own slender means to depend on for every expense of his enterprise, can do little for zoology; but, even as it was, *I had frequent occasion to lament the necessity of throwing away new and interesting objects, especially fishes, because no museum had furnished me with spirits and cases to preserve them in.* It is not fair to impose this expense on the zeal of the traveller who contributes his (*sic*) services gratuitously. *I have a few more zoological notices to submit, however, and expect to add some new fishes to the 2500 already known and described.*
>
> *I shall endeavour to follow the ichthyological system of Cuvier,* the most natural, though the most difficult to class by[k]. [...] *I will first notice those which appear to be rare, or distinct from any already known, in most instances giving the native name.*
>
> [k] "La classe des poissons est de toutes, celle qui offre le plus de difficultés quand on veut la subdiviser en ordres, d'après des caractères fixes et sensibles. Après bein (*sic*) des efforts, je me suis déterminé pour la distribution suivante, qui dans quelques cas pêche (*sic*) contre la précision, mais qui a l'avantage de ne point couper les familles naturelles." CUVIER, *Règne Animal*, Tome II, p.110.
>
> (*EM*, 121–22, emphasis added)

English audiences in 1825 were assumed to have had a reading knowledge of French to understand note 'k'. Both Anglophone and Francophone readers of the *Excursions* then beheld 'those (Fishes) which appear to be rare, or distinct from any already known' to European knowledge as classified by Cuvier's 'system'. The sum of '2500' further reveals Sarah's insider knowledge of the species coverage for Cuvier's 22 volume *Histoire naturelle des poissons* endorsed in the Prospectus that he published for it in 1826. The serious blow in our introduction that was the loss of irreplaceable fish specimens in Sarah's barrels now finds sharper magnification. Her *Excursions* salvaged something from these losses on Edward's account: her 'Fig. 1', 'The Lepidopus, or Hake of the Tagus', recorded the transcultural fields in which the Bowdichs had set out to contribute and to make their name. Their various 'Portuguese' port stops *en route* to Sierra Leone – Lisbon, Madeira, Cape Verde, the Gambia – were integral to their larger comparative West African marine natural history endeavours. Sarah's barrels contained the substantiating evidence – and potentially the type specimens on which the description and name of a new species is based – of fish discoveries 'distinct from any already known' (to European science) that she alone had carefully collected, selected and preserved from aquatic field research in locations that are also marine interzones. Because Sarah was very unusually the expert 'I' preparing her specimen barrels, hers were also the means to recover their contents for science. Her descriptions and figures of West African Fishes in the published *Excursions* (*EM and EMFr*) offer the clearest testimony to hers as their first account for European science, including her record of their given 'native name' and Latin name. This chapter illuminates the fish descriptions in the French *Excursions* of 1826 for the first time precisely for their foremost international scientific significance. Although it was endorsed upon publication, Sarah's work on Fishes is almost completely overlooked as a foundational contribution to modern West African ichthyology in its transnational histories.

'I shall endeavour to follow the ichthyological system of Cuvier ...'

Sarah's 'I' here unusually promotes her primary agency in applying Cuvier's new 'system' of fish classification (in place in his 1817 *Règne animal*) in 1823–1824 as integral to her pioneering perspectives in the field in (tropical) West Africa. In also attesting to her training at the forefronts of international French ichthyology at the Paris Muséum, this 'I' clearly met the highest scientific classification standards (irrespective of gender) for her resulting advances to the world coverage of Cuvier's new multi-volume *Histoire naturelle des poissons*. Sarah, however, did more than collect and describe; she also named West African Fishes for European science in the 'zoological notices' of Chapter 5 of the *Excursions* and its 'zoological appendix'. These contributions have too rarely been recognised by scientists and historians of science, because ingrained bias sets and governs the international rules for species nomenclature as discussed below. The lights of the 1826 French *Excursions* will more prominently confirm in this chapter Sarah's rightfully first place in the history of fish studies in West Africa, which she also first systematised for Cuvier's *Histoire naturelle des poissons* (1828–1848). It underpins today's international scientific databases for fish studies, such as the international English FishBase.org and the French-language CLOFFA.[6]

If we compare the inside title page of the English *Excursions* of 1825 (see Figure 0.1 in the Introduction) with that of the first for the two-volume French *Excursions* (see Figure 1.1), the differences in the latter's small print provide distinguishing evidence of Sarah's importance in international French Muséum natural history. The additional information also renders erroneous the attributions of authorship in French library cataloguing to 'feu T. E. (Bowdich)' bracketed in the indicative Lyon Library copy, rather than to 'Mme Bowdich' (his widow) in the same large font and typesetting.[7] As recounted in this book's acknowledgements, I first called up the two-volume 1826 *Excursions* intrigued as to why the BNF catalogue had classified it under 'Cuvier' (as its 'author'), when he never went to Madeira. Little did I expect to find his name in this smallest typeface below a '*Mme* Bowdich', or to see Humboldt's in apposition as the authors of 'Notes' for what is clearly a woman's scientific work when they were the foremost experts in their scientific fields. Of additional scientific import is the imprint, 'F. G. Levrault', in similar small type below 'Cuvier'. I knew this official publisher for the Paris Muséum also printed Cuvier's multi-volume works. From its inside title page information alone, the 1826 *Excursions* had to be significant in French book history both for its new scientific contents and as the work by a clearly remarkable woman author for the time. As I have demonstrated for the 'Notes' by Humboldt as in fact an extensive epilogue commentary on the comparative geology of the French *Excursions*,[8] Cuvier's are

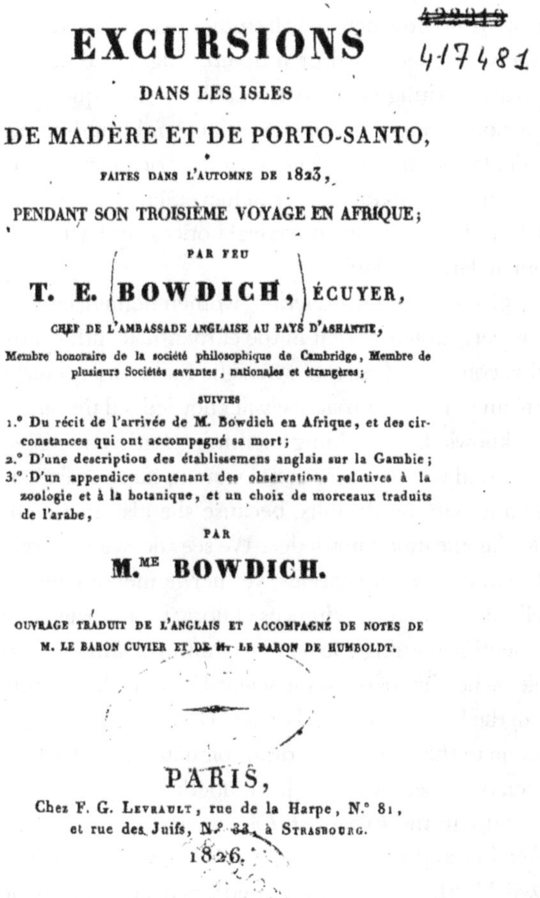

EXCURSIONS

DANS LES ISLES

DE MADÈRE ET DE PORTO-SANTO,

FAITES DANS L'AUTOMNE DE 1823,

PENDANT SON TROISIÈME VOYAGE EN AFRIQUE;

PAR FEU

T. E. BOWDICH, ÉCUYER,

CHEF DE L'AMBASSADE ANGLAISE AU PAYS D'ASHANTIE,

Membre honoraire de la société philosophique de Cambridge, Membre de plusieurs Sociétés savantes, nationales et étrangères;

SUIVIES

1.° Du récit de l'arrivée de M. Bowdich en Afrique, et des circonstances qui ont accompagné sa mort;
2.° D'une description des établissemens anglais sur la Gambie;
3.° D'un appendice contenant des observations relatives à la zoologie et à la botanique, et un choix de morceaux traduits de l'arabe,

PAR

M.ME BOWDICH.

OUVRAGE TRADUIT DE L'ANGLAIS ET ACCOMPAGNÉ DE NOTES DE M. LE BARON CUVIER ET DE M. LE BARON DE HUMBOLDT.

PARIS,

Chez F. G. Levrault, rue de la Harpe, N.° 81, et rue des Juifs, N.° 33, à Strasbourg.

1826.

Figure 1.1 Inside title page of the French *Excursions.*

also no footnotes. All clearly signed 'Cuv', they are integral to volume one (*EMFr*) in key insertions in the Madeira report – mainly Chapter 5 – and in the final zoological appendix. All concern Fishes. The latest scientific validations by both Barons therefore make the French *Excursions* quantifiably larger and qualitatively more momentous. If their scientific endorsements are unavailable to readers only of the 1825 English *Excursions*, and to Anglophone critical legacies including modern ichthyology in international English in consequence, the greater is the resounding silence concerning the 1826 version for its untold legacies in Francophone history of nineteenth-century natural science. We investigate three in this chapter's study of Cuvier's 'Notes' that illuminate the pioneering contributions of 'Mme Bowdich' to fish studies. An alert is therefore in order in what follows as potentially dry, technical or uninteresting, since descriptions of Fishes are rare in scientific travel writing of the period and its subsequent study. Fish nomenclatures are also notoriously alien, multiple, slippery and mind-numbing. To understand Sarah's expert articulations and naming of West African species new to European science is, however, to marvel doubly at her achievements. They accord both with Paris Muséum standards in the 1820s for 'zoological notices' and with the International Code of Zoological Nomenclature today.

For the reasons given in the introduction, women nomenclaturists and their naming legacies for science were largely absent in the early nineteenth-century period except on two very unusual accounts. A first was when the male 'discoverer' – the patronymic is the default of Linnean naming – proactively acknowledged the female provenance of the new finding and its knowledge by adding a feminine Latin ending ('-ae') to the identificatory surname.[9] A second was an exceptional woman naming for science hidden behind the (male) patronymic naming defaults, because she also had command of Latin and the latest scientific classification knowledge. We see the evidence on the inside title page above that Sarah had both qualifications, yet her name and authority as 'S. Bowdich' on her plates befell the automatic prejudices of library cataloguers: 'Bowdich' could only be 'T. Edward'. Scientific naming conventions, however, already accommodate the rare instances of double ownership of the same scientific name, for example Johann Reinhold Forster (1729–1798) the father of Georg Forster (1754–1794), through addition of the first letter of the forename to the surname to disambiguate them. Such precedent for Sarah's case has in fact been practised by David J. Mabberley in the one study to date of Sarah's contributions to botany in the *Excursions* (*EM*).[10] Mabberley distinguished the volume's new species by location and hence name: those for Madeira attributed to Edward he qualified as 'Bowdich'; 'the seven validly published new names' for the Cape Verde Islands and the Gambia ascribed to Sarah he qualified as 'S. Bowdich', to acknowledge and distinguish her equally primary contributions to knowledge.

Encapsulated here are the problems and lessons of recording and acknowledging the exceptional women who named new species in the nineteenth century, in order that hers as the scientific name is ensured in subsequent scientific legacies. Current disciplinary parameters and training then pose additional problems for (scientific) naming rigour. Very few specialists in a particular branch of science, with ichthyology as indicative, are also experts in the history of its development as largely the domain of cultural historians and historians of science, themselves rarely science-trained.[11] Comparative knowledge of the given 'science' and its cultures (plural) of production is rarer still, particularly for Fishes: they transcend national-territorial and national-linguistic demarcation. Even more problematic are the rules to promote stability of species naming that govern the International Code of Zoological Nomenclature, with the multiplicity of fish names being particularly problematic. It therefore (over)promotes (re)naming after 1899 in modern fish taxonomies, and the 50-year rule thereafter, because recent naming is assumed as the more accurate. A 'senior synonym' (older historical name) is therefore doubly liable to becoming invalid. When not used after 1899 it is determined a *nomina oblita*, because it is already replaceable or replaced by its 'junior synonym'. This latter is then the more likely to have been widely validated in the last 50 years. 'Widely' is quantified in the International Code by usage in at least twenty-five (expert) publications by more than ten different authors in a span of not less than ten years (and when few scientific papers reference those older than twenty years in print). Moreover in the history of modern systematic ichthyology, expertise has increasingly focused over time on a particular fish species, rather than on its instance within larger fish ecologies. Implementations of the International Code are therefore particularly detrimental to the very few women contributor-nomenclaturists in ichthyology period, but especially for this sub-discipline as a late entrant in zoology before 1899. The first special issue on women in ichthyology in 1994 is therefore indicative of the scientific naming biases above of the International Code. In its overdue acknowledgement of major twentieth-century women leaders in ichthyology in 'ET, Ro and Genie' (Ethelwynn Trewavas, Rosemary Lowe-McConnell and Eugenie Clark),[12] the specialist ichthyologist contributors were unaware of pre-1899 'foremothers', including in other international (non-Anglophone) histories and cultures of ichthyology. For scholars of Victorian Studies and Travel Writing, however, Mary Kingsley (1862–1900) is the heralded first woman scientific traveller and collector of Fishes in West Africa.[13]

The 1899 watershed and 50-year rules of precedence and precedent in the International Code therefore ignore the all-important historical conditions that determine (inter)national scientific naming and discipline formation. These conditions are crystallised in Sarah's unprecedented 'I' above: 'I have a few more zoological notices to submit, however, and expect to add some new fishes to the 2500 already known and described'. The history of systematic ichthyology in English as the current international vernacular of science is itself a form of 'junior synonym' classifying: it replaces French as the 'senior' international vernacular of science throughout the nineteenth century. As we see below, where fish specialists have acknowledged the *Excursions* (*EM*) for its discovery and Latin naming of fish species qualified by 'Bowdich, 1825', they never register the 1826 translation as its larger confirmatory supplement that dates Sarah's foundational West African contributions signed '(Bowdich, 1825)' *stricto sensu* within the founding legacies of modern French-international ichthyology itself no less. For example, Paolo Parenti's important renewed interest in the (1825 English) *Excursions* in 2019 proposes for Fishes qualified by 'Bowdich, 1825' three instances of *nomina oblita* (forgotten name) and four instances where the younger synonyms are their *nomina protecta* (name granted protection) by operation of both the 1899 date and 50-year rules.[14] In effect, these designations ultimately dismiss earlier, and more importantly parallel, ichthyology traditions that the 1826 *Excursions* discloses. By tracking instead Sarah's astonishing contributions to knowledge of West African Fishes within (international) French and English scientific heritages for modern ichthyology, this chapter can also make the strongest case by the already biased 1899 and 50-year rules of the International Code for '(Bowdich, 1825)' more properly to be revised to '(S. Bowdich, 1825)' in at least three instances. Moreover, major fish databases such as FishBase and CLOFFA would also be enriched by Sarah's rare first recording of their indigenous names. One example for reasons of space suffices to demonstrate Sarah's 'notice (of) those which appear to be rare, or distinct from any already known, in most instances giving the native name' (*EM*, 122):

> Figure 37 is also a new genus, for which I have preserved the native name Seleima, formed of the Portuguese pronoun *se*, and a corruption of the noun *leme*, a helm, and to which I have added aurata, as a specific appellation, from the golden hue given by the 8 orange stripes. It belongs to the second tribe, of the fourth family of Cuvier's division, Acanthopterygiens. (*EM*, 238)

Fish classifications and nomenclatures indeed appear mind-numbing unless understood as the major conduits for important and fascinating (re)discovery of first-namer provenance. That Sarah's collection and scientific naming here was acknowledged by expert contemporaries importantly establishes the legacies of key successors, by means of whom her enduring place can be reconstructed through key evidence. It is set out in two two tables below, each according with 50-year rule datelines, to demonstrate the continuing range and reach of Sarah's work in French and English. In Table 1.1 are the species (plural) today that Sarah first named (in bold) for science in her double-language *Excursions*. Table 1.2 then highlights (in bold) the West African range of her original naming as still viable. The names of key ichthyologists in the discussion are the recognisable authors in the tables.

The Natural Histories of Fishes: 'French'

The multifarious problems of (in)accurate historical fish nomenclature were precisely the challenge that Cuvier tackled in the early 1820s, and sought to resolve in his definitive *Histoire naturelle des poissons* (1828–1848). It undertook to spring-clean the erroneous, duplicate or multiple identifications from earlier records (in many languages), to adjudicate on the precedence of names, and to safeguard unconfirmed species for future verification. Cuvier's Muséum methods for his new history as well as science of modern ichthyology constituted volume 1 of his *Histoire naturelle des poissons* (1828). It accounted for the many precedents in comparative knowledge of longer intercultural heritages (in Europe, the Middle and Far East, the Americas), in order to establish his own. Indeed, his personal library prepared the Bowdichs' work on 'new' West African Fishes.[15] If Cuverian accuracy and intercultural legitimacy underpin the higher methodological stakes for their findings, Sarah further availed herself of the same resources in Paris and its Muséum, to validate her new West African fish descriptions ahead of publication of the *Excursions* (*EM*) as Cuvier was composing his 'Notes' for its French translation (*EMFr*). The loss of her barrels of preserved fish specimens had denied Sarah her potential *holotype*, or type specimen, to verify her 'zoological record'. Nonetheless, her description, drawing and provenance information established its *neotype*, that is 'a specimen selected to replace a type specimen that has been lost or destroyed' (Collins). We return below to the crucial place of her neotype(s).

The publication of the French *Excursions* with Levrault not only defined its importance, but also secured its wider scientific dissemination through Cuvier's auspices at the Paris Muséum. Its *Bulletin des Sciences Naturelles et de Géologie* reviewed all new works and discoveries of specialist and general interest to science. In 1824 (volume 3) de Férussac, the Muséum's specialist alongside Lamarck on molluscs, reported Sowerby's

Table 1.1 The Species Today first qualified by 'Bowdich, 1825'

Reference	Species
Blanc M. and M. L. Bauchot, 'Révision des Thalassoma (Poissons Téléostéens Labridae) de l'Est Atlantique', *Bulletin du Muséum National d'Histoire Naturelle* 32: 1 (1960), 88–96.	***Julis Squami marginatus***
Collingnon, Jean, 'La systématique des Sciaenidés de l'Atlantique oriental', *Bull Inst. Océanogr.* 1155 (16 October 1959), 1–11.	***Otilithus dux***
Edwards, Alasdair, 'A new Damselfish, *Chromis Lubbocki* (Teleostei: Pomacentridae) from the Cape Verde Archipelago with Notes on Other Eastern Atlantic Pomacentrids', *Zoologische Mededelingen* 60: 12 (25 July 1986), 181–207.	***Chromis Triacantha***
Fowler, Henry Weed, 'New, Rare or Little-Known Scombroids', *Proceedings of the Academy of Natural Sciences of Philadelphia* 57 (1905), 56–88.	***Seriola picturata***
MacGregor, John S., 'Synopsis on the Biology of the Jack Mackerel (Trachurus Symmetricus)', *Archiv für Fischereiwissenschaft*, Special Scientific Report No 526 (April 1966), 1–16.	***Trachurus picturatus***
Merella, Paolo, Francesc Alemany and Amàlia Grau, 'New Data on the Occurrence of *Pontinus kuhlii* (Bowdich, 1825) (Osteichthyes: Scorpaenidae) in the Western Mediterranean', *Scientia Marina* 62: 1–2 (1997), 177–79.	***Pontinus kuhlii***
Parenti, Paulo, 'On the status of some nominal species of fishes described by Sarah Lee Bowdich (*sic*) in the account "Excursions in Madeira and Porto Santo during the autumn of 1823"', *Boletim Museu do História Natura do Funchal* lxix (2019): 5–12 on seven nominal species (declaring three *nomina oblita*, and younger synonyms for the four others as *nomina protecta*).	
Reintjes, John W., 'Annotated Bibliography on Biology of Menhadens and Menhadenlike Fishes of the World', *Fishery Bulletin* 63: 3 (1964), 531–49.	***Clupea fimbriata***
Trewavas E. and S. E. Ingham, 'A key to the species of Mugildae (Pisces) in the Northeastern Atlantic and Mediterranean, with Explanatory Notes', *Journal of Zoology* 167: 1 (May 1972), 15–29.	***Chelon bispinosus***
Whitehead, P. J. P., 'The West African Shad, *Ethmalosa fimbriata* (Bowdich, 1825): Synonymy, Neotype', *Journal of Natural History* 4 (1967), 585–93.	***Clupea fimbriata***
Wirtz, Peter, Ronald Fricke and Manuel José Biscoito, 'The Coastal Fishes of Madeira Island—new Records and an Annotated Checklist', *Zootaxa* (2008), 1–26.	***Trachurus picturatus***

receipt of T. Edward Bowdich's work on various shell discoveries from Porto Santo.[16] This *Bulletin* also reported the imminent publication of the longer work containing it, 'par les soins de l'intéressante veuve de ce célèbre voyageur' (under the auspices of the interesting widow of this famed traveller, 93). In volume 5 for early 1825, and specifically under the rubric of 'la Botanique' (Botany), a long review of the *Excursions* by Antoine Guillemin (1796–1842) clearly attributed the plant lists for 'Bona Vista, St-Jago et Banjole' to 'madame Bowdich', long before Mabberley unknowingly did the same in 1978.[17] Volume 6 in the second half of 1825 contained a further description of the *Excursions* by one of Cuvier's disciples, Anselme Gaëtan Desmarest (1784–1838), in this case the contributions to zoological research 'dues au zèle de ces intéressans voyageurs (thanks to the zeal of these interesting travellers)'. Like Guillemin, Desmarest clearly demarcated the attribution of the discoveries. Those for Madeira, especially fish species described in Chapter 5 of the *Excursions*, are the work of 'Monsieur', whereas those in the zoological appendix pertaining to Cape Verde and the Gambia are by 'Madame':

Les poissons nouveaux, ou regardés comme tels, par M. Bowdich, sont les suivans: *Physis furcatis, Labeo sparoides, Smaris Royerii, Seriola picturata, Scorpoena Ruhlii, Serranus rufus, Choettodon Leachii, Zeus Childrenii* ; les caractères de chacun d'eux sont suffisamment développés pour qu'on puisse les reconnaître, et les quatre premiers sont figurés par madame Bowdich. [...]

Un appendice a pour objet de rassembler toutes les notions recueillies sur les productions naturelles de cette partie de l'Afrique [la Gambie], et les îles de San-Jago et Bona-Vista. Madame Bowdich y place des descriptions assez détaillées des espèces, qu'elle et M. Bowdich y ont recueillies ou observées, et parmi lesquelles sont surtout des poissons dont elle a donné des figures lithographiées. Malheureusement l'exécution de ces figures ne nous paraît pas suffisamment approcher la perfection nécessaire pour bien représenter des animaux, dont les caractères spécifiques deviennent chaque jour d'autant plus difficile à fixer qu'on en découvre un plus grand nombre.

Ces poissons nouveaux sont ceux que madame Bowdich a nommées: AMORPHOCEPHALUS granulatus, SELEIMA aurata, MUGIL bispinosis, BODIANUS maculatus, PRISTOPOMA humilis, DAISTON speciosus, DENTEX unisponosis, SCIEOENA elongata, CLUPEA fimbriata, BALISTES dradiata, DENTEX diplodon, LABRUS jagonensis, TETRAODON loevissimus, LICHIA tetracantham, PIMELODUS gambensis, ANOMALODON incisus, CHROMIS triachantha, JULIS squamimarginatus, SCIOENA Dux.

(The new fish, or those regarded as such, by Mr Bowdich, are the following: *Physis furcatis, Labeo sparoides, Smaris Royerii, Seriola picturata, Scorpoena Ruhlii, Serranus rufus, Choettodon Leachii, Zeus Childrenii*; the characteristics of each are well-enough developed to identify them, and the first four are drawn by Mrs Bowdich. [...] An appendix gathers together the collections and natural productions of this part of Africa [The Gambia], the islands of St. Iago and Bona-Vista. Mrs Bowdich includes in it quite detailed descriptions of the species that she and Mr Bowdich collected or observed, and among which are above all the fishes for which she has provided lithograph illustrations. Unfortunately, the execution of these figures seems to us not to approach the necessary perfection for true representation of animals, whose specific characteristics become all the more difficult with each day to fix, until found in greater number. These new fishes are those that Mrs Bowdich has named: AMORPHOCEPHALUS granulatus, SELEIMA aurata, MUGIL bispinosis, BODIANUS maculatus, PRISTOPOMA humilis, DAISTON speciosus, DENTEX unisponosis, SCIEOENA elongata, CLUPEA fimbriata, BALISTES dradiata, DENTEX diplodon, LABRUS jagonensis, TETRAODON loevissimus, LICHIA tetracantham, PIMELODUS gambensis, ANOMALODON incisus, CHROMIS triachantha, JULIS squami-marginatus, SCIOENA Dux.)[18]

Cuvier went further in his 'Notes': his *imprimatur* endorsed the (new ichthyological) science of many of her contributions.[19] But his role as expert respondent and first commentator was primarily to verify and validate the accuracy and identification of her named species. For the Fishes of Madeira and Porto Santo described in Chapter 5 of the *Excursions*, the fuller scientific description for each – fins, ray counts, particular features, colours – was already consigned by Sarah to a numbered footnote on the page. Cuvier's supplementary 'Notes' only corrected technicalities within the main textual description – 'ventral' for the erroneous 'pectoral' fins for '*physcis furcatus*, fig. 28' and '*Labeo sparoïdes*', fig. 29 (*EMFr*, 190–91) – and nomenclature. For example, 'Bowdich' uses *Labeo* for cyprinid fishes. For the *chixarra* or *Seriola picturata* (*EMFr*, 192, to which we return), *plectropomes* (*EMFr* 193) and *Scorpoena kuhlii* (*EMFr*, 193), Cuvier added further specialist remarks concerning comparative anatomical features that challenged the real differences of the last-named from another *Scorpoena*. If Cuvier's 'Notes' for *Labeo* and *Seriola picturata* directly attributed incorrect observations to Edward ('*M.* Bowdich'), no intervention marked 'CUV' made a definitive judgement on the finding as new for Western ichthyology.

In the zoological appendix for the Gambia, however, the many more 'Notes' signed 'CUV' appeared as a separate paragraph, added in smaller font below the 'zoological notice' in question. As also the authority on Birds before undertaking his *Histoire naturelle des poissons,* Cuvier's many 'Notes' both corrected the ornithological information (in similar vein to the fish in *EM* Chapter 5), and more clearly stipulated where a bird species, for example for the '*Coccothraustes,* Cuv.' (*EMFr*, 352–53), appeared new. Indeed, the significance of nineteen birds among others was further underscored in the Muséum's *Bulletin des Sciences Naturelles et de Géologie* for 1827 by 'S. G. L.' (S. G. Luroth). Strikingly, he attributed the final zoology appendix squarely to 'Madame Bowdich', including the identification and naming of the birds as well as the fish within it. The principal object of Luroth's *Bulletin* review was thus to tabulate the discoveries of the *Excursions* for international science in two columns. The left constituted the names given by 'Mme Bowdich'; the right their respective adjudications by Cuvier's 'Notes' in the *Excursions. Bulletin* readers could be in no doubt. Three bird species were new (*Astur, Lanius Nº1, Muscicapa Nº2*), with a question mark also against a fourth, the *Coccothraustes* (as an 'Espèce nouvelle de *Carduelis*?'). Ornithology specialists will also benefit from undertaking a much fuller investigation of the *Excursions* (*EM*, *EMFr*) to verify Sarah's new contributions to this domain of West African natural history.

Unlike Cuvier's 'Notes' for the Fishes of Madeira in Chapter 5, however, those for the zoological appendix of the *Excursions* (*EMFr*, 361–71) more overtly flagged species that appeared new. Although he attributed two fishes to 'M. Bowdich' (*Balistes radiata* and *Pimelodus gambiensis*, *EMFr,* 362–3 and 363–4 respectively) to correct 'him' on certain points, Cuvier added comments only to a further *seven* of the nineteen fishes in total that have descriptions with figures in this appendix (all ascribed to *Madame* Bowdich). Luroth's *Bulletin* report of 1827 once more facilitated understanding of their significance, because he listed the fish under the birds in his two respective columns described above, and reiterated the endorsements contained in Cuvier's 'Notes' for *nine* Fishes. Of these, *Balistes radiata* is an 'Espèce bien distincte' (a clearly distinct species); *Pimelodus gambiensis* an 'Espèce nouvelle' (a new species) and *Dentex unispinosus* questionably a 'Type d'un nouveau sous-genre?' (a type of new sub-species?) (127). By the most conservative count based on this evidence, Sarah's contributions by specialist report on the final zoological appendix of the *Excursions* (*EM*; *EMFr*) added at least three birds, and at least three Fishes to West African species known to European science.

Yet what of the ten Fishes in Sarah's appendix that failed to be noteworthy in Cuvier's expert reconsideration as the world authority in ichthyology in the period? By his lights, it would be too easy to consign these to oblivion (*nomina oblata*), together with their first (female, non-French) recorder-nomenclaturist. To take them seriously instead is to reconsider their differently expert (first-ranking) science in the West African field for the period, and hence their challenge to the perennial equation of scientific authority with 'experts' in national institutes and museum collections, as Luroth's list with double-column headers and evaluation epitomised. The scientific 'weight' for the Fishes in the *Excursions* was tipped heavily right to Cuvier's adjudications. Even if we read these important *Bulletin* articles of 1824–1827 as a surprising acknowledgement for the time that French (field and museum) science could be conducted by 'Mme' as well as by 'M.', the worthiest pupil in the field would never equal her mentor-adjudicator, or his inestimable banks of Muséum knowledge. Focus instead on the ten Fishes that Cuvier disregarded, yet 'Madame Bowdich' picked out as 'new' or 'rare' in her words above (among others she could have collected for want of barrels), highlights the 'until found in greater number' in the *Bulletin* report translated above. Cuvier had simply not seen enough West African specimens to gauge the fuller importance of Sarah's contributions and naming of Fishes from this region. Indeed, if three out of the nine re-collated by Luroth were new, so could be a further three of the ten Cuvier was silent about. Even one among these would prove the excellence of Sarah's training in his 'system', her expertise in applying it in new fields and hence contribution of greater potential importance, because it was independently identified by an expert authority in her own right and with none of Cuvier's many Muséum advantages.

Among the Fishes without a 'Note' by Cuvier is 'Clupea fimbriata, new species, Bowd.' ('fig. 44' in the *Excursions* in Figure 1.2). Sarah's description and drawing for this 'new species' in her report clearly designate its provenance and explain its nomenclature:

> I have given it this specific name, because every scale is fringed, which makes the fish have a very peculiar appearance. The dorsal fin has 16 rays, the ventral 19, and the pectoral 5; the back is of a brilliant azure, the sides are of a pale yellow, and the belly is silvery; the caudal and anal fins are of a deep yellow. Found at Porta Praya' (*EM*, 234).

Figure 1.2 '*Clupea fimbriata*, Bowdich pictured in Figure 44'. Courtesy of the University of St Andrews Libraries and Museums, *Excursions in Madeira and Porto Santo*, rP702.M16B7.

Its importance in subsequent indicative *French* history of ichthyology challenges the 1899 cut-off and 50-year rule of the International Code, because this fish is recognised through Sarah's naming of it as 'fringed' (*fimbriata*) in the scientific taxonomies by which it has been widely known with and without '(Bowdich, 1825)' across the West African region, including (River) Gambian, and before Whitehead's key study in 1967 to which we return.[20] As among the most common fish in West African waters (see Table 1.2), it is known as the Bonga (or Bonga Shad): 'The predominant fishing method is gill net encirclement of Bonga (*Ethmalosa fimbriata, Clupeidae*). Bonga makes up, on average 73% of total artisanal marine landings [...] and is the cheapest and most affordable fish for the Gambian.'[21] Because it also ranges widely from the marine to freshwater reaches of the extensive River Gambia ria estuary, it is a frequent subject of contemporary scientific studies monitoring the levels of salinity in this important catchment, further impacted by hydroelectric and other schemes in neighbouring Senegal.[22] I could have added the references cited in notes 21 and 22 to Table 1.2, but its larger point is to contextualise the crucial 'S. Bowdich' in its concluding record of 2005 by Durand et al. Here is the clear evidence in their naming of '*S[arah]* Bowdich' as of profound two-hundred-year importance for West African ichthyology, and for its (history of) knowledge. Her larger significance behind it is then also in the lineage from 'Cuvier' to 'Cadenat' in Table 1.2, which I now unlock.[23] Let us recall the absence of Cuvier's 'Note' for it in 1826. Yet it is included in the *Histoire naturelle des poissons* in the penultimate volume of 1847 under the name 'Ethmalosa Dorsalis'. This overwriting of 'Bowdich' and of 'fimbriata', however, could not eradicate specialist French knowledge of first provenance in the *Excursions* (*EM*) for this fish, and of 'S.' to qualify 'Bowdich'. For example, Gérald Belloc's report of the important 1937 French scientific expedition of the 'Président Théodore-Tissier' to survey the fish populations of West Africa, and the resulting (anonymous) report that

Table 1.2 The *West African* '*Clupea fimbriata* Bowdich', 1825 (Porta Praya), 1950–2005

Cadenat, Jean, *Poissons de mer du* ***Sénégal*** (Dakar: IFAN, 1950).

Mainguy P. and M. Doutre, 'Variations annuelles de la teneur en matières grasses de trois clupéides du **Sénégal** (*Ethmalosa Fimbriata* Bowdich, *Sardinella Eba* C.V., *Sardinella Aurita* C.V.', *rev. Trav. Inst. Pêches Marit.* 22 no. 3 (1958): 303–21.

Guyonnet, Benjamin, Catherine Aliaume, Jean-Jacques Albaret et al, 'Biology of *Ethmalosa fimbriata* (Bowdich) and fish diversity in the Ebrie Lagoon (**Ivory Coast**), a multi-polluted environment', *ICES Journal of Marine Science* 60 (2003): 259–67.

Ama-Abasi, Daniel, Sieghard Holzloehner, Udeme Enin, 'The Dynamics of the Exploited Population of *Ethmalosa fimbriata* (Bowdich, 1825, Clupeidae) in the **Cross River Estuary** and Adjacent **Gulf of Guinea'**, *Fisheries Research* 68 (2004): 225–35.

Fafioye O. O. and Oluajo, O. A., 'Length-Weight Relationships of Five Fish Species in Epe Lagoon, **Nigeria** ['*Ethamalosa fimbriata* Bowdich' is the fifth species], *African Journal of Biotechnology* 4, no. 7 (2005): 749–51.

Durand J.-D., M. Tine, J. Panfili, I. T. Thiaw, R. Laë, 'Impact of Glaciations and Geographic Distance on the Genetic Structure of a **Tropical Estuarine** Fish, *Ethmalosa fimbriata* (*Clupeidae*, S. Bowdich, 1825)', *Molecular Phylogenetics and Evolution* 36 (2005): 277–87.

included seven from the *Excursions* (*EM*).[24] These clearly register the unbroken French-international importance of Sarah's work, but cannot explain it.

Cuvier had several important disciples in ichthyology, whose different French legacies directly inform current scientific research on world fish, including electronic database resources such as FishBase and CLOFFA. The first is national-institutional in Achilles Valenciennes (1794–1865), Cuvier's direct successor at the Paris Muséum. As Cuvier's former principal *aide naturaliste*, then co-author of the *Histoire naturelle des poissons*, Valenciennes undertook its completion after 1832 upon Cuvier's death, but not the full coverage of its scope. Its 22 volumes exclude several large fish groups including 'gadiformes, apodes, pleuronectes, plectognathes, ganoïdes, sélaciens' as Théodore Monod noted.[25] The important survey by Marie-Louise Bauchot, Jacques Daget and Roland Bauchot pointed out that cartilaginous fishes and cyclostomes (lampreys) are also absent.[26] As David Starr Jordan reported in 1902, successors to Valenciennes, such as Auguste Duméril (1812–1870), completed the work on sharks and ganoids.[27]

The Swiss-born Louis Agassiz (1807–1873) epitomises the second disciple heritage, the Paris 'Muséum' translocated. After his Paris training, he made major contributions in Neuchâtel, *inter alia* on fossil fish and the life cycle of salmonid Fishes, before he emigrated in 1847 to the United States, to re-create Paris practices as director of the Lawrence Scientific School at Harvard. Among his disciples in ichthyology and systematics was David Starr Jordan (1851–1931), author of the 1902 history of US ichthyology and of *The Genera of Fishes: A Classification of Fishes* in 1917.[28] In the latter, Jordan reviewed the new fish discoveries of 'T. Edward Bowdich' in the 1825 *Excursions* and, like Cuvier, 'missed' '*Clupea fimbriata*' (118–19). Henry Weed Fowler (1878–1965), one of Jordan's students, then published 'The Fishes of the United States Eclipse Expedition to West Africa' in 1919.[29] Fowler referenced the *Excursions* (*EM*) for '*Trachurus Picturatus* (Bowdich)' in its fifth chapter and '*Anomalodon* Bowdich' from the 'zoological appendix' (*EM*), but assumed both to be the work of Edward (198–99 and 211, respectively).

Henri Milne Edwards (1800–1885) exemplifies the third and overtly 'field' heritage: his work on marine invertebrates at the Muséum was furthered by examining them in their natural habitats. His new priority of field study importantly led to the establishment by Victor Coste (1807–1873) in 1850 of France's first marine station at Concarneau,[30] followed by Roscoff and Banjuls-sur-mer among many others also in France's colonies. For Harry W. Paul, the creation of marine biology laboratories in France 'may have been the most significant development in nineteenth-century French biology because of their importance as centers of research and research for a whole generation of scientists'.[31] Ichthyologist Jean Abel Gruvel (1870–1941) was among those in the period of France's major 'scramble' for Africa from 1880 (to rival British presence). His fourth mission to examine fisheries in West Africa in 1908 is reported by Maurice Zimmermann.[32] Gruvel's interests at the Paris Muséum's laboratory dedicated to 'Pêches Outre-mer' (fish from overseas) expanded further in 1922 through his new assistant, Théodore Monod (1902–2000). The latter's undertaking of a mission to study Mauritanian fish and crustaceans turned into a lifework devoted among others to Fishes of tropical West Africa, including as the first director in 1938 of IFAN (Institut Français d'Afrique Noire, now Institut Fondamental d'Afrique Noire) in Dakar. It became the 'mother' institute for others in the region.[33] Among Monod's legacies are his work on the systematics of *Clupeidae* in 1961[34] and, more importantly, through his two disciple-assistants, Jean Cadenat (1908–1992) and Jacques Daget (1919–2009). Monod sent the first to Goree, to set up a marine laboratory there. Daget established the Mali IFAN and network of IFAN institutes, later to succeed Monod as director both of the Dakar IFAN and then Paris Muséum. If Africa has a 'natural wealth of fish biodiversity', it is not then the case in light of these three major French expert ichthyology traditions that it 'has a dire poverty of functional systematic scientific institutions and resources'.[35]

This brief history of scientific legacy[36] for French-international ichthyology reinstates Sarah's name directly and indirectly as Cuvier's only female disciple in the later work of key ichthyologists in all three 'French' heritage traditions. To be alert to the particular importance of correct citation of the *first-naming* of species new to Western knowledge is to rediscover the creator of original taxonomy, 'S. Bowdich', and her fish name created in the field in 1825. Tracing key heritages also newly informs expert dissemination of ichthyology as a scientific discipline institutionally, textually and interpersonally via key mentors and peers.[37] All play pivotal roles in understanding the further transmission (gains and losses) of scientific names, and these in the intercultural heritages of *fauna* (and *flora*) species. This contextual plurality was Cuvier's main historical point in volume 1 of his magisterial *Histoire naturelle des poissons* to build his discipline-redefining work in modern ichthyology. If its species richly inform electronic databases subsequently, the first of its 22 volumes became available in English translation only in 1999.[38] Sarah's note 'k' differently assumed French as the international vernacular of science to be widely understandable to her English readers in 1825. It is notable, therefore, that FishBase relies mainly on Anglophone provenance metadata, especially the work of Whitehead, for '*Ethmalosa fimbriata* (Bowdich, 1825)'/the 'Bonga Shad'. The identificatory photo by 'Durand, J.-D.' for this fish in FishBase is silent on his attribution in 2005 of it to 'S. Bowdich' (Table 1.2).[39] By contrast, the information for the same fish in CLOFFA (see Appendix 2) more closely follows Cuvier's *historical* intentions by enriching and verifying the confluence of experts responsible for upholding (or replacing) its original naming. Our tables now make some names among them identifiable and personable for their hands in retransmitting Sarah's naming of *Ethmalosa fimbriata*, with or without '(Bowdich)' or even '(S. Bowdich)'. But this provenance data cannot explain, or re-create, the complex transmission of expert knowledge of, and interest arising from, this fish – 'fide Whitehead, 1967' highlighted in Appendix 2 – in the extraordinary woman who first defined and named it.

In the twenty-first century, research engines and searchable databases facilitate in seconds the painstaking research that earlier taxonomists such as David Starr Jordon undertook in specialist archives and collections. Choice of search word(s) in strings of potentially synonymous or competing search terms becomes more important. '(Bowdich) and fish' will not take the researcher to the backstory above for *Ethmalosa fimbriata*, or another fish in the *Excursions*, and not to a study (such as Luroth's) examining all the West African Fishes in this work, or in one or other of its 'waters' (Madeira and Porto Santo; Cape Verde and the Gambia). A further difficulty arises when a subsequent correction/junior synonym replaces the name that Sarah originally designated, so that her first identification is lost retrospectively. Researcher's happenstance informed by comparative book history research as here can lead to evidence that remakes otherwise missing links. I can add two more Fishes to the list of Sarah Bowdich's named contributions to the ichthyology of West Africa that Cuvier 'missed'.

The first emerged during my archival work to find other studies before this chapter that had engaged specifically with the history of ichthyology of the Gambia, and potentially Sarah's place recorded within it. The extensive BNF collections turned up the (undigitised) work of the Swede G. S. O. Svensson in 1934, *Fresh-Water Fishes from the Gambia River*.[40] A specialist in lungfishes, Svenssen undertook an expedition in 1931 to the River Gambia, research for which required knowledge of earlier historical as well as scientific studies (and mirroring a century later the Bowdich-Mollien-Falconbridge preparation for Sierra Leone). Svensson could find only the *Excursions* (*EM*, which does not include any lungfish), before the work in 1898–1899 of a further English zoologist and specialist in African lungfish, J. S. Budgett. Svenneson's breadth of interest, however, meant that he also recorded the Gambia's 'Sciaenidae' (88-89), and in particular *Corvina dux*:

> This species was described by Cuvier and Valenciennes, 1830 under the name *Corvina nigrita* and it has been called so by all later authors, as far as it has been known. It was however, described and pictured by Mrs S. Bowdich under the name of *Sciaena Dux* already in 1825. Her description is by no means complete nor satisfactory, but together with the picture given by her, quite enough details are given to justify the species. (99)

Had Svenneson not then added '(S. Bowdich, 1825)' and a fuller taxonomy note, I as a non-expert would not have been alerted to the junior synonym, from which to widen verification of her original nomenclature, *Sciaena dux*. Why Cuvier and Valenciennes in 1830 then replaced her earlier name when they knew her work *(EMFr)* is a question I have asked elsewhere regarding the ensuing (lack of) transmission for her later *Fresh-Water Fishes of Great Britain* (1828-1838).[41] The CLOFFA database again provides fuller information on the taxonomy of this Fish, but the *Sciaenidae* are a large family to search. Moreover, this fish has a further new nomenclature thanks to the work of the same E. Trewevas ('ET' above) in 1962: *Pseudotolithus (Fonticulus) elongatus* (Bowdich, 1825).[42] She (Ethelwynn) makes no mention of her ichthyological 'foremother'. The longer and larger transnational evidence here again directly challenges the 1899 and 50-year rules of the International Code when strictly applied.

A similar fate of renaming and overwriting befalls a third fish that Cuvier disregarded for comment in Sarah's zoological appendix, *Labrus jagonensis* (Figure 1.3). I had already begun searching Portuguese ichthyology records when I visited Lisbon in May 2013. A major exhibition, '360° Ciência Descoberta', at the Fundaçâo Calouste Gulbenkian disclosed it in a beautifully-lit display case of vials containing fish in spirits, allowing their '360° scientific discovery'. It stood out for its label in legible brown ink, attracting my excited attention: 'Museu Bocage. Labrus jagonensis Bowd. Ilha de Cabo Verde Inn (*sic*) Lowe' (from the collections of the Museum of Natural History of the University of Lisbon). The exhibition catalogue also confirmed the provenance of the scientific nomenclature as 'Bowdich'.[43] Although beautifully preserved in spirits, it lacked the vibrancy of Sarah's account:

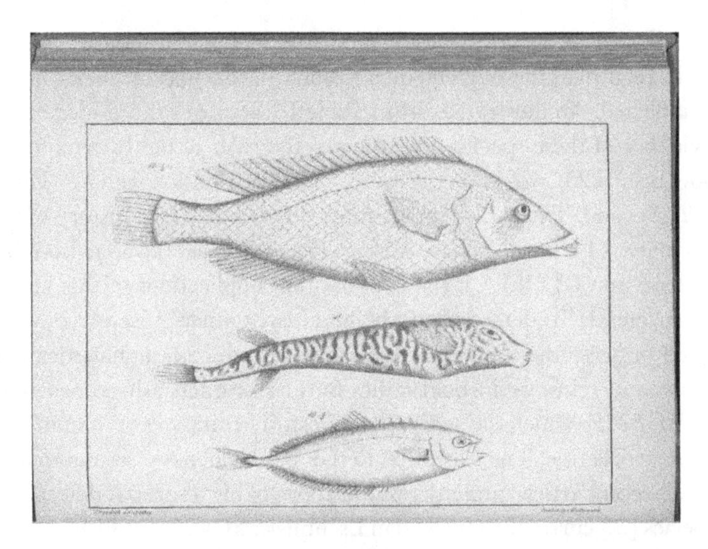

Figure 1.3 'Labrus *jagonensis*, Bowdich pictured in Figure 47'. Courtesy of the University of St Andrews Libraries and Museums, *Excursions in Madeira and Porto Santo*, rP702.M16B7.

> *Labrus jagonensis,* espèce nouvelle; Bowd., fig. 47:
>
> Four large teeth project form the front of the upper jaw, behind which is a row *en velours*; the dorsal fin has 25 rays, the pectoral 18, the ventral 8, the anal 14 and the caudal 12; the preoperculum is radiated, and the operculum deeply scalloped; the whole fish is a brilliant red. Found in Porta Praya and in the Gambia. (*EM,* 234 French in the original; *EMFr*, 364)

How could a non-ichthyologist connect this name with its official successor in the World Register of Marine Species, *Bodianus speciosus* (Bowdich, 1825)?[44] In this database, the 'original description' is allegedly 'not documented'. Yet the *Excursions* (and their plates) are accessible in online readable form in two international languages. In the many important contexts of the French West African IFAN, Jean Cadenat provides the further expert intermediations (in French) required to make the link, in a description and figure drawing in his work heading Table 1.2.[45] It long precedes the two recent magisterial studies in English, by Paolo Parenti and J. E. Randall in 2000 and Martin F. Gomon in 2006, who do not cite it.[46] If the onwards validity of *Bodianus speciosus* (Bowdich) is confirmed by Parenti and Randall, the search tag '(Bowdich, 1825)' is a needle in electronic database haystacks, qualifying many different species in various states of naming decay, unless comparative historians of ichthyology track their correct provenance.

The Natural Histories of Fishes: 'English'

The precedence and verification of the first name(r) for the thing in the face of codes, conventions and international languages of scientific nomenclatures are key issues in the specialist work above in overtly French-language legacies. That they reference the *Excursions* without Cuvier's 'Notes' in 1826 underscores the latter's major loss for science in the expert (French international vernacular) reception and classification history of the nineteen '(Bowdich, 1825)' Fishes of West Africa. These fish data in themselves, in their notice in English as the (current) international vernacular of science in the last century, are unaffected. An illustrative list in Table 1.1 arranges this notice first to present multiple evidence of these species according to the rules of the International Code so that '(S. Bowdich, 1825)' would be the more correct attribution (and *pace* Durand, 2005 in Table 1.2). Second, Table 1.1 demonstrates why comparative history of ichthyology taxonomies matters for Sarah's West African Fishes in several waters both behind and within FishBase and CLOFFA if the conservation implications of this knowledge are to be (re)discovered.[47] To demonstrate the number in other guise of '(Bowdich, 1825)' species, I had to order the important fish name information alphabetically by expert article author, and reinstate in bold italics Sarah's original fish nomenclature in the *Excursions* (*EM, EMFr*) which the accepted modern-scientific species name 'translates' in the respective article title. The exception to the 'Bowdich' name as automatically missing is Parenti's recent article (and my résumé above of his results). It determines '[s]even nominal species [...] currently unplaced in Eschmeyer's Catalog of Fishes on line (Fricke *et al.*, 2019)', and qualified (in the same source) as 'not accompanied by type material' (6). In 1967, Whitehead understood the profound detriment to the science and *history* of ichthyology of missing holotypes, not only for knowledge of the first/earlier names for the fish in question but also of its nomenclaturist(s). Sarah had no preserved fish specimen in the tragedy of her barrels formally to verify her zoological record of '*Clupea fimbriata*' as '(Bowdich, 1825)' or her other Fishes. The nub of Whitehead's work 'to discuss and stabilize the nomenclature of this important species' (585) was to call for the acceptance of *Ethmalosa fimbiata* (Bowdich, 1825) as a 'neotype' (587) amid other rival synonyms. He verified Sarah's description and drawing against the only possible West African marine clupeid genera – *Sardinella, Ethmalosa, Sardina* and *Ilisha* (*EM*, 586) – and against work by later expert predecessors in fish taxonomy to conclude that 'In spite of the poor but quite adequate figure given by Bowdich (1825, fig. 44 [...]), and the relative paucity of alternative West African clupeids with which this figure could have been identified, the Bowdich name was ignored or missed [...] for over a century, until Fowler (1936: 177) showed that it applied to the present species' (587). Whitehead's re-evaluation of the evidence in 1967 depended extensively on ichthyology book history and museum collections to re-describe the 'neotype for *Clupea Fimbriata* Bowdich, 1825' (590) for posterity. The result for the science and history of ichthyology is recorded as 'fide Whitehead' in Appendix 2. Yet his revised description also overwrote Sarah's important identificatory 'fringed' of the scales and nuanced colours above: '*Colour :* upper ⅓ of body slate-coloured, remainder of flanks and sides of head silver-gold. A very faint dark spot in humeral region. Tips of anterior dorsal rays dark brown, but otherwise all fins hyaline' (592). Whitehead's due scientific reclamation of the Bonga Shad as *Ethmalosa fimbriata* (Bowdich, 1825) also rests on the problematic claim overturned clearly above that it had been overlooked for over a century. The River Gambia, like ichthyology of the region, is a confluence of (inter)national interests. Table 1.2 maps the French-international resilience and larger acknowledgement of '*Ethmalosa fimbriata* (Bowdich, 1825)' across its many West African habitats, which Whitehead and other earlier (and later) Anglophone experts miss. Moreover, neither Whitehead nor most recently Parenti (and all the other indicative experts between) discuss the material, intellectual and (inter)cultural importance of the scientific qualifier '(Bowdich, 1825)'. It is surely worthy of further comment on Parenti's part that 'Sarah Bowdich is the authorship (*sic*) of 27 nominal species including eight species currently recognized as valid' (6)? The larger credit is Durand's correct reattribution in 2005 of 'S. Bowdich' to this fish, to acknowledge its first (fore-) name for ichthyology and hence for its history. But Whitehead's arguments for acceptance of *Ethmalosa fimbiata* (Bowdich, 1825) as a 'neotype' (587) also pertain to at least eight of Sarah's other West African Fishes based on their extant recognition by world ichthyology today.

In 1905, Fowler (like Cuvier before him) questioned how to designate fish that are 'New, Rare or Little-Known'. This chapter has asked the same questions of Sarah as representative of key figures hidden in plain sight in the histories (plural) of natural history because outside, yet working between the bars of its national-institutional codes for (proper) science conduct for naming, collection and publication. She and her *Excursions* are no footnotes in the intercultural history of comparative natural history (and modern ichthyology). Rather, the optics precisely of Cuvier's 'Notes' and the French receptions of the *Excursions* testify to the very remarkable 'S(arah) Bowdich)' in his recognition of the international (French) scientific work of her zoological appendix. If his was her training and preparation for it, his 'Notes' (like Humboldt's) also demonstrate his unusual promotion and mentoring for his times of 'outsiders' – foreigners, women, nonconformists – to Muséum science as at least his peers in their work 'overseas'. Our chapter study of Sarah in the contexts of Cuvier's 'Notes' can then go further, to reinstate her naming and descriptions as second to none for West African fish study in the 1820s and still standing out today. As 'Seleima aurata' or 'Ethmalosa fimbriata' illustrate, she named them first for ichthyology *in the field* for their new contributions to world Muséum collection resources. Second, she did so in 1825 by applying Cuvier's new classifications that define his *Histoire naturelle des poissons* and modern ichthyology in the nineteenth century, but before his multi-volume work appeared. In these, as two of her pioneering perspectives for ichthyology are the qualitative intercultural measures of her expertise. Her foundational naming and collection of West African species in the field included long-standing knowledge of their local names.

In the two tables above, none of the collated articles or experts addressed the 'gaps' or 'mismatches' in scientific nomenclature that are the result of major national geopolitical realities not factored in to the 'Statute of Limitation' Article 23 (b) of the International Code of Zoological Nomenclature. Sarah's field contributions in 1823–1824 are indicative. The Canaries were under Spanish rule, the Azores and Cape Verde archipelagos under Portuguese dominion, and Madeira was returned to Portugal after occupation by Britain during the Napoleonic Wars. Travellers with Spanish authorisation could only stop at the Canaries: those with Portuguese (or British) authorisation could not disembark there. Important comparative field studies of *flora* and *fauna* in their larger shared region could not be undertaken. Sarah's *Excursions* are the more remarkable counter-example of comparative achievement, because for the Gambia she could also draw on her earlier encounters with Madeira, Cape Verde and 'Ashantee'. Geopolitical factors and national rivalries also account for historical gaps and *inaccuracies* in subsequent scientific recording, rediscovery, renaming and re-evaluation of species, particularly for taxonomy. Major electronic databases could do much more to curate and contextualise 'senior synonyms' and their often several 'junior synonym' replacements in the nineteenth- and twentieth-century international vernaculars of science. They bypass colonial geopolitics contributing to the already exclusionary tests on time (1899, the 50-year rule) to locate key fish names that are otherwise expunged from allegedly international science databases.

Intellectual and cultural history of Madeira in 1823 then also rather beautifully makes the case for intercultural Fish knowledge of common as well as 'rare' Fishes. MacGregor in Table 1.1 underpins the importance of one example, the 'Blue Jack Mackerel'. It takes on entirely different hues, including historical, when its various scientific and other vernacular names appear together as '(Bowdich, 1825)' (Figure 1.4). In the settings of Funchal's busy working fish market ('Mercado dos Lavradores'), locals do not need to know its Portuguese name, 'carapau negrão': only its vital statistics in Portuguese grace this identificatory ornamental tile decorating the walls. 'Chicharro' is the vernacular Spanish name for the 'chinchard du large' in French and 'blaue Bastardmakrele' in German. In the *Excursions*, Sarah names the fish '*chinchara* or *chixarra*, fig. 27, a new species of *seriola*' (*EM*, 123; '*chinchara* ou *chixarra*, fig. 27, nouvelle espèce de *seriola*', *EMFr*, 191–92). Moreover, her black-and-white outline drawing labelled 'fig. 27' opposite the description in the English edition of the *Excursions* is unmistakable in the Mercado tile illustration. Visitors to the Funchal market from any nationality are thus much more likely to identify the thing among the catches on the market stalls from the picture, even if they do not know the English, Spanish or even Latin name(s) for it. They will see 'Bowdich' clearly on the tile, where history of science cannot. For ichthyologists such as Reintjes or Wirtz (Table 1.1) seeking to compile and update annotated checklists, the task of assimilating and assessing current and previous names for a species is already easier thanks to the work of earlier specialists – such as Blanc and Bauchot, Fowler and Trewevas – but also more difficult, depending on the bases (and databases) that define

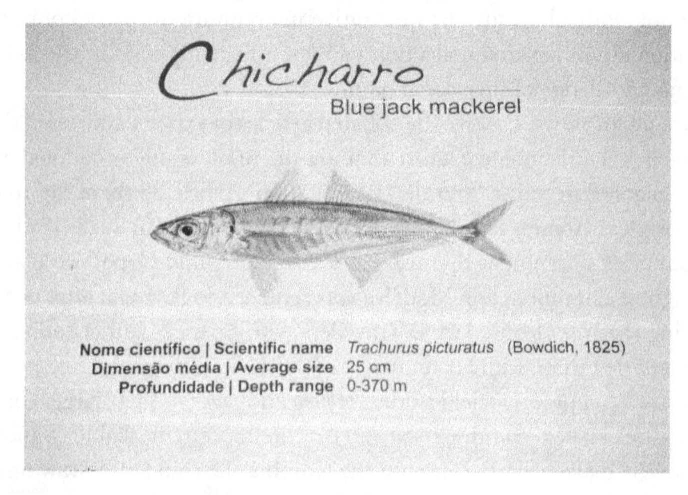

Figure 1.4 Illustration in ornamental tilework, 'Mercado dos Lavradores' (Fish Market), Funchal. Copyright Neil Mackenzie, May 2014.

'first' contributions to science. Are the major historical and archival studies undertaken by previous expert ichthyologists, such as Fowler more or less reliable than recent reappraisals, such as by Wirtz, where eyewitness accounts from diving trips verify older textual and collections-based evidence? How do common names in other languages testify to longer precedent in comparative knowledge traditions, where specimen and pictorial corroborations are absent? Indeed, the terms for classification of a species as 'new', 'rare' or 'common' depend upon a range and precision of geographical, habitat and bathometric definitions (given in the Mercado tile in Portuguese and English) that are comparably imprecise in the titles in both tables above. As Wirtz notes, 'Madeira' and the Madeiran archipelago are not the same entities.[48] Table 1.2 extends their Macaronesian reach and West African limits. Sarah's provenance 'data' for each fish – as collected in St. Iago, Porto Praya in Cape Verde etc. – in the zoology appendix in 1825 is therefore invaluable to modern environmental fish biologists as well as ichthyologist-taxonomists.

This chapter has now unpacked what Sarah *did* in her *Excursions* despite the appalling loss of her barrels. At least three Fishes in her 'zoology appendix' without Cuvier's *imprimatur* demonstrate first-ranking science irrespective of gender. The corollary is then more interesting. The work of 'M. Bowdich' for the Fishes of Madeira in Chapter 5 of the *Excursions* may also have been more properly the work of 'Mme'. Writerly proprieties prevented Sarah from making her case more overtly except in the 'continuation' of her narrative for publication, including its zoological appendix. Her voice is clear, despite protocol for female authority in the opening remarks framing the Bona Vista and the final Gambian chapter:

> I feel so great a repugnance to appear before the public, and so great a distaste to those subjects in which I have lost my guide and instructor, that the present narrative will labour under many disadvantages, besides those which may arise from incapability. It is but justice, however, to those who felt interested for Mr. Bowdich in his public character [...] and to *those who make the cause of science their own*, to relate the circumstances of his last voyage, with their fatal result.
>
> I particularly lament, that, *contrary to his usual custom, his notes were very few, and those so obscurely written, that even I, who am so accustomed to decipher his memoranda, can derive but little assistance from them*: therefore, that I may not injure a reputation which stood so fair with the learned and the good, I must request my readers *to consider me as responsible for every error*. (*EM*, 173–74, emphasis added, 'Relation', *EMFr*, 275).

The lesson of this chapter for our remaining study is that we should everywhere assume in Sarah's work her responsibility for every new and interesting fact that advances 'the cause of science'. Ichthyology today has at least eight species of West African Fishes to acknowledge and recognise in the foundational work of *S.* Bowdich, 1825 in their fields.

Notes

1 Anna Maria Falconbridge, *Two Voyages to Sierra Leone during the Years 1791-2-3*. London: Printed for the author and sold by different booksellers throughout the Kingdom, 1794.
2 See Christopher Fyfe. 2004. 'Falconbridge [*née* Horwood], Anna Maria'. Oxford Dictionary of National Biography. Oxford: Oxford University Press.

3 The prefatory 'I' in *Elements of Conchology*, iv covers Sarah as well as Edward in the account of her access to the Muséum shell collections, Cuvier's private library and freedoms to take specimens home to draw.
4 For the history of malacology, and for the Madeiran archipelago, L. M. Cook, R. A. D. Cameron and L. A. Lace claim the work by R. T. Lowe (1830) as the 'first' to examine the remarkable variety of snail species on Madeira, despite cursory acknowledgement of the earlier *Excursions*. ('Land Snails of Eastern Madeira: Speciation, Persistence and Colonization', *Proceedings of the Royal Society of London. Series B* 239 (1990): 35–79.) Yet the third chapter – 'Visit to Porto Santo' (*E*, 72–101; *FrE*, 114–88) – and the final section of the zoology appendix for 'St-Jago and the Gambia' (*E*, 239–43; *FrE*, 373–77) clearly specify the Bowdichs' extensive contributions to knowledge, regarding both fossil and living species. Edward's excursion to Porto Santo importantly applied Cuvier's understandings of fossil shells – especially marine species found on current landmass – to determine geological dating. This work was reported (in French) by 'F'[de Férussac], 'Description et Figures de Plusieurs Hélices, découvertes à Porto Santo par T. Edward Bowdich Esq, par G. B. Sowerby', *Bulletin des sciences naturelles et de géologie* 3, no. 74 (1824): 92–93.
5 For a study of its double instructions on taxidermy and on what scientific travellers were tasked to collect, see Orr, 'The Stuff of Translation'.
6 FishBase: the FishBase Database at https://www.fishbase.org
7 The Bibliothèque de Lyon print and open access editions catalogue 'M. T. E. Bowdich' as the author, https://numelyo.bm-lyon.fr/f_view/BML:BML_00GOO0100137001100306807. The 'Album' of plates (interleaved in the one-volume 1825 English edition) constituting the second volume ascribes it to 'M. Bowdich', despite Sarah's name being clearly on the plates themselves. See the Bibliothèque Nationale de France online Gallica edition to access the second volume at https://gallica.bnf.fr/ark:/12148/bpt6k9600570n?rk=64378;0.
8 See Orr, 'New Observations'.
9 See the important inclusion of 'Cumming*ae*' in nomenclatures by Agassiz of new fossil fishes discussed by Mary Orr, 'Collecting Women in Geology: Opening the International Case of a Scottish "Cabinétière, Eliza Gordon Cumming (c 1798–1842)", in Cynthia V. Burek and Bettie M. Higgs, eds., *Celebrating 100 Years of Female Fellowship of the Geological Society: Discovering Forgotten Histories*. London: Geological Society, Special Publications 506, 2021.
10 David J. Mabberley, 'Edward and Sarah Bowdich's Names of Macaronesian and African Plants, with Notes on those of Robert Brown', *Botanica Macaronesica* 6 (1978): 53–66 (60). Mabberley draws no wider conclusions from this fact or for the history of botany, given his evidence that Robert Brown failed to recognise the Bowdichs' botanical contributions. Mabberley's study also exemplifies how important critical work 'disappears' when published in a niche journal with a small print. However, for those who care to look and cite in languages other than English, it is available online open access.
11 See the similar point made for geological papers by Colin J. R. Braithewaite, '(i)t is important to remember that older publications may yet contain data of importance. We should know why, and by whom, theories were first conceived and the evidence adduced to support them'. 'Transactions and Neglected Data', *Scottish Journal of Geology* 47, no. 2 (2011): 179–88 (179).
12 Eugene K. Balon, Michael N. Bruton and David L. G. Noakes, eds., 'Women in Ichthyology: An Anthology in Honour of ET, Ro and Genie', *Environmental Biology of Fishes* 41, no. 4 (1994): 7–125. The editors' decision to refer to important women ichthyologists by diminutive, pet or nicknames would not find equivalents for male counterparts.
13 See, indicatively, Ulrike Brisson, 'Fish and Fetish: Mary Kingsley's Studies of Fetish in West Africa', *Journal of Narrative Theory* 35, no. 3 (2005): 326–40. http://www.jstor.org/stable/30225805.
14 Paulo Parenti, 'On the Status of Some Nominal Species of Fishes Described by Sarah Lee Bowdich (*sic*) in the Account "Excursions in Madeira and Porto Santo During the Autumn of 1823"', *Boletim Museu do História Natura do Funchal* lxix (2019): 5–12.
15 See note 3.
16 See note 4.
17 Antoine Guillemin, *Bulletin des sciences naturelles et de géologie* 5 (1825): 347–50.
18 A. G. Desmarest, *Bulletin des sciences naturelles et de géologie* 6 (1825): 396–98 (397). The translation is mine.

19 The French *Excursions* omit the prefatory sections of the English text, clearly indicating Sarah's scientific hand in the volume.
20 Peter J. P. Whitehead, 'The West African Shad, *Ethmalosa fimbriata* (Bowdich, 1825): Synonymy, Neotype', *Journal of Natural History* 4 (1967): 585–93.
21 M. Njie and H. Mikkola, 'A Fisheries Co-Managements Case Study from The Gambia', *The ICLARM Quarterly* 24, nos. 3–4 (July–Dec. 2001): 40–49.
22 See Jean-Jacques Albaret, Monique Simier, Famara Smbou Darboe, Jean-Marc Ecoutin, Jean Raffray and Luis Tito de Morais, 'Fish Diversity and Distribution in the Gambia Estuary, West Africa, in Relation to Environmental Variables', *Aquatic Living Resources* 17 (2004): 35–46; Monique Simier, Charlene Laurent, Jean-Marc Ecoutin and Jean-Jacques Albaret, 'The Gambia River Estuary: A Reference Point for Estuarine Fish Assemblages Studies in West Africa', *Estuarine, Coastal and Shelf Science* 69 (2006): 615–28; Vasilis Louca, Steve W. Lindsay, Silas Majambere and Martyn C Lucas, 'Fish Community Characteristics of the Lower Gambia River Floodplains: A Study in the Last Major Undisturbed West African River', *Freshwater Biology* 54 (2009): 254–71.
23 Ichthyology works hard on its accessibility through article titles and wording. The French homophone 'cadenas' – padlock in English – for 'Cadenat' inspires my argument.
24 Gérard Belloc, 'Rapport general sur la cinquième croisière du navire "Président-Théodore-Tissier"', *Revue des Travaux de l'Institut des Pêches Maritimes* 10, no. 3 (1937): 269–325 at https://archimer.ifremer.fr/doc/00000/5755/. The anonymous 'Recherches systématiques sur les poissons de la côte occidentale d'Afrique. Liste des poissons littoraux recoltés par le navire "Président-Théodore-Tissier" au cours de sa cinquième croisière' is in the *Revue des Travaux de l'Institut des Pêches Maritimes* 10, no. 4 (1937): 425–564 at 325 at https://archimer.ifremer.fr/doc/00000/5775.
25 Théodore Monod, *Achille Valenciennes et l'Histoire naturelle des poissons*. Paris and Dakar: IFAN, 1964, 25.
26 Marie-Louise Bauchot, Jacques Daget and Roland Bauchot, *L'Ichthyologie en France au début du XIXe siècle*: L'Histoire naturelle des poissons *de Cuvier et Valenciennes*. Paris: Muséum national d'Histoire naturelle, 1990, 19–22.
27 Auguste Duméril, *Histoire naturelle des poissons ou Ichthyologie générale*. 2 vols. Paris: Librairie Encyclopédique de Roret, vol. 1, 1865; vol. 2, 1870. See David Starr Jordan, 'The History of Ichthyology', *Science* 16, no. 398 (Aug. 15 1902): 241–258 (249).
28 David Starr Jordan, *The Genera of Fishes: A Classification of Fishes*. Repr. Stanford: Stanford University Press, 1963.
29 Henry Weed Fowler, 'The Fishes of the United States Eclipse Expedition to West Africa', *Proceedings of the United States National Museum* 56 (1919): 195–292.
30 See Yves Le Gal, '2009: Le laboratoire de biologie marine de Concarneau a 150 ans', *La Lettre du Collège de France* 26 (June 2009): 49–52.
31 Harry W. Paul, *From Knowledge to Power: the Rise of the Science Empire in France, 1860-1939*. Cambridge: Cambridge University Press, 1985, 103–17 (116).
32 Maurice Zimmermann, 'Afrique Occidentale. Mission Gruvel et Chudeau', *Annales de Géographie* 18, no. 99 (1909): 278–80.
33 See Jean-Claude Hureau, 'Un exceptionnel naturaliste éclectique', in 'Théodore Monod: un homme curieux', *Autres Temps. Cahiers d'éthique sociale et politque* 70, no. 1 (2001): 25–38.
34 Théodore Monod, '*Brevorha* Gill, 1861 et *Ethmalosa* Regan 1917', *Bull. IFAN* 23 (1961): 506–417.
35 P. H. Skelton and E. R. Swartz, 'Walking the Tightrope: Trends in African Freshwater Systematic Ichthology', *Journal of Fish Biology* 79 (2011): 1413–35 (1415).
36 In historiography, the reconstruction of the lineages and networks of a 'great person' is known as prosopography.
37 See Orr, 'Women Peers'.
38 See the *Historical Portrait of the Progress of Ichthyology: From Its Origins to Our Own Time by Georges Cuvier*, ed. Theodore W. Pietsch, trans. Abby J. Simpson. Baltimore: Johns Hopkins University Press, 1995.
39 See https://www.FishBase.org/summary/SpeciesSummary.php?id=1594.
40 G. S. O. Svensson in 1934, *Fresh-Water Fishes from the Gambia River*. Stockholm: Almquist & Wiksells Boktrycheri A-B, 1934.

41 See Orr, 'Fish', 206 and 240.

42 http://vmcloffa-dev.mpl.ird.fr/table/taxon2/view?idtaxon:int=2838.

43 'VL16: Colacção de Peixes Africanos. *Labrus ti iagonensis* (Bowdich 1825) Ie Cabo Verde Inv, MNHNC–MB06-004694)', in Henrique Leitão, Teresa Nobre de Carvalho, Joaquim Alves Gaspar and Antonio Sánchez, eds., *360º Ciência Descoberta*. Lisbon: Fundacão Calouste Gulbenkian, 2013. Curious reader-visitors can even glimpse this '360º Ciência Descoberto' vial on the exhibition film, available on 'Youtube', https://www.youtube.com/watch?v=HZwX2VdmtPE (at 4 minutes 38/39 seconds). The Rev. R. T. Lowe (1802–1874) was the later-published authority on the fish of Madeira in his *A Synopsis of the Fishes of Madeira; with the principal Synonyms, Portuguese Names and Characters of the new Genera and Species, Corr. Memb. Of the Zool. Soc.* Communicated 28 March, 1837.

44 See http://www.marinespecies.org/aphia.php?p=taxdetails&id=303639.

45 Jean Cadenat, *Poissons de mer du Sénégal*. Dakar: IFAN, 1950, 252–55 and Fig. 150 (190).

46 Paolo Parenti and John E. Randall, 'An Annotated Checklist of the Species of the Labroid Fish Families Labridae and Scaridae', *Ichthyological Bulletin of the J.L.B. Smith Institute of Ichthyology* 68 (2000) at https://tspace.library.utoronto.ca/html/1807/23206/fb00001.html and Martin F. Gomon, 'A Revision of the Labrid Fish Genus *Bodianus* with Descriptions of Eight New Species', *Records of the Australian Museum, Supplement* 30 (2006): 1–333 (92–93) also with colour plates.

47 See the indicative work in historical baseline research by Emily S. Klein and Ruth H. Thurstan, 'Acknowledging Long-Term Ecological Change: The Problem of Shifting Baselines', in Kathleen Schwerdtner Máñez and Bo Poulsen, eds., *Perspectives on Oceans Past: A Handbook of Marine Environment History*. Dortrecht: Springer, 2016, 11–29. https://doi.org/10.1007/978-94-017-7496-3_2 and Loren McClenachan, Francesco Ferretti and Julia K. Baum, 'From Archives to Conservation: Why Historical Data are Needed to Set Baselines for Marine Animals and Ecosystems', *Conservation Letters* 5, no. 5 (2012): 249–59.

48 Both Wirtz and Whitehead point out that previous experts often failed to recognise that the Fishes listed in the *Excursions* were not all 'Madeiran'.

Chapter Two

A FIRST NATURAL HISTORY OF FISHES ILLUSTRATED FROM THE LIFE IN *THE FRESH-WATER FISHES OF GREAT BRITAIN* (1828–1838)

The first chapter revealed the magnitude of Sarah's foundational contributions to knowledge and naming of West African natural history new to European science in the *Excursions* (*EM*; *EMFr*). The publication of her achievements under her own name on both sides of the Channel also set her pioneering work in the field at the forefront of the new specialist Paris Muséum discipline of world ichthyology, alongside botany, ornithology and malacology. The imprimatur of both Cuvier and Humboldt on the 1826 edition only further affirmed Sarah's applications of their highest scientific standards for the period. The *Excursions* are then the more towering in their testimony to her overcoming of almost insurmountable personal and scientific loss. If her husband-'instructor' (*EM*, 173) in scientific field exploration overseas and her fish specimens were irreplaceable, the published text more clearly exposed Sarah's own irreplaceable independent scientific acumen. The strengths of her earlier self-determinations in natural history addressed in the introduction (pp. 14–15) and glimpsed in Chapter 1 in her preparation in 1824–1825 of the *Excursions* using Paris Muséum resources (p. 34) had once again overcome her untimely personal, financial and scientific straits. In little more than one year after her return to London, the proceeds of the *Excursions* could repay her 'sympathizers' and further support her family. But the same dilemma that Sarah faced in 1824 confronted her anew. How could she continue her independent work in natural history as a widow with three young children to support?

Sarah's solution was simple, because unwavering in her remarkable 'continuance of the voyage' of natural history endeavour in the words of the inside title page of the *Excursions* (Figure 0.1 and Figure 1.1). Her resolve also crystallised its larger venture. Her inventive and pragmatic new departure in 1824 was to make London (Britain) the 'overseas' for her ongoing French natural history work. Sarah was already employed in Banks's Library as Cuvier's *aide naturaliste* in all but name, to verify fish drawings for his *Histoire naturelle des poissons*, as his peer in French ichthyology.[1] As a Briton and a woman in its domain, she was also unprecedented among its Muséum experts and overseas 'correspondents'. To invert gain (London) for loss (Paris, the Gambia/Sierra Leone) in this way was also an entrepreneurial, creative and strategic move for a nonconformist widow making her living by her pen and pencil in Britain in 1826. Indeed Sarah had everything to gain, because her pioneering natural history knowledge and experience in Paris, Madeira and the Gambia was now in print in her own name on both sides of the Channel. Sarah could no longer investigate other West African *fauna*, but she could apply her pioneering perspectives in their fields to British contexts, especially regarding Fishes, since she had no rival models in new (French) ichthyology in Britain to predefine her work. Sarah's 'continuance' of it in London was already in place through her regular collaborations with Cuvier and her other Paris Muséum friends of natural history.[2] This chapter will demonstrate that Sarah's canvassing of Cuvier in her *Fresh-Water Fishes of Great Britain* (1828–1839) provided different opportunities to set its bar by his definitive, 22 volume Muséum-based *Histoire naturelle des poissons* (1828–1848). She could do so only through what she had learned in the field in West Africa.

The multiply-trained focus of Sarah's zoological appendix and accompanying Fish drawings in the *Excursions* (*EM, EMFr*) therefore constitute the step-change in her achievements for (French) ichthyology that is the 10-year project of the *Fresh-Water Fishes of Great Britain.* To take forward discipline-defining scientific work building upon one's own, however, required not only Sarah's (self-)critical and newly comparative perspective but also her larger understanding of the scientific field(s) in question. The London that was Sarah's new 'overseas' of Paris and West Africa was the new outlet for her pioneering comparative field perspectives in *The Freshwater Fishes of Great Britain.* But I could only articulate this pivotal 'London' for it through imagining myself in Sarah's shoes in 1826. How could she best her contributions in the *Excursions* anew was the question I also asked of my extensive, even definitive, critical study of her *Fresh-Water Fishes of Great Britain* in French ichthyological contexts?[3] I will therefore not rehearse my multiple supporting evidence, but return to what was 'standout' in Table 2.1 about the *Fresh-Water*

Table 2.1 The *Fresh-Water Fishes of Great Britain* as 'standout'

Sarah's Firsts in Ichthyology:

- the first (woman) to undertake a detailed, specialist, study of Britain's most important, and also rare, fresh-water fish, and to have this work internationally recognised;
- the first (woman) to describe fresh-water fish in text and image only from living specimens in the field (and to establish this criterion for 'serious' ichthyology);
- the first (woman) to use and apply Cuvier's classification system, even before its application in his *Histoire naturelle des poissons*, ichthyology's discipline-defining work and methodology;
- the first (British/ European woman) to describe and name some twenty new West African fish in the field, with these endorsed by Cuvier, a world expert on fish;
- the first (British woman) to have this West African contribution published in French as the then international vernacular of ichthyology; (See Chapter 1)
- the first (woman) to publish specialist works on (British) ichthyology targeted specifically at juveniles;
- the first (woman) 'populariser' of ichthyology to open the field to the 'Amateur Naturalist' who could be female as well as male;
- the first woman in the history of (European and North American) ichthyology of '[M]any Accomplishments,' and thus the eminent precursor of Rosa Eigenmann, and her many important twentieth-century successors;
- a list of firsts […] to elucidate Sarah's authority as an internationally renowned (British) ichthyologist first, and as the first (British) woman in its larger history second.

Fishes of Great Britain in my earlier concluding list of Sarah's firsts for ichthyology of the period.[4] The 'London' above then articulates her new 'angles' of vision (to borrow my own article title keyword), of which we discover two in this chapter. How Sarah came to undertake her second major contribution to natural history is revelatory first of how she undertook it to completion and to publication: it was not Cuvier's idea.[5] When seen in the same form as her *Excursions* in different milieux, however, Sarah's project turned into a further expedition in pioneering field *recovery* that concomitantly supported her now London-based work *en famille*. Second, the chapter newly investigates how Sarah's work on *British-European* freshwater fish species was taken up in France after its important mention by Cuvier in his Prospectus of 1826 and volume 1 of his *Histoire naturelle des poissons* (1828).[6] I had already tracked the occlusions of the *Fresh-Water Fishes* in nineteenth-century British ichthyology, and the possible reasons for them.[7] Was the same fate for Sarah's 'standout' *Fresh-Water Fishes* replicated in France, especially after Cuvier's death in 1832 (and volume 8 of his *Histoire*)?

New Departures in 1820s' Britain: Text with Illustration and Scientific Description

A wealth of scholarship on Britain's burgeoning nineteenth-century print cultures and their diversity of periodical presses highlights the importance of these media in the Victorian period for the dissemination and popularisation of science.[8] The lacuna, as Jonathan R. Topham has noted, is that 'the agency of early nineteenth-century publishers has barely begun to be addressed'.[9] In addressing it, Topham then overlooked travel writing of the 1820s as an important 'scientific' genre. He also did not account for the diversity of British publishers as a key factor in the science publishing markets of the period. As our introduction observed (p. 16), Britain had no national natural science institutes or museums during 1825–1826. It therefore had no official scientific publishing house like F. G. Levrault (for the Paris Muséum), although organs such as the Royal Society had long published *Transactions*. This lack of a national scientific publisher in the 1820s and 1830s was no disadvantage,[10] because many British publishers had the expertise to produce the quality lithograph and/or colour illustrations that guaranteed scientific publication success. The Bowdichs had already published with the most renowned, including John Murray, J. Smith (in Paris and then Cambridge), Longman, Hurst, Rees, Orme and Brown, and George B. Whittaker for the English *Excursions* (see Appendix 1A). Topham's 'agency' for Britain's publishers of natural history therefore lay in their entrepreneurial diversity, largely because demand from readers and authors was multiple, and without set (official institutional) models. Publishers in the 1820s therefore formed and informed their amateur (that is, interested specialist) scientific readerships in different formats, including by subscription, to safeguard the future of their expanding markets. Indeed, the niche for books about natural history and its particular branches only became more viable in the 1830s after British readers had grown their appetite for informed knowledge through cultural periodicals and magazines by subscription. *The Edinburgh Review* (founded in 1802) is indicative, and it served as the model for *The* (London-based) *Magazine of Natural History*. It began publication only in 1828, with 'Mrs Bowdich' among its article contributors (see Appendix 1B). *The Naturalist's Library* (Edinburgh, Lizars) enjoyed enormous successes from 1833 to 1843, in part because the costs for its prized illustrations were already factored into the more prohibitive pricing otherwise for quality printing and binding. The quality plates of the series became an additional attraction for wealthier reader-collectors, and for potential authors and artists to make their name and livings from natural history publication.

Sarah's sex and class in 1825–1826 therefore presented no major obstacles to her securing of a prospective British publisher for her ichthyology project, since none of those cited above had a dedicated natural history list (with concomitant societal bars). She was also in the unusual position of having already published with London's most established. Her larger problem as a serious potential 'science' author was that such a role did not properly exist as the means to earn a living: it was other than the publication pursuit of the leisured (gentleman) naturalist.[11] Apart from the subscription list for *ad hoc* publication, there was no system for the brokering and financing of (illustrated) publications commissioned from authors able to advance new knowledge of natural history for amateur naturalist readerships. Later models for writers of science as 'professionals' or 'popularisers' in Victorian Britain are therefore anachronistic in the earlier nineteenth century. British publishing houses of this period with the print technologies to produce illustrated formats were particularly alert to the entrepreneurial advantages of creating and feeding new demand for scientific subject-matter and its knowledge. Target markets were fashionable literary and artistic society in London, Edinburgh, Manchester and other cities. Publishers were key figures in these society networks, brokering recommendations and transactions between patrons of the fine arts, prospective authors, magazine editors and illustrators.

Two such publisher brokers are pertinent here. The first was the German, Rudolph Ackermann (1764–1834), who set up a lithographic press in 1795 in the Strand, moving to various addresses there as his business interests took inspiration from continental innovations and expanded into 'The Repository of Arts'. His publishing house, renowned for its high-quality colour illustrations, also manufactured and supplied specialist art materials. Additionally, it hosted a drawing school, a gallery, a library and social space for wealthy collectors to meet, the venue forming an integrated emporium for influencing fashionable taste. In 1809 Ackermann launched the monthly magazine, *The Repository of Arts, Literature, Commerce, Manufactures, Fashion and Politics*.[12] In targeting the most discerning reader of taste of both sexes, it paved the way for Ackermann's launch in 1823 of the first British literary annual and Gift Book, *Forget Me Not*, commissioning new illustrated article-length work from the best authors, poets and illustrators.[13] To assure the quality, originality and diversity of appropriate subject-matter, Ackermann drew for suitable contacts on various contemporary magazine editors. Among their number is our second broker, William Jerdan (1782–1869). He edited and commissioned new work for *The Literary Gazette; A Weekly Journal of Literature, Science and the Fine Arts* from 1817–1850.[14] Its title encapsulates the place of burgeoning scientific interests spearheaded by literary travels, and by paintings of foreign landscapes and the natural world. Both men were in Sarah's particular debt, because they had both sought to alleviate hers upon the news of T. Edward Bowdich's premature death. Our proof is a notice in the *Somerset*

House Gazette, and Literary Museum (of 22 March 1824, 399), reprinting the letter she had sent before setting sail from Bathurst, 'in which she announces her bereavement [and] depicts her situation':

> 'I am about to try your friendship, dear Sir, in a thousand ways, for I am now alone in the world, widowed and unprotected. Your friend expired on the 10th of January [...]. No Will has been left, therefore it is my duty to administer to his effects [...]. I hope, looking to my own support, through *S* and *M* to get employed in different works in natural history, setting up as an artist in that line. God grant me success, I am not fit for anything else [...].'
>
> Under these circumstances, therefore, some of the friends of the late Mr. Bowdich have concluded to appeal, on behalf of his widow and children, to the good feeling of those who can appreciate the disinterested devotion of life and talent to a noble object, or who, having the interests of Science and Literature at heart, recognize in the circumstances of Mr. Bowdich's death, and the consequent situation of his family, a claim upon their liberality. [...]
>
> Subscriptions received by Charles Konig, Esq. British Museum; J. G. Sowerby, Esq. 156, Regent Street, John Tomkins, Esq. South-Sea House, Messrs Coutts & Co. Strand; Messrs Ladbroke & Co. Bank Buildings; Messrs. Longman & Co. Paternoster Row, and Mr. Ackermann, Strand.

A further 'Address' in the same issue clarified the missing name, Jerdan, behind those 'having the interests of Science and Literature at heart': 'It is fortunate that the interesting case of this lady and her infant family, has been adopted by the extensive influence of the Literary Gazette, and by the personal exertions of Mr. Ackermann' (414). Ackermann's subsequent commissioning of *Forget Me Not* stories from Sarah Bowdich further facilitated his 'works of mercy' (above). Jerdan, however, played a less fortunate part in seeking to employ Sarah as her letter to *Somerset House Gazette* had proposed in the 'setting up as an artist' in the 'line' of natural history.

'God Grant Me Success, I Am Not Fit for Anything Else' (But Pioneering Natural History)

When a rare copy of *The Fresh-Water Fishes of Great Britain* came up for auction in 2013 at Christie's, its website provided the following information about its provenance: 'instigated in the early 1820s by Lord de Tabey and the publisher William Pickering who planned a volume on *British Ichthyology* to be edited by Jerdan. A prospectus was issued'.[15] A further copy of the latter appeared with a fuller account in 1852–1853, when Jerdan published his four-volume *Autobiography*.[16] Among its cameo portraits of the many artists, writers, collectors and gentleman patrons of the arts and sciences that Jerdan had counted as his society friends, contacts and even business partners was one in particular (in Volume 4 Chapter 8, 'My own Life'), Sir John Fleming Leicester, first Baron Lord de Tabey (1762–1827). He was a renowned collector and patron of English art, as well as an accomplished amateur artist. His English art projects were to include a *British Ichthyology*, illustrated with specialist colour drawings. Jerdan's responsibility in the venture was to write the introduction and descriptions. When Lord de Tabey became seriously ill in

1825, he could no longer proceed financially or artistically. Jerdan was left to find alternative artists and financial backers for the project, because he was unwilling both to shoulder the costs (due to his many other commitments and ventures) and to relinquish the project were someone else to take responsibility for it. To this end, Jerdan set out his version of the 'plan' for the *Ichthyology* in a four-page account following the reproduction of the original 'Prospectus' (see Appendix 3).[17]

Jerdan's *Autobiography* thus provides a very detailed blueprint, to record his original and future interests (as the potential editor), and for the project to be 'carried into effect by some fond disciple of Izaac Walton, in conjunction with a spirited publisher' (145). Walton was the author of *The Compleat Angler* (1653) as a precedent-setting manual for sport fishing. History puts Jerdan in less generous light than this 'spirited publisher'. Knowing of Sarah's many suitable talents and her financial straits, Jerdan enlisted her to the project, but then left her to make good the substantial costs of its production ('estimated at 200 guineas' in Appendix 3). The evidence, without naming Jerdan, only came into the public domain in 1888. On 31 December 1887 *The Field* made report of 'A Rare Book' sold by auction 'at the rooms of Sotheby, Wilkinson and Hodge' assuming that 'Mr Bowdich' had to have been the author/ity of a work about fish. On 14 January 1888 *The Field* then printed the letter by 'Tedlie Hutchison Hale (formerly Bowdich)' to establish her mother's authorship of *The Fresh-Water Fishes of Great Britain*:

> I hasten to supply the information regarding my parents [...] glad, in any way, to be the means of keeping a venerable name from oblivion. When Mrs Bowdich returned from Africa a widow in 1824, she was asked to publish a work on the fresh-water fishes of Great Britain, pledging herself to draw the specimens as they came fresh from the water. Those who, at first induced her to undertake this work, soon found that it would, financially, be a losing concern, and left her to bear the burden alone. My mother felt pledged to her fifty subscribers, but, having three children to support by her pen and pencil, could not afford to devote all her time to this one work, which accounts for the length of time it was in completion. I believe the published price for a number containing four original drawings, besides letter-press, was £2.2s.[18]

Ackermann again stepped in, to produce Sarah's quality publication of the work in Gift Book format in 11 numbers between 1828 and 1838, with a twelfth unfinished except in outline. In this period, however, her 'burden' in producing them was not only financial. On various occasions, she had also to enter into delicate correspondence with those of her subscribers paying 52 guineas each for the complete set who, like the Duke of Sussex, delayed or forgot payments.[19] Furthermore, she had the logistical demands and costs of site visits to support, in order to capture her Fishes in the appropriate season with the aid of local fishermen. From her master drawings made *in situ,* Sarah then faced the barely imaginable greater 'burden'. She reproduced her original, scale – that is no pun – watercolour drawings of each fish fifty times for the four Fishes per number, a task she then repeated every year for 11 years. Jerdan was not among the list of subscribers to *The Fresh-Water Fishes* (see Appendix 4). However, his *Literary Gazette* reviewed its unusual achievements positively, and with feeling (Jerdan may have been the anonymous author given the original prospectus) in its second number in 1829:

> When it is considered *how laborious a task it must be to produce the multiplied drawings* which illustrate this beautiful and interesting publication, it is not surprising that so long a period has elapsed since we announced the appearance of the first No. The present No., like its predecessor, does Mrs. Bowdich's taste and industry the highest credit. The drawings, which are the perch, the chub, the common eel, and the minnow, have a brilliance and a truth in them *that we have rarely seen equalled, and never surpassed; and the descriptions, both scientific and general, are very satisfactory*. (Sat. 13 June, No. 647, 395, emphasis added)

The material rarity of Sarah's extraordinary endeavour lies in its small print run and hence limited availability for inspection, in that fewer than a third of the original 50 subscriber copies are held in world libraries rather than in private collections. The even greater rarity, however, is the artistic-scientific quality of the drawings described contemporaneously and definitively above as 'rarely seen equalled and never surpassed'. Sarah's *Fresh-Water Fishes* preceded similar projects on British Fishes undertaken by Jenyns, Yarrell and others, and overlaps with the first decade of production of Cuvier's multi-volume *Histoire naturelle des poissons* (1828–1848) with plates completing the set. Are hers the pioneering perspectives in *The Fresh-Water Fishes of Great Britain*' that remain 'never surpassed' also in retrospect, when compared with these named competitors and their successors? Answers to this question lie in the closer re-examination of Sarah's work for what the viewer above terms its 'brilliance' and 'truth', that is, its eminent scientific qualities in image and text. Their integrity, we will argue, heightens Sarah's unrivalled natural scientific expertise, making of her drawings and descriptions exemplar measures for all future European scientific knowledge of Fishes, freshwater or marine.

'Brilliance'

The exceptional artistry of Sarah's illustrations for *The Fresh-Water Fishes of Great Britain* was unknown to art history and history of science until the literary historian, Barbara Gates, drew it to critical attention in 1998: 'What made [Sarah Bowdich's] forty-four watercolors so remarkable was her experimental use of gold and silver foil to give them a sense of the metallic iridescence of fish scales'.[20] Subsequent commentator-specialists in art history and fish illustration such as Huon Mallalieu and Christine E. Jackson[21] have echoed but not cited Gates's further underscoring of Sarah's pioneering work 'from life'.[22] Sarah has thus become more a collected scientific illustrator worthy of mention,[23] than she is celebrated as a trailblazer in scientific illustration as integral to identificatory description, especially in light of key developments taking place in European ichthyology in the late 1820s.[24] Sarah's plan in the preface of *The Fresh-Water Fishes of Great Britain* is therefore significant, both to distinguish it from Jerdan's initial venture (see Appendix 3), and to capture what was innovative in her approach by comparison with Cuvier's (illustrated) *Histoire Naturelle des Poissons*. We should now be suitably alert to Sarah's characteristic understatement:

> A work of the following description requires but little preface: at the same time I am desirous of offering my readers a few words on the plan I have adopted, and the endeavours I have made to ensure accuracy.

> *My object has been to give, rather a correct representation of the individual Fish than to form a picture; and by so doing, I trust I have satisfied the Naturalist, without offending the Amateur.*
>
> In my classification I have been kindly assisted by Baron Cuvier, whose system I have adopted, and who has given me the nomenclature he intends using in his forthcoming great Work on Ichthyology. The regular series of the Families, however, has been interrupted for the sake of variety in each Number, and those least interesting to the eye are mingled with their more beautiful companions. When the Work is completed, the Drawings may be easily unsewn, and classically arranged, according to the references given in the text. Another consideration has been, the time and labour required, as far as it affects the appearance of the Numbers at reasonable intervals. To ensure this, the large and small have been thrown together, that each set may bear its true proportion and be published at regular intervals,
>
> I have hitherto been particularly fortunate in procuring good specimens, and have been aided by friends and strangers with equal zeal. *Every drawing has been taken from the living Fish immediately it came from the water it inhabited, so that no tint has been lost or deadened, either by changing the quality of the element, or by exposure to the atmosphere.** (*FWF*, iii, emphasis added)

Where Sarah appeared to adopt the Leicester-Jerdan emphasis on offering a 'correct' representation, her specific qualifications of the word demonstrate the difference of her approach. It was primarily to satisfy the 'Naturalist' without excluding the interests of the 'Amateur'. The mention of Cuvier in the following sentence heading her main paragraph (outlining the practical and scientific rationale of the project) aligned 'Naturalist' with 'Cuvier'. Her aim was to meet the standard of the foremost expert in world ichthyology even before the first volume of his definitive *Histoire Naturelle des Poissons* was published. We know from Cuvier's 'Prospectus' for it (of 1826) that a key task had fallen to 'Madame Bowdich'.[25] Only she had the expert knowledge for Cuvier's purposes to make quality 'copies', that is, correct scientific representations of the fish drawings in the Library of Joseph Banks originally made by Forster and Parkinson on Cook's Voyages. Her resulting reproductions then enabled Cuvier to classify the fish in question for insertion into relevant volumes of his (future) work.

As each of Sarah's descriptions with accompanying colour plate in her *Fresh-Water Fishes* fully endorses, a principal object for modern fish taxonomy is the correct identification of the number and positioning of fins with their ray counts, and of other important secondary features differentiating similar species (see Figure 2.1). Variation in shape or colour may also distinguish fish of the same species at specific stages of their life cycle, or in different rivers. As the Rud also illustrates very clearly here, identificatory ray counts given in the text (immediately below the header treating the species nomenclatures in Latin and European vernaculars) for the dorsal, pectoral ventral, anal and caudal fins can be checked for accuracy against the image. The succinctness of the text concerning the secondary distinguishing features for this fish ahead of its fuller description again makes the viewer pay greater attention to the image: 'No barbs; irides yellow; body flattened, and sloping suddenly towards the tail.' The main description, filling in the colours, then endorses the 'scientific' points that make it also acceptable to the 'Amateur' as well as to the angler:

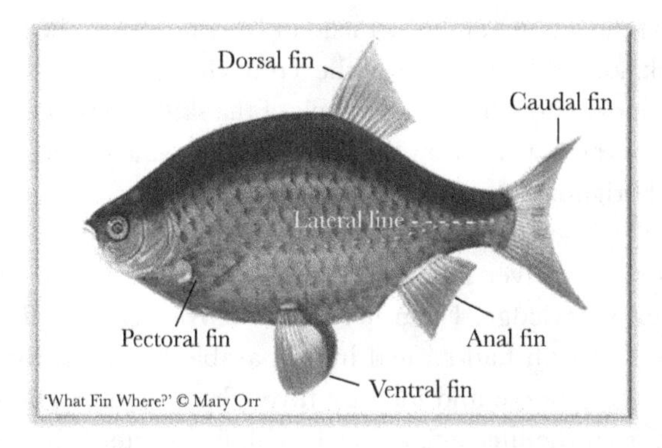

Figure 2.1 'Fin Arts in Ichthyological Description'. My overlay adaptation of Sarah's drawing of the Rud acknowledges the image originally hosted on the now defunct Philadelphia Academy of Natural Sciences – Digital Collections – 'Drawn from the Deep – Bowdich' illustration (untitled) http://www.ansp.org/museum/digital_collections/fish/bowdich.php (last accessed 21/5/2008).

> *The Rud resembles both the Roach and the Bream in shape, but its colouring is more splendid than that of either of these fishes.* The back is olive, and the side blue; towards the belly it is of a bright yellow, *and in some waters* this almost supersedes the blue, and in the sum gives the fish a golden hue. The fins are frequently all red, and red marks are found upon the body; but in the annexed drawing the dorsal and pectorals are of a yellowish brown. The caudal fin is sloped like a crescent, the head is small, and the scales are large. *It is by no means a common fish in England, though abundant in Pomerania and Brandenburgh (sic). It prefers ponds and still waters, and is chiefly to be met with in Oxfordshire, Northamptonshire, Lincolnshire, and in the Trent.* The specimen here presented was caught in the Stour, close to Sudbury, in Suffolk; *but the Rud does not frequent all parts of that river.* It spawns about the month of April; but Bloch says that it does not deposit its eggs all at once, its flesh (when the fish is large), is said to be firm and of good flavour. (No. 21, no pagination, emphasis added)[26]

The same economy of inter-reflecting image and text conveys further specialist (European) identificatory information here for armchair collectors or Museum ichthyologists, but concomitantly provides key field identification for naturalists and anglers of the given fish's particular habitats and habits. Sarah's remarkable precision, concision and detail in her description as inter-reflected in the 'brilliance' of the watercolour image make her identificatory study of each Fish a 'textbook' representation. Chapter 8 will return to this important feature of her work. Hers is a masterclass in (Cuvierian) comparative anatomy of each Fish subject, based upon her expert scientific observations of them in and out of the water in the field (and in the Paris Muséum collections). Her drawings of Britain's Fishes from the life also take fullest cognisance of freshwater habitat *distinctions* in the phrase, 'by [not] changing the quality of the element'. Each drawing would have entailed having several specimens for the given species kept alive in their own river water in buckets, to prolong the time in which their true, that is underwater, life colours could be captured for science.[27] This technical practice translated the instruction 'how' (quoted above from her Preface) 'no tint has been lost or deadened either by changing the quality of the element, or by exposure to the atmosphere.*' The qualifying asterisk notes '*The colours of the Trout change directly after they leave the stream; but I was lucky enough to avail myself of the skill of a friend, who supplied me with a succession of them as I sat on the bank, and by which I secured the tints in all their delicacy and brightness (p. iv).'

The 'brilliance' of Sarah's (scientific) 'textbook' plates shines not only in her new application of gold and silver foils but also in the foils of her own unprecedented comparative 'field' knowledge. From her work on West African Fishes in preparation for her barrels, Sarah had gained incomparable knowledge in hot climates of what immediately to observe and identify through ray counts and colouration, the better to capture from the life each of her British freshwater fish specimens. Sarah's imaging to scale 'in the flesh' to the highest scientific colour specifications clearly rivalled equivalent (Muséum) fish preservation in spirits, a technique in which she was also adept as we saw in Chapter 1. But Sarah's plates go further. They also translated her new quality measures for correct scientific field illustration – scale, colour, fin-ray detail – back into recent Paris Muséum science format, of which she had intimate insider knowledge. To prepare the header information for each British Fish (to verify its British naming among its scientific and vernacular nomenclatures in previous *European* authorities), Sarah had consulted all the works in Cuvier's extensive ichthyology library.[28] Although few contained sketches or outline illustrations, Sarah's plates followed the 'schema' conventions for (Muséum) fish illustration also to be deployed in Cuvier's *Histoire naturelle des poissons*. The standard delineation with the caudal fin to the right enhanced clear figuration of outline shape and anatomical form, including the number and positioning of the fins, and of the lateral line (see Figure 2.1). Sarah's 'basic' figures for the *Excursions* in Chapter 1 retrospectively reveal this Muséum model. By contrast, woodcut illustrations in British ichthyology books, such as H. Cholmondeley Pennell's *The Angler-Naturalist* (1863) – where an impressionistic river or shoreline backdrop for 'The River Bullhead, or Miller's Thumb' is indicative – assert the angler's rather than the naturalist's identifications and interests in its form.[29]

From its first fascicule (1828) onwards, Sarah's *Fresh-Water Fishes* also brought to Paris Muséum house style for ichthyology illustration her greater authenticating hand in the field. None of her British or European ichthyology predecessors provided 'correct' colour illustrations of her order above 'taken from the living Fish immediately it came from the water it inhabited'. None likewise provided eyewitness detail of 'the living Fish' specimen to verify other distinguishing details such as the sizes or shapes of the scales on different parts of the body, again so visible in the Rud or in Sarah's naming of 'Clupea fimbriata, new species, Bowd.' for its 'fringed' scales discussed in Chapter 1. Her 'correct representation' not only surpassed the best experts in the past but also more remarkably Cuvier's 'forthcoming great Work on Ichthyology', accompanied by colour plates made by Muséum specialists. His 24-page Prospectus published in 1826 clarified his methodology and criteria to establish what would be subject-defining about the *Histoire naturelle des poissons*, including its plates. It would examine

> l'organisation de chaque espèce à l'extérieure et à l'intérieure; rapprocher les espèces qui ne diffèrent que par la grandeur, les couleurs, les proportions; en former de petits groupes que l'on rapproche eux-mêmes entre eux d'après l'ensemble de leur conformation, et remonter ainsi à des groupes de plus en plus généraux, que l'on distribue toujours d'après les mêmes règles [...].
>
> Des desseins faits sous nos yeux par MM. Werner et Laurillard, dont les talents sont déjà bien connus du public, suppléent à ce que la parole ne peut exprimer. Il y en aura au moins un pour chaque groupe [...] et l'on les multipliera lorsque les formes singulières le demanderont. [...]
>
> [the organisation of each species both externally and internally; to align species which differ only in size, colours and proportions; to form small groups of these to align among them those in accordance with the whole of their kind, and thus to establish more general groups which we distribute according to the same rules [...].
>
> Drawings made under our instruction by Messrs Werner and Laurillard, already well known to the public for their talents, supplement what words cannot express. There will be at least one drawing for each group [...], more when singular forms require it.[30]

The immediate distinctiveness of Sarah's 'representations' from field observation also permitted her expert adjudication on the confusing identifications of predecessor experts concerning the 'size, colours and proportions' of otherwise similar fishes, for example the Trout, by offering scale-model figures as 'textbook' descriptions: No. 1 Stockbridge Trout; No. 14 Rickmansworth Trout; No. 38 Great Grey Trout. These and all Sarah's watercolour illustrations for *The Fresh-Water Fishes* surpassed the international standards and conventions for accuracy in scientific fish illustration that Cuvier was setting for the *Histoire naturelle des poissons*, as its manuscripts everywhere show. Cuvier's many pencil notes corrected wash colouration and shading on proof copies of the illustrations made by Werner and Laurillard above from specimens – dried or in spirits – in the Muséum collections. Sarah knew them and their work personally, having also worked in the Muséum laboratory dedicated to scientific illustration. Her own style and stylistic development is therefore the more fascinating, and by comparison with her seemingly rudimentary outline drawings of West African Fishes in the *Excursions*. The Rud above is indicative in its all-important detail above, 'irides yellow'. If repetition of 50 times for each Fish enhanced Sarah's observational techniques coordinating expert hand and eye for subsequent work in the field (and ensuing numbers), her particular skill in observational scientific representation is Fish authenticity *by eye*. The Rud's is clearly yellow in the image; its individualised eye is an essential part of her 'correct representation', whereas Werner, Laurillard and other Muséum illustrators mostly draw the same (signature) 'eye' for all species, irrespective of the size, shape or colour of the original in reality, because they worked from preserved specimens: the eye socket was hollow, or completed with an artificial glass eye. But Sarah's authentic eye colouration also creates a further illuminating, and enlivening effect for the whole: she mentions the Rud's overall 'golden hue'. Basic computer graphic features allowed me to set a blue frame around this reproduction to draw out the multiple shading of the body above the lateral line, but I could also experiment with yellow, orange, red, green and black surrounds. Each differently drew out the variations in colouration as the frame colours drew attention to the 'brilliance' of her description and iridescent effects of her shading. The yellow around the eye of the Rud contrastively and comparatively illuminates the body shading below its lateral line, and the yellow-to-orange-reds of its ventral, dorsal and caudal fins.

For the art connoisseur with little interest in the new ichthyological knowledge shining through the beauty of Sarah's 'still lives', Ackermann's endorsement of her work in deluxe illustrated Gift Book format ranked *The Fresh-Water Fishes of Great Britain* as the acme of superior taste, collectability and superlative artistic production. Her 'taste and industry' (*Literary Gazette* above) in pioneering a new science of applying foils – literal and metaphorical – to the study of fish blew out of the water by comparison what Cuvier's magisterial, illustrated *Histoire naturelle des poissons* would achieve. The highest new standards for scientific illustration of Fishes from life in the combined expertise of Sarah's rendering of scientific and artistic observation remain an astonishing achievement in their time and still today. In Lord Tabey's prescient words for the original venture (see Appendix 3) which would also have fallen far short of Sarah's *Fresh-Water Fishes*, 'nothing can be had *excellent* and *cheap*.' Production effort and publication costs made Sarah's unique venture impossible to copy or easily replicate, although the Cornish ichthyologist and fine watercolourist, Jonathan Couch (1789–1870), also captured fishes 'from life' in his privately published four-volume *A History of the Fishes of the British Islands* which appeared after Sarah's death.[31] The same challenge of showing fish in their 'true' colours remains today, despite major advances in underwater photography. The quality of the water, the camera or the publisher's digital reproduction often dims results, or obscures overall clarity of the observational and scientific detail that Sarah's 'correct representations' so successfully achieved. Once seen, like the Fish in question, her illustrations and scientific descriptions are indeed 'forget-me-nots', the most illuminating (scientific) *aide mémoire* for the (expert) Naturalist, the Amateur and the Angler. Brilliance is verifiable scientific truth in the field, and in the museum laboratory.

'Truth'

The Fresh-Water Fishes of Great Britain also reflects by adoption a key practice in the zoological appendix of the *Excursions*, namely Sarah's meticulousness regarding Fish provenance information. It is exemplified in the Rud above, and given in almost all cases because the river of origin matters for identificatory ichthyology, as the Trout distinctions above exemplify. I earlier footnoted the environmental interests of such provenance 'data' in her work for comparative river quality, fish stocks and pollution today, but did not investigate them further.[32] In Table 2.2, I again enlist the larger perspectives of this 'map' by reformatting my footnote on Sarah's provenance information, to draw new attention to the pioneering perspectives it reveals of the *Fresh-Water Fishes* as a whole. My new caption, clarification by dateline rather than number of the serial publication, and grey background highlighting together illuminate the extent and unprecedented reach of Sarah's British 'excursions in forty-four Fishes', to capture her specimens from life for larger knowledge of British and *European* freshwater ranges. In effect, Sarah mounted

Table 2.2 Provenance data: Sarah's Tour of Britain in 44 Freshwater Fishes

For 1828:	the Trout	('Fordwich in the river Stour');
	the Carp	('a pond belonging to Mrs Marryat, of Wimbledon House, in Surry (*sic*)');
	the Roach	('Bures St Mary's in Suffolk [...] on the banks of the Stour');
	the Bleak	('Henley on Thames');
For 1829:	the Perch	('in the mill-pool at Henley-on-Thames');
	the Chub	('the river Lea, near Ware, in Hertfordshire');
	common Eel	('the Thames');
	the Minnow	('Henley');
For 1830:	the Barbel	('the Thames at Henley');
	the Pope	('same gravel pool at Henley on Thames, from which the Perch, Drawing V, was taken');
	the Dace	('the Thames, between Kew and Richmond bridges');
	the Loach	('the New River, near Broxburn, in Hertfordshire');
For 1831:	the Tench	(no provenance);
	Rickmansworth Trout	('between Rickmansworth and Stockbridge');
	the Gudgeon	('the river Stour, near Sudbury, Suffolk');
	the Lamprey	('Mortlake on the banks of the Thames');
For 1832:	the Pike	('Trimlee, in the neighbourhood of Ipswich');
	the Bream	('the Thames');
	the Thames Shad	('the Thames');
	Stickleback – two species	('Carshalton out of the Wandle');
For 1833:	the Rud	('the Stour, close to Sudbury');
	Glut Eel	(no provenance);
	the Crucian Carp	(no provenance);
	the Bull-head	(no provenance 'in small rivers, and almost clear brooks');
For 1834:	the Flounder	(the 'Thames Flounder');
	the Lamprey	('Little Bench, close to the city of Gloucester');
	the Shad	('the Severn');
	Grig Eel	('the Thames, but more especially in Berkshire and Oxfordshire');
For 1835:	the Severn Salmon	(the 'Severn');
	the Barbolt	('some miles up the Orwell, in Suffolk');
	the Smelt	('Rochester');
	the Pride	('the Chelt [...] common in the small rivers of Gloucestershire and Oxfordshire');
For 1836:	the Grayling	('the Wye, which flows through Bakewell, in Derbyshire');
	the Gwyniad, or Schelly	('Tower Hallsteds, on the borders of [...] Ullswater');
	the Northern Charr	('the lake of Coniston');
	the Azurine or Blue Roach	('Houghton, close to the seat of Lord Derby');
For 1837:	the Parr or Samlet	('The Skegger has now almost entirely disappeared from the Thames, in consequence of the various manufactures on the banks of the river');
	the Great Grey Trout	('the lake' – Ullswater);
	the Graining	('streams in Lancashire');
	the Snig Eel	('the Avon in Hampshire');
For 1838:	the Bull Trout	('the Tweed, close to the town of Berwick');
	the Vendace	('Castle Loch');
	the Parr	('the Whitadder, one of the tributaries of the Tweed');
	the Bream-flat	('Bures on the borders of the Stour')'.

a new independent expedition for (French) ichthyology that creatively, centrally and concomitantly accommodated in its strategic itinerary of London's important rivers – in grey background highlighting – her support for her family now of secondary-school age.

Between 1828 and 1838, the larger fact of this 'map' is that only a third of Sarah's indicative British Fishes were taken from the Thames and its tributaries as her 'local' river with easiest access. Her collecting took her to Kent (Fordwich is down river from Canterbury), to Suffolk (close to where she grew up), Hertfordshire, the Test Valley (between Rickmansworth and Stockbridge), to Gloucestershire and the Severn (nearer Bowdich relations in Bristol), and much further afield for Fishes in Nos. 36–38 in Derbyshire, the Lake District, Lancashire, Hampshire, Berwickshire and her one trip to Scotland, to Loch Maben (Dumfrieshire). These 'excursions' represented significant organisational undertakings in terms of landowner and angler contacts, journeying and accommodation costs and other financial outlay. In Sarah's 'plan' above, 'variety' in each number of four Fishes in terms of interest (colour, rarity, size, etc.) also now reveals how it defrayed her many 'costs' of labour from the life – in the field, in her London home – to make her 50 correct representations of each new Fish.

In its wider British cultural interest, Sarah's *Fresh-Water Fishes* should therefore have attracted the greater attention both of wealthy connoisseurs and of those seeking to make their names in British Ichthyology. As I have shown, however, Sarah's contemporaries also working with Cuvier's new classifications in the 1830s and 1840s – Robert Hamilton, William Jardine, Leonard Jenyns, William Swainson, James Wilson and William Yarrell – displayed a startling disregard concerning the existence, let alone excellence of her work.[33] To add to its/her firsts in Table 2.1, here was the first dedicated study of the *freshwater* fish species according to Cuvier's latest 'system' of classification in any region of the globe, and the first to identify their number and variety in Britain (excluding Ireland) from a thoroughgoing eyewitness inspection that also offered comparative European information (see the Rud once more). Almost as an aside in his Prospectus of 1826 for the *Histoire Naturelle des Poissons,* Cuvier had remarked upon the significant gap in knowledge that Sarah's *Fresh-Water Fishes of Great Britain* not only swiftly filled but also significantly set new standards for: 'Nous nous sommes particulièrement attachés à nous procurer les poissons d'eau douce de l'Europe, d'ordinaire si négligés dans les cabinets' (we have been particularly keen to procure European freshwater fishes, which are usually so neglected in collections).[34] Various descriptions therefore demonstrate Sarah's early understanding of the science of fish colouration as determined by the reproductive lifecycle. Three descriptions are indicative of the powers of Sarah's specialist eye and knowledge as 'satisfactory' at once to 'the Naturalist' and to the expert Amateur/Angler:

> The Chub alters very much at, and just before, the spawning time. The green of its head, the gold of its scales, and particularly the red of its lower fins, all become more bright and decided, and it is then a handsome fish.' (*FWFGB,* No. 6, Chub n. p.)

> Age and difference of waters and food, have a great effect on the colours of this fish; some of them are of a bright golden tinge, and others of a dark and dusky hue. The males are said to be paler than the females. I have heard from two witnesses that at Glassenbury, in Kent,

> near Cranbrook, the ponds contain Tench with blood-red spots on their sides, but I have never met with any, or been told of any elsewhere. I have searched through all the principal works on Ichthyology, and, except the variety called by Bloch Cyprinus tinca aurata, which has black spots, I do not find authority for any thing (*sic*) of the kind. (*FWFGB*, No. 13, Tench n. p.)

> The colouring of the annexed drawing is far from shewing the Charr in the state in which it is most pleasing to the eye, but it is that which marks it when best for the table; at other times the rose turns to a brilliant orange, and all the colours are heightened. […] The different hues which the Charr assumes has given rise to a great confusion in the nomenclature of its varieties; it would seem, however, that Mr Agassiz and Mr Yarrell have reduced those of Great Britain to two species. (*FWFGB*, No. 35, Northern Charr n. p.)

It is then not without irony that of Sarah's 50 British subscribers (see Appendix 4) only Sir Humphry Davy (1778–1829), a keen fisherman and published expert on Salmon in 1828,[35] had the necessary acumen to appreciate Sarah's path-breaking work. He did not live to see all its numbers. Sir William Jardine (1800–1874), resident at Castle Loch near Loch Maben and the author of *Illustrations of British Salmonidae* (1839–1841) was also not among her subscribers, although Roderick Murchison (1792–1871) was, and the likely '*M*' in her letter of 1824 (*S* being Sowerby in all probability, but not a subscriber). In all three excerpts, Sarah's remarkable economy and precision of detail provide invaluable field observations for the most informed. Moreover, the last example demonstrates that Sarah kept abreast of the latest contributions to international freshwater ichthyology by Yarrell in Britain, and Agassiz for Switzerland, especially after Cuvier's death in 1832 (when she also lost direct access to his library). In French history of ichthyology, only Marie-Louise Bauchot has traced Sarah's name for her artistic involvements within Cuvier's *Histoire naturelle des poissons* as the 'copyist' of the Forster and Parkinson drawings made on Cook's voyages, but not as the author of the *Fresh-Water Fishes of Great Britain* cited within it.[36] The final part of this chapter can now fill the 'gap' in Cuvier's 'Prospectus' for the first time, by highlighting the many references to Sarah's work on British/European freshwater Fishes in its later volumes.

To return to the three ichthyological legacies of Chapter 1, Achilles Valenciennes was the all-important intermediator for the future of Sarah's work in British ichthyology both before and more crucially after Cuvier's death. In 1829, the *Histoire naturelle des poissons* saw publication of volumes 3 (completion of the Perch family of Fishes) and four bony Fishes of the Scorpaenidae family (such as Mullets, Scorpion fish, Bullheads and Gasterosteidae such as Sticklebacks). Since members of these families are represented extensively across the globe (in mainly marine, but also in brackish and freshwater habitats), they presented a monumental (re)classification challenge for Cuvier. In the Preface to Volume 4, he reported how he met it by sending Valenciennes and Laurillard to major specialist collections – the first to London and Berlin, the second to Nice for nine months – to research, verify and make drawings of the Fishes in question (v–vii). While Sarah was not specifically mentioned here among the 'amis' (friends) in London of the *Histoire naturelle des poissons* project, her work was directly included, both through references to (Georg) Forster's drawings in Banks's Library (for New Zealand Fishes in these families) and a 'beautiful' drawing of a Madeiran scorpaenoid Fish 'par Mme Bowdich' (298). By early 1829, Sarah had not collected her 'correct representation' of the 'Bull-head' (No. 24), meaning that her description could not be referenced in its correct place – by Order and then Family – in Cuvier's 1829 volume. From volume 9 in 1833, Valenciennes assumed the full responsibility for Cuvier's 'plan' and maintained his connections with 'London' correspondents, including Sarah's continuing work copying drawings in Banks's library.

In terms of key Families that include various freshwater species, Cyprinoid Fishes had to wait until volume 16 (1842) for specialist treatment, by which time Sarah had completed the full set of her 'representations' and numbers of *The Fresh-Water Fishes of Great Britain.* Although there is no library catalogue trace in France of copies of her work, the evidence that Valenciennes consulted it emerges in the *Histoire naturelle des poissons* from volume 16 as the first to cite it extensively. The first reference pertains to the 'Carpe Carassin' or Crucian Carp (*Cyprinus carassius,* Bloch), particularly regarding its colours:

> Pennant avait aussi connaissance de ce poisson sous le nom de *Crucian carp.* [...] Depuis Pennant, les zoologistes anglais Turton, Flemming (*sic*), Jennyns (*sic*), Yarell (*sic*) l'inscrivent dans leurs ouvrages. Mme Lee (Bowdich) en a donné une représentation dans son bel ouvrage des poissons d'eau douce de la Grande-Bretagne. Il me paraît que les couleurs argentées dont elle l'a ornée sont un peu brillantes, ainsi que le rouge des nageoires. (87)
>
> [Pennant also knew of this fish by the name of 'Crucian Carp' [...]. Since Pennant, the English zoologists Turton, Flemming (*sic*), Jennyns (*sic*), Yarell (*sic*) inscribed it into their works. Mrs Lee (Bowdich) has given a representation of it in her beautiful work on the freshwater fishes of Great Britain. It seems to me that the silvery colours with which she has decorated it are a little bright, as also is the red of the fins.]

The 'brightness' that Valenciennes questioned should not invalidate Sarah's 'brilliance'. He would not have seen specimens captured immediately after they had been taken from the waters in which they were living. In Chapter 11 of the same volume (on Gudgeon and Tench), Valenciennes commented upon the 'excellentes figures' (excellent drawings) of both Fishes in 'Lee, *Fresh-water fish. of Brit*' (309 and 331). The same artist seemingly attracted his contradictory views.

In volume 17 of the *Histoire Naturelle des Poissons* (1844) dedicated to 'Ables' or White Fishes, the description and figure by 'Lee' are similarly noted for the Bream (not fascicule 10 as stated, but No. 18, (22)) and the Rud (again wrongly noted as No 31 instead of as above No. 21 (118)). In the same volume, Valenciennes also singled out for the Dace (la Vandoise) the work of 'M. (*sic*) Bowdich' and 'l'expression de son élégant et habile pinceau' (Mr. (*sic*) Bowdich and the expression of 'his' elegant and expert paintbrush) (211). Sarah's work is similarly cited for the Bleak (282; *FWFGB*, No. 4) and the 'Minow' (*sic*) (373, *FWFGB*, No. 8). Valenciennes however failed to notice 'comme une espèce distincte' (as a distinct species) (217) in *Leuciscus lancanstriensis*, Shaw the Graining as also represented in Sarah's penultimate fascicule (*FWFGB*, No. 39). Perhaps the more fascinating and precious aside concerns Valenciennes' explicit acknowledgement of *her* continuing work for this volume of the *Histoire naturelle des poissons* as its London collaborator. Sarah was his major expert authority through her several important drawings of Chinese White Fishes in Banks's Library collections, to enable him to distinguish between very similar species (257, 360, 362, 363), including the Silver Carp (*Leuciscus molitrix,* nob.). While Sarah clearly had not made these drawings 'from life' to render their colours and proportions 'correctly', her expertise and eye for the important differences within this Family for her *Fresh-Water Fishes* lent Valenciennes her rare subject expertise that enabled his.

The final reference to her work on British freshwater fish then occurs in volume 21 of the *Histoire naturelle des poissons* (1848), treating Salmonid Fishes. It concerned the 'Northern Charr' (see above): 'il existe dans le recueil de Madame Bowdich une brillante représentation du Charr des contrées septentrionales de l'Angleterre' (242; *FWFGB*, No. 35) (In the collection by Mrs Bowdich, there is a brilliant representation of the Northern Charr of England). In the same volume, Valenciennes then surprisingly failed to note her equally 'brilliant' representations of two rare species of British *Coregoni*, the (Welsh) Gwyniad (*FWFGB*, No. 24) and the Vendace of Loch Maben in Scotland (*FWFGB*, No. 42). 'Brilliance' and 'Truth' however sum up Sarah Bowdich's remarkable number of listings in the *Histoire naturelle des poissons* as a 'zoologist' of the first order of international importance with figures such as Agassiz, despite working outside the ranks of Muséum science.

In Sarah's preface to her *Fresh-Water Fishes*, Cuvier's 'great work on Ichthyology' was the measure for her new scientific knowledge and classification of fish. By the same measure, the 'brilliance' and 'truth' (quality and expertise) of her application and execution to recover British-European freshwater species for science in 1828–1839 could not be better illuminated than in the many citations by Valenciennes from 'le recueil de Madame Bowdich' above in respective volumes of the *Histoire naturelle des poissons*. It remained the definitive work for world ichthyology as the 'FishBase' or 'CLOFFA' of its time discussed in Chapter 1 until at least the end of the nineteenth century. Too few ichthyologists subsequently have paid attention to repeated (expert) mentions hidden in plain sight within its foundational volumes. As the Red List of endangered native species grows longer, Sarah's phenomenal and pioneering study of British Ichthyology speaks almost more strongly today about conservation of specific river habitats and the collection of faunal heritages for their widest appreciation by expert, amateur and angler alike in cognate habitats of the world. Sarah's clearest unwritten statement in the complete set of her published numbers is the testimony of their result: to leave the laboratory is to enter the interconnecting rivers of (freshwater) ichthyology so that the many differences (diversity) of *local* colours in fish populations are better acknowledged.

Capturing Fishers from the life was not the only mark of the new comprehensiveness of Sarah's *Fresh-Water Fishes.* In its inclusion and arbitrations on British cartilaginous Fishes that are Eels and Lampreys, it surpasses the *Histoire naturelle des poissons,* which contains no volume on them. Moreover, and unlike the works by her male peers in Cuvier, Valenciennes and Yarrell, Sarah's larger legacy in her Gift Book to ichthyology was its appeal for the importance and significance of next-generational national and international freshwater habitats, including their scientific beauty and truth. Her list of subscribers is then worthy of notice, not only as representative of the great, good and affluent (the landed gentry and bankers) but also of the wider audiences Sarah had set out to captivate. Almost a fifth are women, clearly engaged in their own 'scientific' interests, such as the geologist and fossil collector Barbara Rawdon Hastings (1810–1858) listed as 'Baroness Grey de Ruthyn', or because they were widely cultured, such as 'Miss Waddington' one of the sisters of George Waddington, traveller, theologian and scholar. A further significant group were gentlemen scientists, such as Herschel and Murchison, and in J. G. Children representative figures building the collections of the British Museum, as well as publishers such as John Murray and illustrators like Dr Frank. In her new departures for (French) ichthyology in London, Sarah's expedition of Britain in 44 freshwater Fishes remains 'standout' not only for its rare scientific acumen in text and image but also for its pioneering comparative field perspectives for ichthyology that provide far-reaching environmental understandings today.

Notes

1 See Orr, 'Women Peers'.

2 In a note to the fourth part of her *Memoirs of Baron Cuvier*, Sarah acknowledges various key 'friends' including 'MM. Desfontaines, de Jussieu, Brongniart, Geoffroy Saint-Hilaire, Frederick Cuvier, Chevreuil, Valenciennes, Deleuze, Laurillard and (the late) MM. Haüy, Latreille, Thouin, Royer, Dufresne, Vanspaendonck and Lucas' (*MBC*, 159–60).

3 See Orr, 'Fish'.

4 Orr, 'Fish', 206 and 235.

5 Orr, 'Fish', 220.

6 Orr, 'Fish', 215, 223–24.

7 Orr, 'Fish', 226–30, 236. See also my expanded study of the (mal)treatment of Sarah's text by Leonard Jenyns, Orr, 'Catalysts', 514–16.

8 See 'ncse: nineteenth-century serials edition'. http://www.ncse.ac.uk/reference/bibliography.html.

9 Jonathan R. Topham, 'Science Publishing and the Reading of Science in Nineteenth-century Britain: A Historiographical Survey and Guide to Resources', *Studies in History and Philosophy of Science, Part A* 31, no. 4 (2000): 559–612. Topham supplies a useful table of (London-based) publishers in 1837 engaged in producing works in 'Medicine & Surgery', 'Natural history, botany, agriculture, &c' and 'Science, including metaphysics, and physics &c.', with their percentage stake in these categories and markets overall.

10 Longman had the largest slice of the trade. See Topham, 'Science Publishing', 34.

11 See David Elliston Allen, *The Naturalist in Britain: A Social History*. Princeton: Princeton University Press, 1994.

12 In effect, it was a 'Great Exhibition' in microcosm of the latest (re)productions long before the world event of this name in 1851.

13 See Appendix 1B for Sarah's many contributions to *Forget Me Not* and other gift books of the period, and Orr, 'Amplifying Women's Intelligence', for a larger discussion both of the genre and Sarah's work within it.

14 See Beverly E. Schneller. 2006. 'Jerdan, William (1782–1869)'. Oxford Dictionary of National Biography. Oxford: Oxford University Press.

15 Orr, 'Fish', 220, note 37.

16 William Jerdan, *The Autobiography of William Jerdan*. 4 Vols. London: Arthur Hall, Vertue Co., 1852–53. It is consultable online at http://www.lordbyron.org/contents.php?doc=WiJerda.1852.Contents.

17 From the electronic edition, 144–47 at http://www.lordbyron.org/contents.php?doc=WiJerda.1852.Contents.

18 Orr, 'Fish', 232. Bibliographical references for *The Field* are in its note 74.

19 Orr, 'Fish', 232.
20 Barbara Gates, *Kindred Nature: Victorian and Edwardian Women Embrace the Living World.* Chicago and London: University of Chicago Press, 1998, 77.
21 Huon Mallalieu, 'The Compleat Ichthyologist', *Country Life*, 26 Feb. 2004, 78, and Christine E. Jackson, *Fish in Art.* London: Reaktion Books, 2012, 216–17.
22 Gates, *Kindred Nature*, 78.
23 See the DSI-Database of Scientific Illustrators: http://www.uni-stuttgart.de/hi/gnt/dsi2/index.php?function=search&table_name=dsi.
24 Orr, 'Fish', is an exception.
25 Orr, 'Fish', 223–24.
26 The colour image of the Rud can be viewed in the recent blog by Helen Carron, '"Rare Book Series". Continuing the Exploration of Women Illustrators' (17 Sept. 2020). https://www.emma.cam.ac.uk/members/blog/?id=472
27 Orr, 'Fish', 234, for the inventive use of a bladder in which to transport live Graining from Lancashire to London.
28 Orr, 'Fish', 222.
29 H. Cholmondeley Pennell, *The Angler-Naturalist: A Popular History of British Fresh-Water Fish. With a Plain Explanation of the Rudiments of Ichthyology.* London: John Van Voorst, 1863, chapter VI, 'The River Bullhead, or Miller's Thumb' (63).
30 Georges Cuvier, *Prospectus: L'Histoire naturelle des poissons.* Reprint. Paris: Imprimerie de Mallet-Bachelier, 1862, 19–20. The translation is mine.
31 Jonathan Couch, *A History of the Fishes of the British Islands.* 4 vols. London: Groombridge and Sons, 1860–65.
32 Orr, 'Fish', 239, note 89.
33 Orr, 'Fish', 226–30.
34 Cuvier, *Prospectus*, 16.
35 Sir Humphry Davy, *Salmonia: or Days of Fly-Fishing.* London: John Murray, 1828.
36 Bauchot, Daget and Bauchot, *L'Ichthyologie en France au début du xixe siècle.*

Chapter Three

A FIRST SCIENTIFIC BIOGRAPHY 'FROM A WOMAN'S PEN': THE *MEMOIRS OF BARON CUVIER* (1833)

When Tedlie Hutchison Hale wrote to *The Field* on 14 January 1888 to keep her mother's 'venerable name from oblivion' as the author of the 'rare' *Fresh-Water Fishes* discussed in Chapter 2 (p. 56), she was also correcting the initial correspondent's larger assumption: a woman's could not have been the pen (or pencil) of a work on Fishes/of science. Sarah's 'venerable name' was therefore doubly important to Tedlie, both as distinct from her father's and with regard to 'Cuvier'. In the next sentence of her letter, she supplied even rarer report of her late mother's own views on her contributions to natural history: 'The work by which my mother most wished to be remembered was her "Memoir (*sic*) of Cuvier," published by Longman, now, unfortunately, out of print.'[1] In light of Chapters 1 and 2, Sarah's 'Rare Book' needs no linking to Cuvier. Her *Memoirs of Baron Cuvier* was no lesser a rarity, however, whether in 1833, 1888 or when digitisation restored it to open access to its original plural readerships, including 'experts' thanks to its modern University Press reedition. As a book 'of enduring scholarly value' in the Cambridge Library Collection's electronic reprint of 2014, this 'biography [...] remained the authoritative work in English on the most distinguished scientist of the age'.[2] Almost more importantly in its day and today, Sarah's work appeared within a year of Cuvier's death in 1833 in three different imprints with major French, United States and English presses recorded in Appendix A. Why had the now open access *Memoirs of Baron Cuvier*/*Mémoires du Baron Georges Cuvier* not attracted more than footnote interest in twenty-first century studies of Cuvier in French and English?[3] To right the potentially larger omissions concerning the French edition for French history of science and scientific biography, my first reappraisal in 2020 of the unprecedented 'Mistress Lee' on its cover put the case for the work's singular importance.[4] Rare was the woman (in France) penning science in 1833 or the (first) biography of an eminent French 'savant' (man of science) such as Cuvier; even rarer that she was British. Rarest of all, was Sarah's skilful reattribution to Cuvier in her *Memoirs*/*Mémoires* of the very paradigm that he had perfected in respect of other major 'savants' (m. pl), the *Éloge scientifique* (official encomium by a peer to the eminent life and major scientific accomplishments of the deceased[5]). That she could then redefine his *Éloge* model in her expansions of it the better to honour him precisely for his (French-)*international* scientific endeavours stemmed from her strategically bifocal viewpoints, crystallised in her choice of title and genre for the 'Memoirs/Mémoires' in light of its treble audiences. In French the noun *mémoire* (m.

and m. pl.) defines the scientific paper delivered or published through learned societies, institute annals and historical reports.[6] The noun *mémoire* (f.) is the faculty of memory by which the past can be remembered and represented. Sarah's *Mémoires* therefore doubly demonstrated her arts as Cuvier's foremost '*mémorialiste*' (historiographer), to commemorate him in and outside France.

By guaranteeing the memory of Cuvier's name for posterity on both sides of the Channel and the Atlantic, Sarah's titular plural *Memoirs* had further definitional traction in English in 1833, as the OED ascertains through key dates in the 1820s:

> **1.** A note, memorandum; a record –1755. † **2.** [...] MEMORANDUM (*rare*). Also *pl.* official reports of business done. – 1829. **3.** *collect. pl.* **a.** A record of events a history treating of matters from the personal knowledge of the writer or the reference to particular sources of information. 1659. **b.** An autobiographical record 1673. **4.** A biography, or biographical notice 1826. **5.** An essay on a learned subject on which the writer has made particular observations. Hence *pl.* the record of the transactions of a learned society. 1680.

Definition four is particularly noteworthy given Sarah's own term in the introduction for her work specifically as a 'biography' (*MBC*, 4). Definitions three (a. and b.) and five further encapsulate the bi-focal purpose of Sarah's 'memoir' project. Her scripting of Cuvier's extraordinary international scientific, legislative and interpersonal reach not only reinvested her mentor's model *Éloge scientifique* with broader generic (international non-French) afterlives. As his biographer, she could also include her own 'life in science' as determined by his in her account. At the same time, however, the tensions of the stylised official (French) format and (Cuvier's) reputation (of which more below) were fully in view in Sarah's rationale for her four-part *Memoirs/Mémoires*, clearly described in its 'plan' (*MBC*, 6) concluding her introduction:

> Unwilling to incur the risk of confusion, by mingling too much anecdote, either with my narrative of events or description of scientific and legislative labours, I have divided the present volume into four parts or portions, that each may bear its own share of detail. The first will give the data of all the important circumstances of the Baron Cuvier's life, in their respective order; the second will contain an account of his various works, as a savant and philosopher; the third will be devoted to his legislative career; and the fourth will be chiefly confined to those anecdotes which will best illustrate his character as a man. In following this method, I may, probably be led into something like repetition; but I hope I shall be excused, if each part shall be found to contain a whole in itself, which facilitates reference. (*MBC*, 6)

The 'Great Life–Great Works' model for the (French) funerary oration, the *Éloge scientifique* and national historiography was then the 'fixity of form', of which more below, that Sarah daringly remade in 1833 with fullest respect to Cuvier through additionally reflecting his public service and private life (parts three and four respectively) in his 'circumstances' and 'works' (parts one and two). Indeed, her very use of the word 'plan' as Cuvier's word in all his major works for their principal design, demonstrates in this one illustration the economy, precision and informed attention to fact that is Sarah's use of scientific language in her 'Introduction'. Her formal tribute to the four main 'domains' of Cuvier's integrated scientific life further vouched for her unusual insider-outsider scientific knowledge of it, including his four-part classification of the Animal Kingdom (his *Règne Animal*, described in part two of the *Memoirs/Mémoires*).

As the pendant to my study of the French edition, this chapter focusses sharper attention on Sarah's remarkable *Memoirs* in its English and American editions for the history of science.[7] Its magisterial blend of extant (French) 'science biography' genres in new intercultural form also discloses the extent of Sarah's (French-international) scientific acumen in her overt acknowledgement in this 'introduction' (for Francophone and Anglophone audiences): hers in 1833 was the first 'biography' of Cuvier full stop ('point final'). If this French period (in both senses) spearheaded my contribution to the history of (women in) French science by making better known the extraordinary pen of 'Mistress Lee', the full stop weight of her work in its three editions is larger. Generically, the *Memoirs/Mémoires* configured in 1833 what would become the 'science biography'.[8] Its forerunner status is therefore coextensively determined by the two major interrelated questions that historians of (women in) science had not asked or addressed (until my 2020 intervention). Sarah herself reformulated them in 1833 in the opening statement of her introduction to the English *Memoirs*:

> Before I enter upon the subject of this volume, I would explain to my readers the motives which have induced me to write it, in order to prevent that appearance of presumption, which may naturally be laid to the charge of an unlearned person, who attempts to write the life of so illustrious a savant. (*MBC*, 3).

In immediately drawing attention to her 'woman's pen' in other words, Sarah preempted the tacit question(s) centrally in her mind, because indubitably in the minds of her readers: why was *she* the writer, when eminently better qualified for the task were the men of science in France (Cuvier's 'savant' peers) or in Britain? Exemplifying the latter was the 'Dr Leach' (William Elford Leach, 1791–1836), whose letter of introduction to Cuvier for the Bowdichs was 'scarcely necessary' (*MBC*, 5). Of principal note here is Sarah's deft rhetorical turning of this major double question as part of her no less magisterial claim to authorship in the twice-repeated 'I' of this sentence. Her opening gambit therefore exemplifies the modesty *topos* (see the introduction, p. 11) of (false) 'presumption' in the face of (false) 'charge', but particularly in respect of the 'naturally' operating Sarah's telling chiasmus: 'an unlearned person' (gender neutral, yet sexed) was the unusual 'peer' of 'so illustrious a savant' (clearly synonymous with Cuvier).[9] In therefore committing to print his life in the advancement of science (to include her own), Sarah's were the largest and highest authorial risks and reputational stakes in 1833, when his French peers and her 'countrymen' in Britain could not. In consequence, Sarah again addressed the(ir) very different reasons to demur (further justifying her *Memoirs/Mémoires*) through direct contrast only with the 'public notices which appeared in England concerning the Baron Cuvier' (*MBC*, 2): 'though all did him the justice of placing him above every other naturalist, not one spoke of his talents as a legislator, and all equally neglected his private character. This and the almost universal incorrectness of detail, no doubt proceeded from ignorance rather than intention.' Her contrary account of avowed 'facts' (*MBC*, 4) as the adept and singularly placed 'unlearned person' (countering such 'incorrectness' and 'ignorance') then mark out Sarah's pioneering contributions as 'a biographer of savants' in her own words (*MBC*, 4). Although she could have no position in official science-making, not only did the *Memoirs/Mémoires* command immediate respect on both sides of the Channel (and Atlantic). Her unprecedented contributions to scientific biography also 'remained the authoritative work in English on the most distinguished scientist of the age' (quoted above) for at least a century, but only until history of science from the 1970s redefined 'authoritative'. The interrelated questions of a woman's pen for science, and on account of a major (inter)national scientist, therefore centrally inform quality judgements of the *Memoirs* for its times and today.

Invaluable Cross-Channel and Transatlantic Critical Insights in 1833

The study (hitherto not undertaken) of three extensive indicative (anonymous) Anglophone reviews upon the appearance of the *Memoirs* in 1833–1834 is revelatory. All three confirm the unquestioned status of Cuvier at the pinnacle of (French-international) natural history upon his death, and the imponderable fact of Mrs Lee's (un)suitability to pen her *Memoirs* of him. In so doing, the reviewers also affirm Sarah's unerring judgement in addressing her task from the outset of her Introduction (quoted above) and as central to her strategy for its four-part form. For the reviewer of *The Monthly Review*

> [i]t would have sounded very strangely indeed throughout Europe, that the fame of one of the most illustrious naturalists of modern times should be left to the guardianship of a lady, who, however endowed with native ability, and however furnished with acquired accomplishments must, from the very necessity of the case, have been wholly, unable duly to estimate the merits of her hero. But when we consider the circumstances under which this biography was composed, we shall see that the whole of the difficulties to which the above statement gives rise vanish like the vapour before the morning sun. The exploits of Cuvier as a man of science were not left to be determined by Mrs. Lee or any other biographer; they were well defined already. What was wanted by the world was some account of the individual, some insight into the husband, into the father [...] Who was to supply the desideratum? [...] All the world will agree that it ought to have been some one (*sic*) who had [...] seen him in the maturity of his experience and genius, and who was known to have enjoyed his confidence to that degree that would constitute in the eyes of the world a guarantee that at least a genuine account could be given of his personal peculiarities, his habits, &c. That Mrs. Lee comes under this description no one will deny who is acquainted with the close intimacy which has subsisted for several years before his death between Cuvier, with his family, and his present biographer.[10]

The reviewer for *The Eclectic Review* clearly took greater notice of the first sentence of Sarah's introduction:

> IF we were disposed to wish that the task of writing the life of this 'illustrious savant' had fallen into other hands, the un-presuming manner in which Mrs. Lee explains the circumstances

> that, in a manner, devolved upon her the honourable office of the biographer, would render it alike ungenerous and unjust to impute presumption to her for undertaking it, or to criticise with fastidiousness her performance. On the contrary, we feel under obligation to her for bringing before the English public this interesting and authentic account of her distinguished friend. Cuvier is a name which has become identified with science, and, like those of Linnæus, Buffon, and Davy, must share in the immortality of the knowledge which he contributed so greatly to advance. But Cuvier was not only distinguished as a man of science: his accomplished mind and estimable character rendered him an ornament of society, the centre of the social circle in which he moved, and the object of affectionate regret and veneration.[11]

In a survey of the *Éloges* on Cuvier that appeared in 1834 (by C. L. Laurillard, G. L. Duvernoy and M. E. Pariset), the reviewer for the New York *Foreign Quarterly Review* notably saw fit to include Mrs Lee's *French Mémoires* among them:

> Mrs. Lee's book, already well known in our own language, is the record of an accomplished friend, who, exhibiting in her appreciation of the writings and public services of Cuvier, a delicacy, a discrimination, an extent of information, and a modesty, most honorable to her sex, has also painted him as he was in private life, and in the bosom of his family, amidst the tranquil occupations of his study, or when sustaining as became him the domestic griefs which in his later years overshadowed him; and she has done this with a fidelity and a pathos to which we think the sympathy and tears of many readers must have borne an unsuspicious testimony. [...] We must let M. Duvernoy describe the minute traits of one whom he most intimately knew. We quote his Notice rather than the more ample account of Mrs. Lee, because no English reader should omit the perusal of one of the most elegant, judicious, and affecting pieces of biography that ever proceeded from a female pen.[12]

That all three review(er)s doubly highlighted her sex – 'Mrs Lee' is a 'lady', hers 'a female pen' furnishing the title for this chapter – with regard to her subject Cuvier as 'man of science' only further confirmed their important contemporary acknowledgement and unanimous adjudication of the authority, worthiness and 'judiciousness' of Sarah's *Memoirs*. In its delivery, the 'authentic', 'accomplished' and 'ample account of Mrs Lee' also vouchsafed against false misapprehensions that this author was 'unknown', or 'unable duly to estimate' her subject (Cuvier, his science). Moreover, the *Foreign Quarterly Review* offers precious comparative evidence that successor *Éloges* by named figures all in Cuvier's closest circles at the Muséum also supplied insider ('intimate') knowledge of him. Sarah's *Memoirs* and contributions to scientific biography cannot then be derivative or rendered second-rate through their potentially gendered focus in the final part on Cuvier's 'private' as opposed to his 'public' spheres of science. All three reviews therefore offer precious insights into the as yet unwritten rules for (scientific) biography of the period, especially in light of the OED's definition four above for the 'Memoir' dated 1826. In unequivocally counting 'Mrs Lee' as foremost among such 'biographers' (without sex), her work was also clearly on par with the best French writers of *Éloges* (*scientifiques*) mentioned above. As *The Monthly Review* additionally underscored: 'The exploits of Cuvier as a man of science were not left to be determined by Mrs. Lee or any other biographer; they were well defined already.' Indeed, Sarah had already stated clearly that 'the labours of M. Cuvier speak for his wonderful mind; and time alone can show, to its full extent, the influence of that mind upon science' (*MBC*, 4). The 'facts' that are illustratively quoted in these reviews from part 1 (Cuvier's CV) and the longest part 2 (description of his many scientific publications) of the *Memoirs* were not in doubt. No review took issue either with Sarah's stated 'English' pitch:

> I applied to the relatives of Baron Cuvier for data. These data were contributed with a readiness which vouched for the sentiments of the family [...]. Recollection crowded upon recollection, anecdote upon anecdote, till, in a short time, it became very difficult to select from the mass. [...] Reflection whispered, that I was able to correct the many errors afloat; that, perhaps, *I was the only one in England,* who, from having been received into the bosom of the family, could personally speak of various circumstances and events; and when I thought of all the affection and kindness I had received, I began to feel that there would be a degree of ingratitude in remaining silent, and determined that I *would, independent of all other publications, attempt to lay open to the English world the noblest part of the gifted individual—his heart.* (*MBC*, 3–4, emphasis added)

In this last as the 'chief purport of the present biography' (*MBC*, 4), the reviewers importantly did not cavil either on the use and substance of the 'anecdotes' that variously demonstrated it in the final part of the *Memoirs*. Rather, *The Eclectic Review* alone questioned why such 'illustrative anecdotes' were not better 'distributed' (rather than confined to part 4):

> The absurd manner (begging Mrs. Lee's pardon) in which the materials of this memoir are distributed into four portions, not consecutive, but synchronical, has separated from the notice of these leading events of his life, some interesting details and illustrative anecdotes which ought to have been incorporated with it. The following additional particulars relating to that part of his life which was spent in Normandy, are supplied by the funeral eulogium delivered by Dr. Pariset. (232)

More careful reading of the opening sentence of Sarah's 'plan' above, however, demonstrates how she had already pre-empted such charges of 'absurd' distribution in her evidence-based 'method' for the whole. It was informed by her knowledge, iteration and extension of Cuvier's French *Éloge* model. Such 'greats' in (French) scientific life and works required no personal life details or 'anecdotes' to explain them. Strikingly, modern historiographers and biographers of Cuvier since the 1970s have mostly charged Sarah's *Memoirs* with the inaccuracies, even polemics, of the 'illustrative anecdotes', to reclassify and footnote her first biography as outdated and misleading thanks to more correct (objective, evidence-based) twentieth-century (modern scientific) standpoints.

History of Science in Comparative View: Science (and) Biography since the 1970s

The fact that history of science is, as Daston states, 'a relatively young discipline [...] coalescing only gradually in the twentieth century' confuses the terms of engagement with Sarah's *Memoirs* that are 'science', 'scientists' and 'science biography', especially when applied to the period before the mid-nineteenth century when 'histories of science had become distinct from scientific publications, although they were still written primarily by scientists, including prominent figures such as William Whewell, Marcellin Berthelot, Ernst Mach, and Pierre Duheim'.[13] Indeed, much cited in British histories of nineteenth-century science is the fact that Whewell coined the word 'scientist' in 1833, because no term in English existed for the French 'homme scientifique'/'savant', whereas clear precedent existed for Whewell in the word 'artist' (and 'naturalist') for persons of equivalent cultural status.[14] Sarah's 'introduction' to her *Memoirs*, however, provides clear counter-evidence that 'savant' and 'man of science' were terms widely circulating in English, as the reviews above of her *Memoirs* in 1834 further endorse. Indeed, the major given for the New York *Foreign Quarterly Review* (as also for the *Quarterly Review* in which Whewell's 'coinage' appeared) that Anglophone historians of science of the period seriously underestimate was that educated 'English' audiences had fluent reading knowledge of French, German and Latin and accessed European literary and science publications. By contrast, French readers – and Cuvier exemplified the case[15] – could not be guaranteed to have the equivalent proficiency to read literary or scientific works published in English. The increasingly national and professional implications that Whewell's anxiety about the term 'scientist' in Britain therefore epitomised were France's international predominance in institutional 'science' (despite defeat after the Napoleonic Wars), whereas (British) 'letters' (ancient and modern) were reputationally secure. If 'Darwin' then became (and remains) the further major reference point for Victorian and more recent British history of science and science biography, the onwards professionalisation of the sciences (and letters) after his death would provoke and further embed what C. P. Snow's much-cited Cambridge Rede Lecture of 1959 termed the 'Two Cultures and the Scientific Revolution'.[16] Specialism in each meant that 'literary intellectuals' and 'natural scientists' in Snow's terms had little common ground. History and Philosophy of Science (HPS) Departments, including that established at Cambridge as HPS in 1972, provided one means to address it, as did pre-emptive literary-critical reappraisals of nineteenth-century science writing, for example the ground-breaking *Darwin's Plots* by (Cambridge academic) Gillian Beer in 1983.[17] As the product of HPS, Dorinda Outram then stands out as Cuvier's major modern biographer since 1984, because she was also versed as an (Anglophone) historian of French science of the 1780–1850 period.[18] Outram was acutely aware of the impasses of scientific biography as exemplified in the 'case' of Cuvier's own *Éloge* as the foremost master of its form. Her work also remains the most extensive survey (and critical bibliography) of 'the extraordinary bibliographical tradition through which we view him' since '[a]lmost every presentation […] since his death in 1832 has been dominated by emphases which were established very soon afterwards'.[19] Top of her list of his (five) earliest major biographers is the 'eulogistic account […] by his friend and protégée, Sarah Lee […] important, for all its apparent naivety, *because for the first time it selects and uses stock incidents from Cuvier's life to argue a definite case for the position of science*'.[20] Outram's key acknowledgement and (three-page) critique in 1976 of Sarah's *Memoirs* therefore served to structure her survey of his biographers and, in consequence, inform her own biography of Cuvier's 'vocation, science and authority'. The facts for Outram that '(Lee's) account was until 1964 the only one readily available in English [...] which also included lengthy discussion or summary of his ideas [...] and of other accounts of his life' also explain how it 'blocked the reader's interest in pursuing alternative viewpoints'.[21] If these viewpoints included 'the political aspect of his career [...] his religious ideas [...] and the way his work related to other contemporary movements of thought', Outram most took issue with how Sarah's *Memoirs* effectively set up their 'facts' as polemic, so as to support 'the "Declinist" debate, [...] to call attention to the comparatively flourishing condition of organized science in France [and] the prestige his reputation had gained in England during his lifetime.'[22] To address Sarah's 'one-sided image of Cuvier' in consequence, Outram took up precisely the 'novelty of her account [...] in its polemical use of the biographical form, and the distortions it introduced into future ideas of Cuvier' by delimiting Sarah's *Memoirs* (and other publications).[23] For example, and despite Outram's unsurpassed 'bibliography' in French and English as a critical tool, she made no reference to the *Memoirs* appearing simultaneously in French (and US) editions, evidence that undermines her case against 'Lee' for an allegedly 'Declinist' stance. Sarah's *Memoirs/Mémoires* could not do otherwise than report her eyewitness encounters with Cuvier's world-leading Muséum science from her sustained Paris (1818–1823) and then London-Paris perspectives (1824–1833). For the future co-editor of the influential *Uneasy Careers and Intimate Lives* of 1987 discussed in our introduction (p. 7, 9, 12), Outram's omission to ask or unpack in 1976 how 'Mrs Lee knew Cuvier in Paris [...] and was an intimate of his family until her own death in 1856',[24] returns us to the impasses and blind spots since the 1970s of science biography and history of science of the nineteenth century especially in the period before 1850 that Sarah was already negotiating from the first line of her introduction to the *Memoirs* in 1833.

Of major note are, moreover, the important birthing and legacies of (national) historiography in France in the 1830s by key political figures such as François Guizot (1797–1874) and Louis Adolphe Thiers (1797–1877), or the writer of the monumental multi-volume *Histoire de France* Jules Michelet (1798–1874), pivoted on the regaining of imperial prestige at home and abroad.[25] It was only in the latter half of the nineteenth century that national biography became an important subset of (national) historiography and was particularly favoured as a genre among new classes of professional writers in France and Britain – historians, arts critics, journalists – seeking to make their own names and reputation out of the life and work(s) of more famous and important persons.[26] Outram's important identification of the 'polemics' of the *Memoirs* therefore lies not in its alleged 'Declinism', but in the impasses for national historiography on both sides of the Channel represented by Cuvier's death. How and why Outram and her successor modern historians and biographers then failed to pose or address the question concerning the unusual (British) 'woman's pen' authoring the first biography of Cuvier is the same question differently put in 1833. The highly-regarded Longman, H. Fournier and Harper all published this 'unlearned' author and her work precisely for the significant scientific and commercial interest of this first 'biography' of Cuvier. Sarah's unprecedented viewpoints in the 'bosom' of Cuvier's circle and his science was for the New York *Foreign Quarterly Review* precisely what made the *Memoirs/Mémoires* 'elegant, judicious and affecting', that is polemically diplomatic. It could not have been otherwise, because this work was concomitantly a first (science) *auto*biography (OED definition 3b above) 'from a woman's pen'.

'Posterity Should Benefit by the Example' (*MBC*, 1)

The unprecedented cross-channel phenomenon that is Sarah's 'biography' of Cuvier offered the first 'portrait' of the acclaimed, and multiply decorated Cuvier in text and image after his death, as the inside title page demonstrates (see Figure 3.1). On the right, it also offered new precedents for the fully-formed 'portrait painter' in the female (auto)biographer at work behind her signed record of a lynchpin figure in national-international science. In already pioneering supra-national historiography from outside in, rather than (self-)promotionally from within the national history at stake, Sarah's unusually bipartisan perspectives, let alone her woman's-eye view, made the *Memoirs* generically significant for its authority in the history of nineteenth-century science and (scientific) biography already in 1833. Sarah cannot be classed as a precursor in such orders of 'professional' biographer, however, because her *Memoirs/Mémoires* was a singular venture in what would be the larger corpus of her science writing after 1833 as subsequent chapters will reveal. Her 'reputation', that is her name and renown in (Cuvier's) natural history endeavour, was therefore wholly at stake concerning her commissioning to pen it, and hence in the balances regarding her future natural history writing to maintain her family.

Where no male biographer need make an apology for existing in print, Sarah's introduction further presented her 'biography' of Cuvier as an *apologia* for her very reluctant yet necessary undertaking. Also covered by her pointed English 'pitch' discussed above

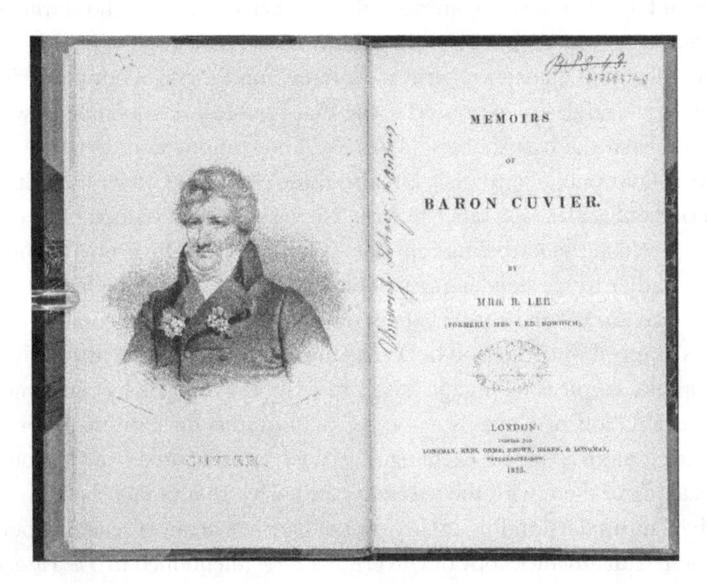

Figure 3.1 'Inside Title Page Portraits'. Courtesy of the University of St Andrews Libraries and Museums, *Memoirs of Baron Cuvier*, sQL31.C9L2.

was her important diplomatic defence of his reputation 'at home' in 1832. Sarah was acutely aware of Cuvier's authority and celebrity-notoriety, both as the peerless exponent of *Éloges scientifiques* making his own daunting for successors to write, and for his pyrrhic victory over his major colleague-antagonist Etienne Geoffroy Saint-Hilaire in their bitter public 'Querelle des Analogues' (Quarrel of the Analogues) of 1830. Where the latter upheld the gradualist 'Transformisme' of all extant and extinct forms from a single 'plan' (before Darwin was working on similar ideas), Cuvier countenanced only 'revolutions' that also entailed extinctions, and un-crossable boundaries between the four domains ('règnes') of the animal kingdom (vertebrates, invertebrates, articulates and radiates) as the 'fixity of form' in his (four-volume) *Règne Animal* (1817; revised edition 1829–1830). In these specific contexts, Sarah's astute 'plan' above to reconfigure the 'Cuvierian' *Éloge* paradigm through its strictest imitation in parts 1 and 2 of her work, also allowed her to address Cuvier's scientific 'fixities of form' by their extensions in parts 3 and 4 of her *Memoirs.* Part 3 then addressed Cuvier's 'talents as a legislator' (his political, public and religious life that Outram deemed 'distorted' above), so as to enlist his holding of many key offices accumulated throughout France's various regime changes, when his personal reputation was increasingly at stake in 1830. New laws that year (the *cumul*) brought extensive curtailment to financial remuneration for multiple public post-holders such as Cuvier. His opponents therefore charged him with manipulating his multiple government posts,[27] to ensure the appointments of members of his extended family and like-minded disciples to key chairs and other positions in France's institutes of science. Sarah made no direct reference in her introduction to the *Memoirs* to these recent French contexts, because its commissioned translation with H. Fournier could take them as read. Rather the turning point (third) paragraph of her justification for penning the (English) *Memoirs* as (her) defence of Cuvier in France and internationally resides in the solicitations (commissioning) of her work; both those who 'seemed to think it a matter of course that I should publish some particulars of my lost friend', and 'one or two influential quarters to write a short memoir for one of our public journals' as part of 'the universal desire expressed to me that I publish the documents which abundantly flowed from the best sources' (*MBC*, 3). These anonymous persuaders were not only 'English' ('our public journals'), but also those closest to Cuvier himself.

The period of Sarah's reluctance to write Cuvier's account is documented in the correspondence (held at the Paris Muséum library) of Dr G. L. Duvernoy, Cuvier's relative and also godfather to his stepdaughter Sophie Duvaucel. One long letter to Duvernoy (in French) from Sarah, and nearly twenty others written by Sophie making mention of her, were exchanged in the period between Cuvier's death in 1832 and early 1833, and enlisted Sophie's auspices to engage Sarah to write the 'biography'. Its longest second part – the evaluation of Cuvier's extensive publications on natural history – caused Sarah the most anxiety in the exchange, because most at stake was her position as a woman speaking of them with the necessary authority (that in fact she had). To resolve any perceived misunderstanding on her part of Cuvier's array of scientific publications, Sarah requested an advance copy of Duvernoy's *Éloge* (mentioned in *The Foreign Quarterly* above), since it was also to elucidate Cuvier's scientific publications through providing an accompanying complete bibliography. Its compiler was none other than Sophie, whose letters to Duvernoy confirmed her instrumental roles in securing Sarah's pen by offering the same direct help to her long-standing British friend. Both Duvernoy's official *Éloge* of Cuvier in 1834 and Sarah's *Memoirs/Mémoires* in 1833 therefore derived from a three-way collaboration principally involving the two women in Paris and London who best knew Cuvier and his science directly. Sarah's published acknowledgement and 'gratitude towards those who have assisted me, either by their notes or their works*' then footnotes at this asterisk only the men: 'Foremost among these [...] Baron Pasquier, M. Laurillard, Dr. Duvernoy, and the Baron de H——' (*MBC*, 6). Sophie's interventions, however, more fully elucidate Sarah's carefully-worded acknowledgement in her introduction (also explaining the 'politics' of her 'English' pitch above): 'I applied to the relatives of Baron Cuvier for data [...] and I seriously applied to the task' (*MBC*, 3). The other side of its coin was Sarah's reinsertion of her autobiographical place thanks to Sophie into her biography of Cuvier, both to justify its writing in the introduction (before its concluding 'plan'), and in the 'illustrative anecdotes' of which Sarah was the observer in part 4. In these ways, and with the scientific material for the second part secured by Sophie's good auspices, Sarah's *Memoirs* potentially melded in 1833 what Thomas Hankins has pinpointed as the major tension of science biography: 'The biographer of a scientist tends to be drawn either to the personal life of his (*sic*) subject or to the technical details of his (*sic*) subject's scientific work. It is difficult to bring these two different aspects together in a harmonious way'.[28]

But Sarah had a further means at her disposal to explain the parts by the whole. From the second sentence-paragraph of her introduction, she owned and expressed her deepest grief and very heavy 'heart' on the loss of Cuvier:

> When death has torn from us those whom we have most loved and revered, and the overwhelming bitterness of grief is past, the first feeling which awakens us from our sorrow is the desire to uphold the memory, and to make known to all men the virtues of the being enshrined in our hearts; a feeling which springs, not only from an honest pride in doing justice to one who is no more, but from a desire that posterity should benefit from the example. (*MBC*, 1–2)

Biography and autobiography had a Janus face. Sarah could make very public her deep personal grief for Cuvier, and her unusual understanding and mastery of such loss in the death of her husband, also directly implicated into the further autobiographical account of her first 'introduction' to Cuvier upon the return of 'Mr Bowdich [...] from his second, and I from my first, voyage to Africa, in the year 1818, and shortly after Mr. Bowdich proceeded to Paris' (*MBC*, 5). The latter's premature death (and that of their first child), and then Cuvier's as their/her mentor and, from 1824 pivotal 'gatekeeper' for her work at the Muséum, propelled the rhetorical and human (intellectual-emotional) challenges of her scientific biography task. The *Foreign Quarterly Review* in particular noted her success in summoning for the reader Cuvier's 'domestic griefs [...] with a fidelity and a pathos to which we think the sympathy and tears of many readers must have borne an unsuspicious testimony'. Sarah had also mourned with Cuvier and his family in 1827 upon the illness and then (premature) death of his last surviving child, her friend Clémentine. It so affected him that he shut his Saturday salon, of which she, her mother and stepsister Sophie (Duvaucel) were the hostesses.[29] To pinpoint this illustration of Cuvier's 'heart' lay not in an 'anecdote' recounted only in the fourth part, but in the unprecedented inclusion in front matter for the *Memoirs* of a facsimile letter by Cuvier (see Figure 3.2), penned before Clémentine died.[30] It encapsulated first-hand his deepest concerns as a father, to speak volumes – the 'pathos' noted by the *Foreign Quarterly Review* – to readers of it beyond his own grave. The second paragraph above of Sarah's introduction, therefore, more powerfully foregrounds the deep shock also of Cuvier's (premature) death and her 'rousing […] from the stunning grief' that enabled the action of her 'woman's pen'. Not only had she 'applied to the relatives of Baron Cuvier for data […] which vouched for the sentiments of the family' (*MBC*, 3). The personal and political diplomacies of her corresponding pen melded their 'data' and 'sentiments' into *Memoirs* that came with her heartfelt eye-witness points of reference. For her cross-channel and transatlantic readers and reviewers in 1834 and today, the result has to be judged by taking seriously her most carefully penned personal 'introduction'.

For historians of nineteenth-century French and British history of science, the year 1833 was differently a watershed for professionalising developments on both sides of the Channel. We noted Whewell's coinage of the 'scientist', to turn gentlemanly pursuits into professional pursuit of larger institutional 'science' as exemplified in Cuvier's work dominating 'Comparative Anatomy' at the Paris *Muséum* for three decades. As

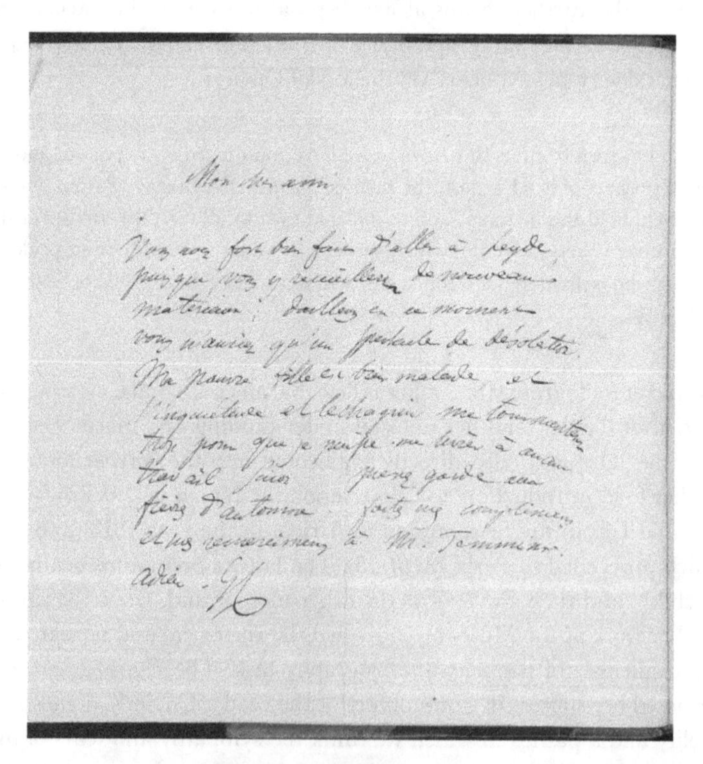

Mon cher ami

Vous avez fort bien fait d'aller à Leyde, puisque vous y recueillerez de nouveaux matériaux; d'ailleurs en ce moment vous n'auriez qu'un spectacle de désolation. Ma pauvre fille est bien malade, et l'inquiétude et le chagrin me tourmentent trop pour que je puisse me livrer à aucun travail. [illegible] prenez garde aux fièvres d'automne. Faites mes complimens et mes remercîmens à Mr. Temminck.

adieu. G C

Figure 3.2 Foldout facsimile of Cuvier's heartfelt letter. Courtesy of the University of St Andrews Libraries and Museums, *Memoirs of Baron Cuvier*, sQL31.C9L2.

Joseph Ben-David for example noted, but did not connect to Cuvier's death France's 'Stagnation and Decline after 1830' and until the Second Empire[31] derived from the (over)centralisation of its sciences in the hands of more or less proactive chairs. The prodigious sum of Cuvier's Muséum and public offices in Part Three of Sarah's *Memoirs*, alongside his extensive publications in Part Two, therefore also pointed to the larger 'vacuum' that their cessation would cause in the 1830s and 1840s as his disciples and antagonists gained ascendancy. With hindsight, her *Mémoires/Memoirs* of 1833 mark this zenith in France as concomitantly the motor revivifying Anglophone natural history endeavour thanks to Cuvier's new classification systems, which Sarah had been among the first to adopt in her (ongoing) *Fresh-Water Fishes of Great Britain*. Her 'biography' of Cuvier could only provide an unusually (bi)partisan view of him in 1833 thanks to her rare outsider-insider encounters with his French-international natural history first, and as a British woman second in his science without a sex (or nationality). These lights also return us to how we reassess Sarah's *Memoirs/Mémoires* for the history of (women in) science and 'science biography' today.

The indicative Anglophone reviews of 1834 above unanimously lauded Sarah as Cuvier's major 'biographer', and the *Memoirs* as in effect '(national) science biography' before the genre became established. To count Sarah in its vanguard is then to rediscover in her *Memoirs*, especially in its working 'method' in her introduction, a masterclass in biography as a diplomatic and rhetorical form. Her pioneering perspectives are then several. In breaking the moulds of the French *Éloge*, and of model 'great life-great works' entries in later national biography including 'science biography', Sarah's approaches challenged the person and the subject of science as uniformly undertaken by 'great national men'. Sarah's integral four-part study of Cuvier's 'domestic' (home and French political) life in his science already pre-empts more recent 'models'.[32] But it was her informed outsider-insider position as his 'biographer' that most challenges 'science biography' and its roster of writers today. The *Memoirs/Mémoires* already spoke with remarkable authority and eloquence on Cuvier's science to its *three* different science publishing 'markets'. They locate the international, rather than national optics for 'important' lives in science, including Sarah's own. Her long-standing insider access to her scientific subject – both the person and the science – guaranteed her access to Cuvier's personal libraries, (unpublished) notes, or other more personal accounts as almost a family member, because she was unusually also a mentee, close associate and correspondent in his public-facing science.[33] Her 'woman's pen' then makes her *Memoirs* the rarest forerunner example for 'international' science biography today in its diplomatic handling of what cannot but must be named, the history of science without sex, nationality or official-institutional stamp.[34]

Sarah left no personal overview of her life in natural history-making, which is why Tedlie's report heading this chapter is so precious. However, the penultimate paragraph of Sarah's Introduction (*MBC*, 5), and her self-insertions into the 'anecdotes' of Cuvier in the final part of the *Memoirs* (for example *MBC*, 154–57; 159–60) leave the most important account.[35] In Chapter 5, we will review her self-reflective scientific notes in different guise, leaving Chapter 9 to unpack the central importance of the science 'anecdote' for Sarah's dextrous pen. Rhetorically and personally, the anecdote in her *Memoirs of Baron Cuvier* offered Sarah the means to justify how and why *she* alone 'in England' could write with authority *the Memoirs of Baron Cuvier* in obligation, tribute and deep grief in 1833 for her mentor and friend in science:

> We [the Bowdichs] became the intimates of the family, with whom, for nearly four years, we were in daily intercourse. We left France with their blessings; and on returning alone to Europe, I was received even as a daughter. My correspondence with M. Cuvier's daughter-in-law, and other branches of the family, has been uninterrupted since that period; I have paid them repeated visits at their own house; and for fourteen years not a single shadow has passed over the warm affection which has characterised our intimacy. (*MBC*, 5)

Here is Sarah's 'science autobiography' in 80 words. The name 'Mrs R. Lee'/'Mistress Lee' on the cover, spine and inside title page (above) of the *Memoirs/Mémoires* in 1833 could then mark a pivotal personal turning point for her name and reputation. As the 'Mrs T. Edward Bowdich' authoring cross-channel works on natural history in the *Excursions* (*EM*, *EMFr*), she could as 'Mrs R. Lee' find new authority as the 'biographer of (foremost) savants'. Yet her personal grief in these double authorial accomplishments made the second here almost the more bittersweet and heart-stopping. Her commemoration of Cuvier in the *Memoirs* spoke to the rationale for its existence through her own life in his science. His death therefore marked the point of no return for his important personal resourcing of her work in the previous seven of the 'fourteen years' above that Sarah recorded. They had been pivotal to the 'continuance' of her narrative in expert natural history-making. The closing question of the Gambian woman's portrait therefore reverberates afresh: 'What will she do now?' Sarah's *Fresh-Water Fishes* was only at its halfway mark, and she needed other outlets to support its completion, having also undertaken the commissioned *Memoirs/Mémoires.* Her asterisked footnote acknowledging those who assisted her to complete it 'either by their notes or their works' (*MBC*, 6) vitally concluded with 'the Baron de H—' who alone in her list wrote no *Éloge* of Cuvier. The terrible 'shadow' (above) of Cuvier's death only better reveals Sarah's parallel harnessing of Humboldt for the *Memoirs* and in her natural history corpus, and his larger parts in her pioneering scientific endeavours and perspectives from the field overseas. As 'Mrs Lee' henceforth, Sarah could set out once more to make her two earlier expeditions to West Africa very much more her own as 'formerly (Mrs) T. Edward Bowdich'. As the unchanged 'Mrs Sarah' she had, through the introduction to the *Memoirs/Mémoires,* also further honed her double 'I's and eyes reporting beyond the bars of science.

Notes

1 See Orr, 'Fish', 232. Bibliographical references for *The Field* are in its note 74.
2 Front matter to Sarah Lee, *Memoirs of Baron Cuvier.* Cambridge: Cambridge University Press, 2014 at https://doi.org/10.1017/CBO9781107444737.
3 See for example Philippe Taquet, *Georges Cuvier.* Paris: Odile Jacob, 2006, and Kathleen Kete, *Making Way for Genius: The Aspiring Self in France from the Old Regime to the New.* New Haven and London: Yale University Press, 2012, chapter 4.
4 See Orr, 'Les *Mémoires du baron Georges Cuvier.*' I also corrected the attribution of the *Mémoires* to Théodore Lacordaire, its translator.

5 See for its embedded traditions in France see Arnaud Saint-Martin, 'Autorité et grandeur savantes à travers les éloges funèbres de l'Académie des sciences à la Belle Époque', *Genèses* 87, no. 2 (2012): 47–68.

6 Cuvier was the major, celebrated memorialist at the Paris Muséum for more than twenty years, both for its many scientific activities – in his annual reports in the *Annales du Muséum* – and for his deceased colleagues: his acclaimed *Éloges* (*scientifiques*) ran to several volumes by his own death in 1832. They included encomiums to eminent foreign as well as French scientific figures, his last being to Lamarck. In French, the noun *mémoire* (f.) is the faculty of memory by which the past can be remembered and represented. Sarah's *Mémoires* therefore also demonstrate her art as a *'mémorialiste'* (historiographer), guaranteeing memory of Cuvier's name for posterity.

7 For an evaluation of the field and its history from 'the mid-nineteenth century' see Lorraine Daston, 'Science, history of': 'By 1900, histories of science had become a genre distinct from science [...] with its own distinctive program of training, institutions (journals, professional societies, university positions), and scholarly standards'. *International Encylopedia of the Social and Behavioural Sciences* 21, second ed. (2015): 241–47 (241).

8 Key Anglophone reference works and contributions on the genre of scientific biography include Michael Shortland and Richard Yeo, eds., *Telling Lives in Science: Essays on Scientific Biography*. Cambridge: Cambridge University Press, 1996; Peter France, 'From Eulogy to Biography: The French Academic Éloge', in Peter France and William St Clair, eds., *Mapping Lives: The Uses of Biography*. Oxford: Oxford University Press, 2002; Thomas Söderqvist, *The History and Politics of Scientific Biography*. Aldershot: Ashgate, 2007, and David R. Oldroyd, ed. and intro. Special Number on 'Biography'. *Earth Sciences History* 13, no. 1 (2013): ii–xiii.

9 For my earlier investigation of Sarah as Cuvier's 'peer', see Orr, 'Women Peers'.

10 'Art. I. *Memoirs of Baron Cuvier.* By Mrs R. Lee', *The Monthly Review* iv (Oct. 1833): 159–78 (159–60). The reviewer's paraphrase of Sarah's introductory account of her first contact with Cuvier prioritises the mediations of 'a very distinguished countryman of her's, the late Dr. Leach, of the British Museum' (160).

11 'Art. III. *Memoirs of Baron Cuvier.* By Mrs R. Lee', *The Eclectic Review* x (Sept. 1833): 228–39 (228–29).

12 'Art. IV. Life and Labors of Cuvier', *The Foreign Quarterly Review* xiv (Dec. 1834): 164–85 (165 and 182).

13 Daston in note 7, 241.

14 For an indicative appraisal, see Richard Holmes, 'In Retrospect: On the Connexion of the Physical Sciences', *Nature* 514 (2014): 432–33. https://doi.org/10.1038/514432a. Whewell's coinage appeared in 1834 in his anonymous review of Mary Somerville's new work on astronomy, 'On the Connexion of the Physical Sciences. By Mrs Somerville', *The Quarterly Review* 51, no. 101 (1834): 54–68. See my recent re-reading of Whewell's reactive anxieties behind his extensive review in Orr, 'Catalysts', in light of Mary Somerville's pivotal place in British science endeavour as a woman with informed 'continental' knowledge.

15 In the final Part of the *Memoirs*, Sarah notes: 'One thing used particularly to annoy him; which was, to find an Englishman who could not speak French' (*MBC*, 165).

16 See C. P. Snow, 'The Rede Lecture, 1959', in Stefan Collini, ed., *C. P. Snow: The Two Cultures* (Canto Classics). Cambridge: Cambridge University Press, 2012, and Collini's important review of it in the contexts of nineteenth- and twentieth-century history of science and cultural history, 1–21.

17 Gillian Beer, *Darwin's Plots: Evolutionary Narrative in Darwin, George Eliot and Nineteenth-Century Fiction*. 3rd edn. Cambridge: Cambridge University Press, 2009.

18 Dorinda Outram, 'Scientific Biography and the Case of Georges Cuvier: With a Critical Bibliography', *History of Science* 14, no. 2 (1 June 1976): 101–37, and 'The Language of Natural Power: the "Eloges" of Georges Cuvier and the Public Language of Nineteenth-Century Science', *History of Science*, 16 no. 3 (1 Sept. 1978): 153–78. These directly inform her *Georges Cuvier: Vocation, Science and Authority in Post-Revolutionary France*. Manchester: Manchester University Press, 1984.

19 Outram, 'Scientific Biography', 101.

20 Outram, 'Scientific Biography', 102, emphasis added to signal Sarah's 'anecdote' materials.

21 Outram, 'Scientific Biography', 102–03.

22 Outram, 'Scientific Biography', 102–03.
23 Outram, 'Scientific Biography', 103–04. It lies outside this chapter to unpack the many misrepresentations of Outram's selective report (104–05) of Lee's work as also informative of Outram's relegations of 'Lee, 1833' to strategic footnotes in her biography of Cuvier cited in note 7.
24 Outram, 'Scientific Biography', 103.
25 For a recent US-focussed appraisal see Lloyd S. Kramer, 'The Declining Study of Nineteenth-Century France', *Central European History* 51, no. 4 (Dec. 2018): 640–45, in which Kramer omits to mention French historiography or its pivotal political figures.
26 The history of the *Oxford Dictionary of National Biography* is a late case in point.
27 Outram's 'Scientific Biography', 105–07 singles out Blainville's attacks as indicative.
28 Thomas L. Hankins, 'In Defence of Biography: The Use of Biography in the History of Science', *History of Science* xvii (1979): 1–16 (2), his statement contextualised by the 'particular problems for the historian of science, because of the great difficulty of integrating science into the rest of human intellectual endeavour. Historians either fight the centrifugal tendency of science to flyoff from the body of history, or they happily fly off with it secure in theirconviction that science has little or nothing to do with the rest of historyanyway'.
29 See Orr, 'Keeping it in the Family: The Extraordinary Case of Cuvier's Daughters', in Cynthia V. Burek and Bettie M. Higgs, eds., *The Role of Women in the History of Geology*. London: Geological Society, Special Publications 281, 2007.
30 The letter reads: 'Mon cher ami, Vous avez fort bien fait d'aller à Leyde, puisque vous y recueillerez de nouveaux matériaux; d'ailleurs en ce moment vous n'auriez qu'un spectacle de désolation. Ma pauvre fille est bien malade, et l'inquiétude et la chagrin me tourmentent trop pour que je puisse me livrer à aucun travail sain (?). Prenez garde aux fièvres d'automne. Faites mes complimens et mes remercimens à M. Temminck. Adieu, GC.' (My dear friend, You have done well to go to Leiden, because you will gather much new material there; besides at this time there is only desolation to see. My poor daughter is very ill, and the anxiety and deep worry so torment me that I can devote myself to no sensible work. Stay clear of the Autumn fevers. Give my compliments and thanks to Mr Temminck. Farewell, GC'., translation mine).
31 The letter is not reproduced in 'google books' versions of the *Memoirs/Mémoires*. The 2014 digitised version in note 2 interleaves it after the 'introduction'. Because the reader has to unfold it in the print copy to read it, the contents are the more poignant. The addressee may have been Valenciennes on a mission to Leiden for Cuvier's *Histoire naturelle des poissons*, or Duvernoy.
32 Joseph Ben-David, 'The Rise and Decline of France as a Scientific Centre', *Minerva* 8, no. 2 (April 1970): 160–79 (172) at https://www.jstor.org/stable/41822018.
33 See Hankins, 'In Defence of Biography' in note 28: 'A fully integrated biography of a scientist which includes not only his personality, but also his scientific work and in the intellectual and social context of his times, is still the best way to get at many of the problems that beset the writing of history of science. [...] science is created by individuals, and however much it may be driven by forces from outside, these forces work through the scientist himself (*sic*). Biography is the literary lens through which we can best view this process' (11–14). See also Anne Jefferson, *Biography and the Question of Literature in France*. Oxford: Oxford University Press, 2007.
34 See Alice Jenkins, 'Writing the Self and Writing Science: Mary Somerville as Autobiographer', in Juliet John and Alice Jenkins, eds., *Rethinking Victorian Culture*. Basingstoke: Macmillan Press Ltd, 2000, 162–78.
35 See Julie Rak, 'Are Memoirs Autobiography? A Consideration of Genre and Public Identity', *Genre* 34 (Fall/Winter 2004): 305–26.

Part Two

HARNESSING HUMBOLDT

Chapter Four

A FIRST (PLANT) GEOGRAPHY OF THE GAMBIA: *EXCURSIONS IN MADEIRA AND PORTO SANTO* (1825)

The introduction drew attention to the visibly minor place of the Gambia on the inside title page of the *Excursions*, as 'supplementary, separated, accidental, niche, unimportant [...] in its interest' (p. 17). Indeed, the full title of the work (*EM, EMFr*) lists the Gambia almost as an afterthought, like the similar place it occupied in Sarah's explanatory preface:

> For the favourable reception of the first part of my book [on Madeira], I feel little or no apprehension. The errors which may have crept in when correcting the press, will justly be laid to my charge, and cannot deteriorate from its excellence. There, indeed, I have not presumed to make the slightest alteration, not even by compressing the Supplement [on the botany of Madeira] into the body of the work [...].
>
> *For the second part* [on the Gambia] *I claim indulgence*, but I do not ask it from the consideration that I am a widow with three orphans to maintain and educate; for, in my opinion, these circumstances form the strongest stimulus to exertion. *I have only to entreat the public to consider, that I make my appearance as an Authoress for the first time, and deprived of the aid which would have ensured me success. Accustomed to submit every word and action to my husband, I now feel a diffidence in my own abilities, which fetters rather than promotes my best endeavours.*
>
> When I recollect the painful struggles, the numerous privations, the years of intense study, which preceded Mr. Bowdich's third voyage to Africa; when I reflect that every hope, every wish, that bound us to Europe was sacrificed; that all personal property, and the greatest bodily and mental exertions were devoted to this one cherished object; *and when I look at the last part of the volume, to which this is the Preface, I feel concerned at the little apparent result. But when I request my readers to bear in mind, that the little that has been done was completed in the short space of a month, I think they will agree with me, that it is a favourable specimen of what might have been effected, had Mr. Bowdich's life been prolonged.*
>
> Although I may deem it necessary to control my feelings in other respects, to the sentiments of gratitude there need be no restraint [...] to those who sympathized with me in my affliction, and met my returning orphans with their bounty [...]. (*EM*, v–vii, emphasis added)

Sarah's carefully penned words concerning the second and 'the last part of the volume' (with no identificatory geography) were one with the open fact of her (sole) authorship, also placed second on the inside cover page. Her authority and significant independent additions to European scientific knowledge of the Gambia were the achievement of 'the short space of a month', undertaken entirely without Edward's interventions. The first chapter explored how Sarah's new 'specimen(s)' indeed in (French) ichthyology contributed 'no little apparent result', because one with this important 'last part' of the *Excursions* particularly in the French edition with Cuvier's 'Notes'. In the opening of its 'Narrative' section ('Relation' in the French edition), Sarah almost protested her self-deprecation too much, however, by underscoring her fullest responsibility for it in the absence of Edward's 'assistance':

> I FEEL so great a repugnance to appear before the public, and so great a distaste to those subjects in which I have lost my guide and instructor, that the present narrative will labour under many disadvantages, besides those which may arise from incapability. It is but justice, however, to those who felt interested for Mr. Bowdich in his public character, without any dearer tie of friendship, and to those who make the cause of science their own, to relate the circumstances of his last voyage, with their fatal result. *I particularly lament, that, contrary to his usual custom, his notes were very few, and those so obscurely written, that even I, who am so accustomed to decipher his memoranda, can derive but little assistance from them:* therefore, that I may not injure a reputation which stood so fair with the learned and the good, I must request my readers to consider me as responsible for every error. (*EM*, 173–74, emphasis added)

This chapter will expose Sarah's singular responsibility for every verity in her 'excursions' in the Gambia. Only through her provision of its all-important 'report' could the terrains of this major river country come to better understanding in Europe on two related counts; its larger African geography and longer West African history. To frame Sarah's pioneering perspectives on both, we first explore her narrative for its important slice of 'British' colonial life in 1823–1824. It accounts for Sarah's (sole) 'continuance' of the Bowdichs' West African project in her concerted Gambian fieldwork 'in the short space of a month' comparatively applying the findings from their recent study of Madeira, and earlier knowledge behind it from their 'Mission to Ashantee'. In short, Sarah's 'narrative' constituted the first European scientific exploration of the Gambia (by a woman), and long before Mary Kingsley visited the region, as Chapter 1 underscored (p. 33). However, it was Sarah's harnessing of Humboldt anew in her report that the second part of the chapter addresses. As I first brought to critical attention, Humboldt's 'Notes' for the French edition of the *Excursion* comprised an extensive epilogue essay.[1] It reviewed the Bowdichs' invaluable new geological study of Madeira (undertaken after their preparatory Paris training under him) to affirm their work as the important connecting piece for recent geology of the archipelago (including von Buch's and his own), and of the wider region. As this chapter now argues, the lights of Humboldt's leading world geology and 'plant geography' determine and most distinguish Sarah's expert fieldwork in the Gambia's 'equinoctial' terrains from those of the 'New Continent' (of the Americas) in which he had famously set foot. Through Humboldtian lenses, Sarah's rich haul (in her Zoological Appendix) of local *fauna,* including Fishes discussed in Chapter 1, and *flora* hitherto undocumented in European natural history and botany, reveal her potentially more significant understandings of the Gambia's ('plant') geographies. Not least in the last section of the chapter, these also display and convey Sarah's extraordinary perspectives on, and accounts of, 'Gambian' peoples. The shortest part of the *Excursions* therefore contains the clearest attestation of Sarah's remarkable scientific and personal character, as captured in profile in the opening 'portrait'. Before she had returned to London in 1824, she had undertaken a pioneering (Humboldtian) survey report of the Gambia, despite being a widow of only one month. How could she do justice to it in print?

As we discovered for her crates in Chapter 1, Sarah's turning of major losses into gains pertained also to Edward's 'obscurely written' results above from his frenetic measuring (along similar lines to his extensive data for Madeira) of the Gambia's bathymetry, altitudes and temperatures in November and December of 1823. They were unusable. Indeed the main step-change, and distinction, in Sarah's pioneering observational purview of the Gambia's natural and human economies as interconnecting nonetheless with the country's geology and geography is glimpsed in her strategic choice of metaphor determining her 'best endeavours'. Sarah's unusual scientific report 'unfettered' her claims to 'diffidence in [her] own abilities' (without Edward), so that she could address the Gambia's (un-measurable) larger African geographies, and hence longer intercontinental cultural economies. They include the country's literal 'fetters' that make the Gambia's seeming geographical (and historical) 'invisibility' particularly invidious. The long-historical practices of known (indigenous) enslavement by the Mandinko, for example, in the 1300s before the Portuguese in 1455 discovered the Gambia's major slave markets, led in the 1600s to Portuguese, Spanish and British commercial interests in the transatlantic trade to provide slaves to work in the mines of Mexico and in the sugar and cotton plantations of the Caribbean. I have already drawn attention to Humboldt's 'passionate abolitionist' interests in American sugar production after 1807 (legal abolition of the Trade) in his epilogue essay for the French edition of the *Excursions*.[2] In therefore harnessing Humboldt's 'unfettered' views of nature, Sarah could make the first report of the Gambia in 1823–1824 for its rich, and interculturally challenging, significance for natural history and for humanity.

A slice of 'British' life in the Gambia in 1823–1824

> THE few general remarks I have to offer upon the settlements of the Gambia, arise from casual observation, and are so trivial, that, if the spot were better known, I should not attempt their publication. Their chief good which I can hope to arise from them, will be that of interesting a future traveller to explore further. (*EM*, 200)

Sarah's seemingly effacing 'few general remarks' on the Gambia turn on her prescience that they represented nonetheless an important first report of the country since recent British resettlement in 1816. Her 'Narrative', 'Description [...] of the English Settlements on the River Gambia' and the 'Appendix' ('containing Zoological and Botanical Descriptions and Translations from the Arabic') in fact total over one hundred pages (*EM*, 173–278), of which the first fifty (*EM*, 173–218) include two remarkable, highly detailed, eye-witness, pen and ink sketches by Sarah of 'Bathurst' and 'Bakkow' made in 1823–1824. The representation of the 'future traveller' for whom she envisaged her study was clearly a 'scientific traveller' (à la Humboldt, or the Bowdichs) capable of joining 'the new settlement [...] about four hundred miles up the river, on an island [...] called McCarthy' (*EM*, 206), rather than a (European) business trader, or even (gentleman) adventurer-sightseer. Importantly, as a white British woman unusually overseas in this 'spot', Sarah documented the fact that she was one of several such 'travellers'. In her note on the same page (*EM*, 200), she documented her awareness of British precedents also qualifying her contemporaries: 'Francis Moore's description of the Gambia in 1738 [...] written with much simplicity [...] enables us to compare the former with the present state of affairs'. Rather more remarkably, this note then concludes with Sarah's critique of British civilising missions (and not African 'backwardness'): 'I am sorry to add, that not the slightest improvement seems to have taken place among the natives, since that period, although we have been in possession of the settlements more than a hundred years'. Therein lay her rather different comparative white British(-woman) perspectives. Sarah was signally aware of not only her male but also her female compatriots in the Gambia.

Until relatively recently in colonial, British Imperial and Commonwealth history, scant attention was paid to the important place in early trading and colonial settlements of missionaries and of non-native women. These different, often non-conformist, minority groups were co-extensive with overseas trading missions, rather than serving differentiating roles and functions.[3] The economic and political administration of a colonial trading post indubitably depended upon the labours of merchants, soldiers, government officials, seamen and carriers in great numbers, but 'civilising' was overtly undertaken by missionaries and involved the interconnected, and often feminised concerns of 'godliness' and 'cleanliness'.[4] These values were inculcated in the indigenous peoples through basic education, which women missionaries often undertook alongside the 'cure' of bodies and souls.[5] Missionary work was therefore among the few suitable occupations that educated and adventurous (European) women could pursue, whether married or unmarried, because it entailed 'care' for the colonisers – nursing those with tropical fevers, educating settlement children – and colonised alike.[6] Arnold Hughes and David Perfect usefully collated in 2006 the key facts already noted by predecessor historians, to which we return, about the intense missionary activity in the Gambia at this time: an Anglican chaplain was first appointed in 1820; the first Wesleyan missionary, John Morgan, arrived in February 1821; William Singleton was the first Quaker in the settlement in January 1821. A footnote about Morgan is a first alert to overtly female as well as indigenous Gambian co-participations in its wholly 'Protestant' yet multi-denominational civilising mission: 'a group of four Europeans and two Gambians educated in England led by Hannah Kilham, did not arrive until December 1823. The mission collapsed in 1824. See Gray, *History*, 311–13.'[7] We glimpsed her in the opening portrait.

Like the small slice that is the Gambia and its River in West-African geography, the tiny window of its history in 1823–1824 containing Hannah Kilham sheds important light on this 'enclave'.[8] Alongside the missionary denominations mentioned in this 2006 footnote (and in the pages from Gray's *History* it cites[9]), the Quakers brought to the Gambia, as to other parts of West Africa, the clearest abolitionist beliefs of all: the equal (theological) value of all human beings in body, mind and spirit, irrespective of race or sex. However, as Alison Twells has explored, the 'collapse' of Kilham's project revolved around the inequalities that she encountered with respect to status: British women and Africans (of both sexes) always had a different footing to the settlement's British men.[10] By focussing on Hannah Kilham's roles as a missionary and educator of Gambian men, Twells (alongside Hughes and Perfect) missed a further fact. Kilham was one of a significant group of European women in the Gambia in 1823–1824. Gray clearly described them only two pages beyond the range cited by Hughes and Perfect immediately above:

> Last of all, but by no means least, mention must be made of the French Sisters of Charity, who in 1823 came at the request of Sir Charles MacCarthy from Goree. In view of the state of the surrounding country their work had necessarily been confined to St. Mary's Island. Despite the many difficulties they did much real philanthropic work amongst the poorer members of the community. Both Mrs Bowdich and Mrs Kilham gave high praise to the Sisters for their ministrations to the sick. Unfortunately the same calamities befel (*sic*) them as befel (*sic*) the reset of the missions. Sister Adèle died and ill-health eventually compelled her companion to leave the colony.[5] [...]
>
> [5] Bowdich, *Madeira and Porto Santo*, p. 203; Biller, *op. cit.* p. 188.[11]

As the only historian of the Gambia to footnote Mrs Bowdich and her '*Madeira and Porto Santo*',[12] Gray's scholarly cross-reference for Kilham, and in the context of other 'Sisters', had also gone unremarked, until Martha Frederiks combed his material afresh in 2003 to extend Gray's mention to include non-British women, but yet again the work of 'Mrs Bowdich' mustered only footnote status:

> (the Cluny sister) Anne Marie Javouhey left the Gambia after about ten weeks [...] The two other sisters, probably Sr. Adèle and Sr. Marcelline, stayed behind in The Gambia, but very little is known about what they did. [...] Hannah Kilham in her diary mentions the funeral of the French Sr. Adèle on 23 December 1823. The name of Sr. Marcelline is mentioned by Mrs Bowdich,[19] who visited The Gambia in 1823.[13]

Sarah's was therefore a singular position in the Gambia within, yet not belonging to any one of the representative 'Christian' (white, female) communities named above busy extending the civilising work of the 'British Settlement' in 1823. Her valuable 'mentions' concerning it then explain why her unusual account has endured as a token footnote within the scholarly footnotes of (among very few) specialist historians of the Gambia. Her account was among the first since Francis Moore's, because her rationale to pen it as 'interesting a future traveller to explore further' reflected his in her note above. Clerk to the Royal African Company, and living on the Gambia River from 1730 to 1735, Moore had viewed it as a 'country much talk'd of, and little known', a 'dark continent' in knowledge that he had set out to fill in his journals and account.[14] His zeal, like Sarah's to correct it identified the Gambia's paradox, repeated with variation down to the twenty-first century. The nation's diminutive geography, limited geopolitical influence in its region, and slight economic value overseas (equating to recent monocultural production of groundnuts[15]) made it of particular interest to non-Gambians, especially Anglophone non-African travellers such as Moore (and the Bowdichs), governors, historians and anthropologists. All sought to put their own stamp on comprehensive, definitive and revisionist accounts of its history.[16] Already in 1964, historian Harry R Gailey Jr. formulated why his work, and subsequent study by qualified external experts, was so necessary:

> [...] books that focused upon the Gambia are either modified travellers' accounts or are outdated.
>
> Thus a vacuum does exist in our knowledge of the Gambia.
>
> Finally the Gambia represents in microcosm most of the problems that confront any new emergent state. [...] The most important aim was to present the political development of the area from a casual appendage of Britain, through the crucial period of the 'scramble' [...] of necessity, therefore, the book can be considered as European oriented. The centrum of power throughout Gambian history for four hundred years has rested outside the territory. The Europeans, in addition, are the only ones to have left considerable records of their contacts with the people. The native Gambians have left few written records, and there have been few detailed anthropological studies undertaken in the Gambia.[17]

Gailey's foremost assumption concerning the 'vacuum [...] in our knowledge of the Gambia' was that Western expertise, including 'detailed anthropological studies', would fill it, when precisely such 'expertise' occluded what was earlier known. Gailey's dismissal of works by earlier travellers and historians is indicative. 'Gambians' are thus homogenised, and assumed to be (altogether) 'unlettered'. Moreover, Gailey must also have been aware of the work of his contemporary, Florence Mahoney, the Gambia's first (woman) historian publishing in English (including a review of his book in 1965).[18] Indeed, her work on the Gambia's mixed-race women such as our book's initial observer was a notable exception. It was the first specialist history since Gray's magisterial *History of the Gambia* (allegedly 'outdated' according to Gailey), to footnote Sarah's observations of them in 1823. A before-and-after approach to key dates – including British Settlement from 1816 – in the Gambia's colonial, and now post-colonial history (including the controversial recent presidency of Yahya Jammeh), has stratified Anglophone non-Gambian, and recently also Gambian, survey accounts, histories and anthropologies. At best, recognition of Sarah's work within them, because already obscured by the title of the *Excursions* explaining the work's absence from history of (women's) travel to the region,[19] has been further reduced to a rare footnote within further footnotes in the history of the Gambia as still 'little known'. This chapter now takes up Francis Moore's formulation, to talk much of Sarah's very differently pioneering geographical and scientific report in 1823–1824 on two intercultural counts, clearly made visible through her Humboldtian scientific worldviews.

'Little Apparent Result': The First (Plant) Geography of the Gambia

The Bowdichs' important comparative geological work in neighbouring Madeira was particularly precious to Humboldt. Because of the Napoleonic Wars, he and Bonpland had had permission only to visit Tenerife en route to the Americas. In consequence, his 'Notes de M. de Humboldt' forming the epilogue for the French *Excursions* (*EMFr*, 426–45) retroactively focus new critical attention to the original (*EM*) concerning its Humboldtian understandings of geology, chemistry and plant distribution. I earlier drew attention to, but did not comment further on this 'notable running thread' in the many footnotes and in text references to 'initial volumes of the *Voyage aux régions équinoxiales du nouveau continent*'.[20] This chapter brings them to main attention for the first time in Appendix 5, and not only because Humboldt's many French publications were referenced and cited throughout the *English Excursions* with no English translations. The range of Humboldt sources in the Madeira section also defined Sarah's scientific exploration practices there, and hence her reapplication of them in her report of the Gambia. Any differences in the latter would also mark her development of them as of further significance.

Sarah's single-handed publication of the first (European) geography of the Gambia since British resettlement of it in 1816 could not emulate, or be judged by the same measures as the Bowdichs' detailed, year-long study of Madeira's geology, principally including extensive measurements of the island's peaks and other phenomena. Although Edward posthumously measured up in Humboldt's 'Notes' of 1826 to the foremost European geographer 'mesurati' in the Barons von Buch and himself,[21] she was not the trained 'measurer' of the Bowdich partnership. But it does not follow that because she did not adopt Humboldt's blueprint of measurement as central to intercontinental scientific exploration, she was no 'scientific' geographer in the Gambia. Rather, as the expert recorder of other no less important botanical and zoological measures,[22] Sarah's work in the Gambia more largely revealed the necessary 'Bonpland' partnership in 'Humboldt''s famous *Voyage*. Her expert attention to *plant* geographies offered no less important measures of larger geological and geographical understanding. Indeed, hers in the Gambia located the very terrains that the Humboldtian model literally and metaphorically overlooked. It is exemplified famously in his iconic views of Ecuador's Chimborazo volcano. The mountain in the work of Humboldt and Bonpland was key to world topographical understanding, with ranges, peaks and cross-sectional plans the means to construct orders of climatological, geographical, botanical and other inter-global sequencing of importance.[23] By complete contrast, the Gambia is universally only some few hundreds of feet (metres, *toises*) above sea level. The topography is not volcanic, and presents little visible 'geognosy' to explain its complex and changing alluvial terrains:

> Numerous creeks intersect the island [of St. Mary's Island on which Banjul is situated], and when the tide retires, leave stagnant pools; the soil, which reflects back the heat with intensity, is in general sandy, with scattered patches of vegetable mould but alluvial and marshy in the neighbourhood of the creeks, the half-dried margins of which exhale a baneful miasma that alone would generate fever, needing no addition to its poisonous effects from the bad quality of the water. The river, in its whole extent, flows through a thickly wooded country, and the mangroves penetrate far into its bed on either side; consequently the return of the tide brings with it a quantity of putrid vegetable matter, which is continually deposited on the banks. (*EM*, 201)

All the 'lofty' subjects and exploits pertaining to the Bowdichs' *Excursions* in Madeira, following Humboldt's priorities in South American 'Equinoctial Regions' were thus necessarily (sea-)levelled in the no less 'equinoctial' Gambia. Its sedimentary *land*-scapes here explain its larger, no less 'Humboldtian', (plant) geography. They also connect to Sarah's one footnote reference to Humboldt in her 'Narrative': 'This completely exemplifies a remark of Baron de Humboldt's, I believe, that "there is more true solitude in sand than in forest"' (*EM*, 191). If the Gambia melds both, the note trigger is a description of the 'barren and uninteresting' Bona Vista (Cape Verde Islands, where the Bowdichs' had stayed briefly to await sea passage that brought them to the Gambia). Of particular note were its springs 'depositing white sediment', the 'porous red clay' jars to hold their water, and the salt collected 'in square, shallow pits' (*EM*, 191). As has only very recently been investigated in this 'offshore', the Gambia 'documents a complex tectono-stratigraphic history' in its (West African/ Atlantic) 'cretaceous continental margin development' when '[s]ubaqueous channel systems (up to 320m wide) meandered through the pro-delta region reaching the palaeo-shelf edge, where it is postulated they initiated early submarine canyonisation of the margin'.[24] In effect, Sarah's description above records an early lithology – the sands, clays and sediments deposited from even more ancient stratigraphy or 'superposition' of rock layers[25] – of the (River) Gambia's much longer (paleo-)geology, and central place on the geological map of the region's major sediment drainages from mainland sub-Saharan Africa.

The coextensive understanding of major water distributions in this shifting sedimentary landscape also integrally informs the Gambia's bio- and human diversity (in today's parlance). Sarah's primary focus on the (River) Gambia's hitherto largely undocumented *flora* and *aquatic fauna* – all in such a small and seemingly homogenous space – and on their 'usages' of the country's axial waters therefore reversed Humboldtian priorities for interpretative 'plant geography'. Where his understood stratigraphical geologies as key to understanding the life forms they support, hers made careful note of the different species living side-by-side in these aquatic and alluvial deposit terrains, because particular land and waterbody features are much more fluid and difficult to demarcate. Specific plants, molluscs and fish are then 'measures' distinguishing (and never entirely corroborating) where the Gambia's intersecting littoral habitats become (more or less) marine rather than estuarine or freshwater, and according to a range of variables including season, river flow or drought. Had Edward's measurements of the distance, height or depth of the River Gambia from mouth to source been completed, these would not have revealed its 'geognosy'. Rather, what seem mere lists in Sarah's 'Appendices' for the Gambia turn into applied Humboldtian 'plant geography' and 'plant geology' maps that include human interventions within them. Two brief examples demonstrate how their particulars interlink the geo- and biodiversity of the Gambia in Sarah's remarkable first 'geography' of it.

In the latter part of the Botanical Appendix, 'Banjole and its Environs' (*EM*, 248–67), Sarah lists 'Oryza mutica[b]' among other indigenous grains (and as food staples). Her note in English is a mine of 'intercultural' information:

[b] The white rice of the Gambia is generally thought to be quite equal to that of Carolina, but in the variety which I examined, the seeds were yellow, flat, and deeply furrowed. When boiled, turns red. (*EM*, 248)

Sarah's early findings have gone without notice in recent scholarship by Clare Madge, for example, corroborating the qualities of this native Gambian 'red' rice, both for its greater drought tolerance than imported white ('Carolina') varieties, and for the extent of women's work in Gambian rice production and food preparation.[26] But Sarah's short note was also significant comparatively in the context of 'Humboldt's' *Voyage* and of the Madeiran report in the *Excursions*, in which Bonpland's *Plantes équinoxiales* was cited (see the first reference in Appendix 5). In it, his important work on (world) grasses also saw them as a major indicator of the development of human agriculture. A long note qualifying 'ornamental grass' cultivated in Madeiran gardens (*EM*, 114) specifically relates to '[t]he rice from our part of the Coast of Africa, is complained of as reddish [...] and grown in quantities in the interior, on the banks of the Adiree or Volta'. Sarah's extensive comparative world knowledge here eminently qualified her to examine the Gambia's various indigenous grains, alongside other food and medicinal plants, and as significant in 1824 for the understanding of the larger region's 'ethno-geography', soil science and sedimentary geology in today's terms.

A second example, Sarah's account of Gambian mangroves, is also highly pertinent to current debates about management of this important 'ecosystem' (now a world SSSI environment), particularly for its oysters and how best to harvest them sustainably.[27] In the concluding pages of the 'Zoological Appendix' (*EM,* 239–243) devoted to molluscs – '[t]he shells which we found at St. Jago and the Gambia [...] and chiefly named after Lamarck' (*EM,* 239) – different oyster species are identified and recorded:

Ostrea—Gam.
"...cristola, Gam.
"...folium. Gam.
"...fucorum, B. V. (*EM*, 240)

Although no trace of these historical ('Lamarckian') names emerges in searches of current (mangrove) oyster names, these clearly indigenous *fauna* and *flora* examples only better reveal Sarah's world (plant) geographical understanding and standing. In her 'notes' to new specimen matter as in her Appendices in *Latin* (as the official language of European science), Sarah was meticulous about official naming according to the most authoritative models – Cuvier for Muséum (world) natural history, Lamarck for (world) shells, Bonpland for (world) botany, Humboldt for (world) geology – to pass her own name in their ranks where this was otherwise problematic (for a woman, a Briton), including when she could also distinguish hers. As we saw in Chapter 1 for Fishes, she was therefore equally meticulous about adding provenance – 'Gam(bia).', 'B(ona). V(ista).' above – as a further signifier of her (new) West African contributions to their European (and American) scientific knowledge. The 'we' in the oyster example above for the Gambia as 'I' (Sarah, not Edward) then also indicates through her many notes and final part of the *Excursions* wherein lay her pioneering perspectives. As detailed in the book's opening portrait, mangrove oyster harvesting was in 1824 and remains today the work of Gambian women. Sarah could therefore advance the causes and knowledge of West Africa's equatorial regions, to single out those of rich *aquatic* interest among geography's loftier and geology's stratified (ad)vantage points, because long-standing human (and women's) settlement history was also more prominently integral to them. Neither Cuvier nor Humboldt made such a case in their work. Moreover, the clearly transatlantic transportations of rice in the 'note' above, and of sugar in Madeira (*EM*, 102–103), also defined parallel movements and translocations of human 'species' in the trafficking of slaves. Sarah's precious eye-witness observations in the Gambia on its *flora* and *fauna* integrally related to its geopolitical relations to French, Senegambian and Goree neighbours regarding trade and continuing (transatlantic) slave trade in particular. Sarah's clear indictments of the multi-ethnic, including indigenous practices despite British controls, also deserve to be much more 'talked of' (Moore, above):

> The river winds very much its course, and during the rains, its water is fresh at Jillafree (or Gillyfree), about twenty-five miles from the mouth, where a factory has been established for the inland trade. Albreda, which I understand is picturesque in its situation, is about a mile from it, belongs to the French, and I do not hesitate to declare, is a known emporium for slaves and smuggling. The Chief, and only authority there, for he is not to be styled Governor or Commandant, received Mr. Bowdich very hospitably [...] Slaves are brought by the concealed agents for the trade to Albreda, where they are secreted by the residents, especially in the houses of the French mulattoes, till a French vessel arrives; a frequent event, as a considerable trade is carried on by means of small craft, between Senegal, Goree, and the River Gambia. It is at Albreda that the bargain is struck; but, as all foreign vessels are subject to examination as they pass and repass the town of Bathurst, they do not ship their live cargo till they reach Salem, situated to the north of the river's mouth; where the slaves, having been marched through the bush or forests, meet their purchasers, and are taken thence to supply any market where they are likely to fetch a good price. Several proofs of this occurred during my residence at Bathurst. (*EM*, 204–205)

If slavery was far from abolished in 1823–1824 a stone's throw from Bathurst, this passage is illustrative of the detail, and other eye-witness verification, authenticating the many intercultural observations that mark out Sarah's Gambian narrative. In a nutshell, and precisely because criticality towards her own 'culture' was also recorded in her work as we see below, here was a forerunner (woman) exponent of 'anthropology' (as opposed to ethnography, and its earlier formation in ethnology) long before the field was constituted.[28] Sarah's pioneering perspectives in its field(s) in her Gambia 'narrative' are summed up in the next subheading. It will investigate two that epitomise the principles of intercultural exchange – and knowledge exchange – at the heart of her work. The first concerns description of the Gambia's many distinct ethnic-linguistic groups from outside in. The second related focus in Sarah's report more unusually concerned the Gambia's majority (and no less literate) Islamic culture, including as a critical reflector already in 1823 of European-British civilising mission.

The Odd (White) Woman Out in the Gambia in 1823–1824: Recovering an Extensive (Intercultural) Report from the Field

Sarah's Unitarian views, 'Paris' scientific training and category of one propelled her interests in the Gambia beyond all usual avenues of European expatriate inquiry, even as the only female member of a very different scientific 'mission' in the country by design and not commission. As the introduction flagged (p. 15), Sarah's learning in Paris of Arabic as the region's *lingua franca* was to access the wealth of local knowledge that her botanical lists for the *Excursions* contain, alongside her 'Translations from the Arabic'.[29] Arabic therefore opened her investigations of the majority indigenous communities around her: 'the state of the surrounding country' comprised the Gambia's largely Muslim populations. Their various ethnic groupings also continued syncretistic, non-Muslim animist practices from prior African as well as Portuguese (Catholic) colonisation and settlement.

The month-long 'survey' that Sarah conducted in the environs of Bathurst offered her particular freedoms as a widow to cross intercultural lines and religious divides. Indeed, Sarah continued unimpeded the Bowdichs' practice proactively to seek out and to enlist indigenous ('native') knowledge: 'The usual means were resorted to, of purchasing the birds, shells &c., brought us by the natives, and every facility was afforded by our countrymen, particularly by the Commandant [Captain Findlay], whose anxiety for the survey seemed to equal Mr. Bowdich's' (*EM*, 197). But Sarah's first report on the 'Population' and 'Account of the manners and costume of the Joloffs and Mandingoes' is then the more striking. Chapter 7 will investigate the importance of her further capture of her subjects in her watercolour plate, 'Costumes of the Gambia' (also illustrating our cover). Here in her particular regard to women subjects, Sarah's descriptions for the period were unusually free of racial (and gendered) prejudice,[30] through her deployment of comparative cultural perspectives and evaluations:

> The mulatto women, who are mostly Joloffs from Goree, are some of them handsome, and pretend to approach nearer to European manners than those of other parts of Africa. [...] They wear pagnes like other natives, and as they are generally tall and gracefully formed, look very elegant. They add a covering to the head, which, if it were not so enormously high, would be pretty; it is an assemblage of several square handkerchiefs, (frequently nine) put on much in the way of those of the French peasantry, but rising in a very high cone at the back of the head, and, on state occasions, ornamented with a broad gold band. They generally wear shoes, and those who go without stockings ornament their ancles (*sic*). (*EM*, 209)

These women are 'handsome', 'tall and gracefully formed', 'elegant' and clearly fully-clothed. Their head-dresses are, if not to Sarah's aesthetic taste, on a scale already of *French* fashion sense, and in their decoration 'ornamented' like the turbans (worn by superior ranks of Gambian men). Unlike their 'French mulatto' counterparts in Albreda in the passage cited above, who were clearly complicit in the slave trade, or later nineteenth-century views of (semi-naked) 'negresses' as in effect prostitutes,[31] the Joloff women from Goree represented an important ethnic sub-group within a multi-ethnic trading community established for some two-hundred years. The history of these Joloff women was only taken seriously in the early twenty-first century in a study by Gambian woman historian, Florence Mahoney, who also referenced Sarah's here as the first report of them.[32] As negotiators ('translators'), and also as wealthy 'business women' in their own right as our opening fictional Gambian woman portrayed, they were key brokers between the various indigenous peoples of the (River) Gambia and white European traders.

Where 'manners and customs' were staples of travel accounts of the period, Sarah's report delves even deeper. Its interest lies in the multiple, intersecting, but also local and very particular practices that existed within larger and longer historical settlement in the Gambia, whether 'Moorish', Portuguese, Mandingo or Joloff. One telling representative example nests within the quotation describing the Joloff women above at the ellipsis. Their contrastingly 'un-European' customs are the object of Sarah's concertedly comparative focus:

> at the same time, they religiously preserve their own superstitions and ceremonies, some of which are disgusting, and others prejudicial: among the latter, is that of shutting themselves up in a room with every crevice stopped, and a large fire burning during child-birth, and neither mother or infant are allowed to breathe the fresh air under a fortnight. This practice is so different from that of other mulatto women that I thought it worth mentioning. (*EM*, 207)

This report could not have been garnered by observation of these birthing rooms solely from their outside, but only through Sarah's conversations with the (Joloff) women informants who had also been inside them. Such intelligence could also not so readily have been gathered by (white European) male observers. While Sarah clearly disapproved – such customs are 'prejudicial', not her more value-laden 'disgusting' – her own experiences of having given birth, and recently in Madeira, make this particular report of childbirth practices personal as well as (participant) observational. Conducted over 150 years before 'women's interest' anthropology in the field also took issue with male-gendered geography, anthropology and ethnography, Sarah's 'mention' because 'different' of Joloff birthing customs is astonishing on its own account as well as in comparative West African terms. To make it, she also had to have been familiar with the birthing practices of other Gambian ('mulatto') women, not least because of their equal interests in hers. Sarah was altogether the odd woman out among her white 'sisters' in the Gambia in 1823, because her own young children clearly marked her out in her daily 'scientific' work. As the opening portrait of this book imagined, they would have been immediately the centre of attention and remark among Sarah's Joloff and other local women informants and, in consequence, her most invaluable intercultural asset for 'scientific report' as this passage shows.

When we then connect Sarah's informed descriptions of Joloffs, Mandingoes and other groups with her 'Botanical list' (which she clarified as 'much less complete than I had expected it would be', *EM*, 266), her 'anthropological' insights appear almost more ground-breaking. Sarah's major contributions to botanical knowledge of the Gambia cannot be judged by the relatively small size of her specimen samples, but rather by her comparative observational fieldwork that challenged previous (Western) authorities on West African floras, which she cites:

> With regard to those which I profess to have determined, I offer them with some degree of confidence, for, since my return, I have re-examined my notes, and the remnants of my specimens, amid the collection in the Jardin du Roi, and have scarcely had a single instance to alter. My books of reference, both for species and localities, have been Persoon and Willdenow. At the end of each name, I have added the country to which the plant has been hitherto supposed to be indigenous, that an idea may be formed of the similitude of vegetation, and I have given the uses made of it by the natives. They were all gathered in a soil more or less sandy, and on a level with the sea [...]
>
> It has been remarked by M. Palisot de Beauvois, in his Flora of Benin and Owaree, that the natives of Africa more frequently make their medicines from Compositae, than any other family. *This is by no means the case with the Joloffs and Mandingoes*; *their remedies seem to be distributed throughout the different families, and the only remarkable circumstance attending them, is the frequency of their antidotes against worms, and lung complaints.* The variation of the climate accounts for the necessity of the latter, but their food, which is chiefly rice and corn, without any great proportion of fruit, does not seem to induce the former disorder. *The guinea worm I believe to be wholly unknown: nor did I see a single instance of enlarged spleen, or elephantiasis, so frequent among the Fantees.* (*EM*, 266–67, emphasis added)

Sarah's 'Bonpland' role in her first (plant) Geography for the Gambia now also clearly reveals its larger place within the Bowdichs' excursions in Madeira. Her herbarium 'remnant' is a plant list (in Latin) of 18 pages, the majority of entries also accompanied by supplementary notes (in English). These display remarkable specialist botanical knowledge that includes impressive 'insider' information about the given plants' differentiated local naming, or culinary and medicinal uses. One example stands for many:

> Hibiscus trionium[g], Hab. in Italia, Africa, &c. [...]
>
> [g] Native name Dummodo. The leaf is boiled with rice, to give it an acid flavour. The Moors make a syrup with it for a cough, and call it Basab. (*EM*, 254).

Sarah's 'notes' therefore clarify her informed fascination in local differences, whether birthing practices or medicinal or other uses of indigenous plants by the Joloff or Mandingo separately, or as practised by 'natives' or 'Moors' (her categories). A key refrain punctuates the extensive notes: 'I was told [...]' (for example, *EM*, 261) or 'is/are said':

> Cassia occidentalis[x], Hab. In America. [...]
>
> [x] This plant seems to be the panacea of the Mandingoes, who call it Bantamara. Its seeds are roasted, and used instead of coffee. The warm baths given for all disorders, have a quantity of these leaves thrown into them. They are said entirely to cure rheumatism, and in all fever cases the bodies of the patients are rubbed with them. (*EM*, 256)

These participant observations, in today's terms, only more clearly illuminate Sarah's pioneering perspectives in the field in 1823.[33] Here was a remarkable early (and woman) expert working in what today is medical anthropology, ethnomedicine and ethnopharmacology.[34] But her means of securing such 'local' knowledge also clarified unequivocally that expert book information ('M. Palisot de Beauvois') was not enough; its application through further observation also entailed a (Bonplandian) 'plant geographical' review.

Supplementary, Separated, Accidental, Niche, Unimportant?

Unlike Kilham, Sarah (and Edward) had not learned the local languages of the Gambia prior to their expedition; it was not their intended destination. While spoken Arabic differs from its written 'standard', Sarah herself noted that '[t]he western dialect [...] approaches nearer to the learned Arabic then the modern Oriental (*EM*, 268).' The 'learned' Arabic was also the Quranic. The short description of 'Bakkow' (Bakau) in Chapter 3 of Sarah's, 'Narrative' cannot then be skim-read as a hasty 'sketch' (Sarah's word below, which her figure might also confirm), undertaken with only one overnight stay. Its 'sea breeze' (healthier air) had made it an important 'annex' to the British hospital in Bathurst – 'a house has been built there for convalescent officers' (*EM*, 213) – but Sarah noted its unsuitability for trade, due to difficulties of access. She could reach it only on horseback by land, with canoe crossings of intervening tidal creeks. The town

> consist[s] of miserable-looking huts, crowded together, filled with smoke, and some not high enough for a middle-sized person [herself?] to stand upright in. The granaries are mingled with the huts [...] and raised on poles, to prevent the encroachments of ants, and other insects. The hall of justice or palaver house, is higher than the others, with two arched entrances, but would not contain more than ten people sitting close together: it is built of the red earth of the neighbourhood, and a passage from the Koran is inscribed over each door. The mosque is one of the worst huts in the town. [...]. (*EM*, 214–15)

> My readers will easily perceive, from the foregoing little sketch, the difference in customs, the striking inferiority of the inhabitants of this part of Africa, to those north and east of the leeward coast. Mr. Bowdich's 'Mission to Ashantee' is a detail of splendour and bravery, accompanied by shrewdness, reflection, and ingenuity, a polish of manner, a taste for arts, and a dexterity of manufacture, shewing an advancement that astonishes us in a people called barbarous.
>
> Whence can this difference arise? Not from their natural productions. The same metals, the same superb vegetation, the same soil, the same climate, exist in both countries. Not from their religion, for what can be more luxurious or splendid than the Musselmen of the East. Not from their greater intercourse with strangers, for there the Mandingoes would have the advantage. (*EM*, 217–18)

Despite Sarah's active sympathies with abolitionist agendas, and Quaker plans precisely to 'cultivate' Bakau,[35] her account pivots more unusually on the question of 'cultivation' in the Gambia's *Muslim* heritages. The town and its surrounding landscape and vegetable productions (mangroves (*EM*, 214), monkey-bread trees (*EM*, 214–216), Run trees (*EM*, 216) and coral trees (*EM*, 216)) interrelate: 'the whole vegetation is very luxuriant, but not owing much to cultivation[u]' (*EM*, 216).[36] Sarah's judgement of the clearly 'inferior' (backward) state of Bacau for comparative West African geographical and historical natural history is by the historical yardsticks not of 'enlightened' Europe, but rather of Asante 'advancement'. The visible discrepancy between West African like and alike posed the larger problem concerning 'cultivation' that European race theory (including Cuvier's) and comparative religion of the period had yet to explain. The open conclusions that Sarah offers in response are even more interestingly pan-African: 'Is it not then a further proof of the Egyptian origin of the Ashantees, suggested by Mr Bowdich in his Essay on their superstitions *&c.*—a fact which would satisfactorily account for their greater progress towards civilisation' (*EM*, 218). The last word of Sarah's 'Narrative' of the Gambia is 'civilisation' in direct connection to West African 'cultivation' for its models that distinguishes it unequivocally from European or, indeed, Christian 'civilising' missions. This rarest account of the Gambia in 1823–1824, because by a white European woman, also provided a rare first 'counter-European' history and (plant) geography of the Gambia.

Sarah's concomitantly 'pro-Muslim' viewpoints in her unusual report – because her Unitarian upbringing respected 'the wisdom of other faith traditions' (Introduction, p. 13) – are, like her plant lists, meticulous concerning their provenance. She recorded and named her main informant, the 'Marrabout' Dongo Kary. The only literate Gambian of her account, he possessed 'English very tolerably' (*EM*, 268). While Sarah's gloss on and translation of Dongo Kary's stories in the final appendix of the *Excursions* provide further fascinating insights into the oral and literary Quranic traditions of his native Senegal, the further interest for European and British history of the Gambia in 1823–1824 is the record of Dongo Kary's response to Sarah (as British, a Christian, a woman), as much as of hers to him:

> The astonishment expressed by the Marrabout at seeing me write, not only my own language, but his also, was very entertaining; as the knowledge of the Moorish females is confined to the repetition of the hymn and common prayer [translated by Sarah, *EM*, 274]; and when I explained the 'hamza' to him, he exclaimed, *as on every other wonderful occasion*, 'white man and woman do every thing (*sic*); your country pass ours.'
>
> It will be seen by the translations that the religious traditions of African Moors are confused and imperfect. Glimpses of the truth are mingled with their own romantic notions, and so long as they ascribe the highest honours to Mohammed, they care little for consistence of circumstance, or connexion of events. The expressions used by our Marrabout were frequently so ludicrous, that we could not avoid smiling, or even laughing, and the seriousness with which he uttered them, added to the effect caused by his having lost an eye, and by his enormous bush of woolly hair, which stuck out from his head in every direction. He would frequently argue with us on the respective merits of our religions, and *I was surprised by his correct acquaintance with the christian (sic) tenets, and his high opinion of their charitable tendency. He invariably confessed the divinity of our Saviour, as a prophet, and placed him in rank next to Mahomet.* (*EM*, 269, emphasis added)

Through the medium of Arabic-English intercultural exchange, Sarah offered her British (and French) readers a view of 'enlightened' western European civilisation in the mirrors of the Gambia's enduring majority Muslim cultures. By downplaying the effect of her roles as 'entertaining' above, Sarah's awareness of her self-positioning in 1823–1824 (as part of the Bowdich couple) in this encounter remains no less striking today. The frequency and delight of these meetings (in the emphasis above) suggest mutual cultural enrichments (plural) that did not separate religion from 'science', discovered reciprocal respect for knowledge (including that of women) and agreed clear differences of belief. Furthermore, this inter-cultural exchange also provides insights into why the intensive British 'missionary' zeal in the Gambia in these crucial years in the 1820s so overwhelmingly 'collapsed'.[37] The given cause in the high mortality rate due to fever among missionaries obfuscates the failures of ignorance (or arrogance). Sarah's singular report of successful interreligious or interfaith dialogue in 1823, when such appellations were only birthed by the 1893 World's Parliament of Religion in Chicago,[38] already demonstrated to Christian missionary societies of all stripes that the Gambia's marabouts and Muslim peoples were no 'dark continent' (Francis Moore's words above) awaiting enlightenment, but were already informed for at least a century about Judeo-Christian beliefs. Sarah's extensive botanical and Arabic appendices for the *Excursions* dispelled with counter evidence the many European myths about backward, ignorant and dirty 'natives'. Where baths, herbal ablutions, purgatives, gargles everywhere punctuate the botanical notes, larger West African intercultural 'cultivation' had in Sarah's narrative of the Gambia critical European consequence.

Sarah's 'little-known' report from the Gambia in 1823–1824 in 'the last part of the volume' has now yielded up some of its best-kept secrets, once it is reviewed through Humboldtian, Bonplantian scientific worldviews and Unitarian perspectives. Despite its apparently 'nothing' to see, this first 'plant geography' of the Gambia connected it interculturally to its wider geological and historical West African region. Sarah's extraordinary 'excursions' from Bathurst in the space of only a month also modelled and extended expert 'Humboldtian' plant geography. The interconnecting range of expertise in *her* 'Notes' to the *Excursions* in aquatic natural and intercultural history included contributions to 'ethnobotany', 'anthropology' and interfaith dialogue as disciplines yet to be born. *Contra* Gailey above, 'modified' travellers' accounts remain key genres for report of scientific exploration. The small 'slice' of Sarah's presence in the Gambia's history in 1823–1824 could not more tellingly uncover the white spaces of European prejudice, ignorance and narrowness of 'Western' view that has largely excluded West African civilisations and optics from intercultural and scientific inquiry, and hence the very rare intermediaries like Sarah who have made them better known.[39]

As imagined by our Joloff woman's opening view by contrast picking Sarah out, this chapter has challenged narratives of the European and British 'civilizing mission' (religious and scientific) through her multiple non-conformisms between its bars. Our study of both Cuvier's 'Notes' in Chapter 1 and Humboldt's 'Notes' in this chapter for their larger methodological importance in the translated *Excursions* (*EMFr*) now also better displays her many 'notes' and appendices in its original (*EM*). If they are integrally 'sedimented' in their purpose, and 'Humboldtian' and 'Bonplandian' in purview, following chapters can now unpack how Sarah's (pioneering) perspectives on the Gambia further enlighten her publications in natural history from 1833. If in Dongo Kary's words 'white man and woman do every thing; your country pass ours', Sarah's newly relevelling report neatly overturns the allegedly men-only worlds that are rugged physical geography overseas, the climbing of new peaks of scientific knowledge of the period and men-only councils for inter-religious dialogue until very recently.

Notes

1 See Orr, 'New Observations'.
2 Orr, 'New Observations', 158.
3 See Porter, '"Cultural Imperialism" and Protestant Missionary Enterprise, 1780–1914', 367–91 as catalyst for subsequent sharper redefinition of the 'civilising' mission, and the place of missionaries (in diverse groupings) within or outside it. A recent example is Bronwen Everill, 'Bridgeheads of Empire? Liberated African Missionaries in West Africa', *The Journal of Imperial and Commonwealth History* 40, no. 5 (2012): 789–805.
4 The issue of moral, social and medical hygiene in colonial settlements was also tied to the mainly non-native female group in ports and capitals: prostitutes. These (white) women had little choice, since they were often transported to colonies for this purpose. Sarah is silent on their existence in the Gambia, but their presence in 'civilising' missions in the later nineteenth century is belatedly attracting research. See Saheed Aderinto, 'Journey to Work: Transnational Prostitution in Colonial British West Africa', *Journal of the History of Sexuality* 24, no. 1 (Jan. 2015): 99–124 and Liat Kozma, 'Prostitution and Colonial Relations', in Magaly Rodríguez García, Lex Heerma van Voss and Elise van Nederveen Meerkerk, eds., *Selling Sex in the City: A Global History of Prostitution, 1600s–2000s*, 730–47. Leiden: Brill, 2017.
5 For a recent study of women missionaries as educators of girls in Sierra Leone in the 1820s see Fiona Leach, 'Resisting Conformity: Anglican Mission Women and the Schooling of Girls in Early Nineteenth-Century West Africa', *History of Education: Journal of the History of Education Society* 41, no. 2 (2012): 133–53.
6 See Elizabeth Provost's valuable work on the subject (including bibliography), 'Assessing Women, Gender, and Empire in Britain's Nineteenth-Century', 765–99.
7 Arnold Hughes and David Perfect, *A Political History of the Gambia: 1816–1994*. Rochester: University of Rochester Press, 2006, 25; note 85 in question, 361.
8 'Enclave' is the word used by R. J. Harrison Church to encapsulate the Gambia in his *West Africa: A Study of the Environment and of Man's Use of it*. 7th edn. London: Longman, 1974, 224–32.
9 J. M. Gray, *A History of the Gambia*. Cambridge: Cambridge University Press, 1940.
10 See Twells, '"So distant and wild a scene"', 301–18.
11 Gray, *A History of the Gambia*, 315. Gray's referencing to 'Bowdich' for this episode is incorrect, and should read '*Madeira and Porto Santo*, 216' ('p. 203' discusses French nuns 'of the order of St. Joseph' involved with Bathurst hospital, and in particular 'Sister Marcelline'). Kilham's stepdaughter, Sarah Biller, authored *The Memoirs of the Late Hannah Kilham*. London: Darnton & Harvey, 1837. For a more recent biography of Kilham see Mora Dickson, *The Powerful Bond: Hannah Kilham, 1774–1832*. London: Dobson, 1980.
12 The index to Gray, *A History of the Gambia*, 500, refers curious readers only to the information cited in note 10. Gray actually footnotes 'Bowdich' on two other occasions (in connection with Kilham). The first (found within the page range quoted by Hughes and Perfect, 313) is 'Bowdich, *Excursions in the Interior* (*sic*) of *Madeira and Porto Santo*, *A History of the Gambia*, 216 and *Stories of Strange Lands*, 134, 135'. The mistaken title indicates a possible mental slip from Gray's consultation of works on Mungo Park, including Bowdich's *An Account of the Discoveries of the Portuguese*, which Gray does not cite. Although the page range given is correct, Gray omits to provide the full reference for *Stories of Strange Lands*, which 'Mrs Bowdich' published as 'Mrs. R. Lee'. His second bibliographical mention is triggered by the following: 'A house, which is now known as Cape House, was erected on the plot of by a number of liberated Africans. As it was not immediately required as a convalescent house, it was handed over

the Society of Friends for their use during the brief period of their labours in the Gambia. [...] The "lighthouse" consisted of a lantern hung on a neighbouring palm tree and does not appear to have lasted very long.[1] [...] [1] [...] Bowdich, *Madeira and Porto Santo*, pp. 213–216'. *A History of the Gambia*, 334. *EM*, 216 again refers to Kilham, but the 'light-house' in this alleged page range appears only as a detail, in Sarah's drawing of Bathurst.

13 Martha T. Frederiks, *'We have toiled all night': Christianity in the Gambia, 1456–2000*. Zoetermeer: Uitgeverij Boekencentrum, 2003, 187. Note 19 in the quoted passage refers the reader to *EMFr*, 319–20.

14 Francis Moore's words in his preface to *Travels into the Inland Parts of Africa: Containing a Description of the Several Nations for the Space of Six Hundred Miles up the River Gambia* ... 2nd edn. London: D. Henry and R. Cave, 1740, v–xi (v). For a discussion of the chronology of this work, see Matthew Hill, 'Towards a Chronology of the Publication of Francis Moore's "Travels into the Inland Parts of Africa ..."', *History in Africa* 19 (1992): 353–68.

15 For a comprehensive history, see Kenneth Swindell and Alieu Jeng, *Migrants, Credit and Climate: The Gambian Groundnut Trade, 1834-1934*. Leiden: Brill, 2006.

16 See respectively Lady Bella Southorn (wife of Sir Thomas Southorn, Governor of the Gambia, 1936–1943). *The Gambia: The Story of the Groundnut Colony*. London: George Allen and Unwin Ltd., 1952; Gray, *A History of the Gambia*; Harry R. Gailey, Jr., *A History of the Gambia*. London: Routledge & Kegan Paul, 1964, and the ethnographical work by Donald R. Wright, *The World and a very Small Place in Africa: A History of Globalization in Niumi, The Gambia*. 2nd edn. Armonk, New York and London: M. E. Sharpe, 2004.

17 Gailey, *A History of the Gambia*, ix–x.

18 Florence Mahoney, 'Review of *A History of the Gambia*, by Harry R. Gailey', *The Journal of African History* 6, no. 3 (1965): 428–29 lauds Gray's work as 'monumental' by comparison with the 'disappointment' of Gailey's. See also Mahoney's *Stories of the Gambia, with Supplement*. Bathurst: Government Printer, 1967; *Creole Saga: The Gambian Liberated African Community in the Nineteenth Century*. Privately printed, *c.* 2006. [BL Yk.2009.b.2027] and *Gambia Studies*. Privately printed, *c.* 2007. [BL Yk.2009.a.7562]. Building on her work is Hasoom Ceesay's *Gambian Women: An Introductory History*. Bundung: Fatoumatta's Print and Communication Centre, 2007.

19 See Deirdre Coleman, ed., *Maiden Voyages and Infant Colonies: Two Women's Travel Narratives of the 1790s*. Leicester: Leicester University Press, 1998, and our discussion in the introduction of Pratt's *Imperial Eyes*, and McEwan's 'Gender, Science and Physical Geography', 215–23.

20 See Orr, 'New Observations', 149.

21 See Daniel Kehlmann, *Measuring the World*. Trans. Carol Brown Janeway. London: Quercus Publishing, 2007.

22 See Orr, 'New Observations'.

23 See Bernard Debarbieux, 'The Various Figures of Mountains in Humboldt's Science and Rhetoric', *Cybergeo*, 2012. https://doi.org/10.4000/cybergeo.25488.

24 See Max Casson, Gérôme Calvès, Mads Huuse, Ben Sayers and Jonathan Redfern, 'Cretaceous Continental Margin Evolution Revealed Using Quantitative Seismic Geomorphology, Offshore Northwest Africa', *Basin Research* 33 (2021): 66–90 (66).

25 See the title and contents of the anonymous translation (by the Bowdichs') in Appendix 1, *A Geognostical Essay on the Superposition of Rocks in both Hemispheres by Alexander von Humboldt* (1823). Where unusual beds of sand are found among layers of sandstones (294–95), these are measured by altitude, not sea-level.

26 Clare Madge, 'Collected Food and Domestic Knowledge in the Gambia, West Africa', *The Geographical Journal* 160, no. 3 (Nov. 1994): 280–94, and her 'Therapeutic Landscapes of the Jola, The Gambia, West Africa', *Health & Place* 4, no. 4 (1998): 293–311.

27 See Britt Crow and Judith Carney, 'Commercializing Nature: Mangrove Conservation and Female Oyster Collectors in the Gambia', *Antipode* 45, no. 2 (2012): 275–93, and the Report in 2021, 'Women Shellfishers and Food Security Project. Participatory Assessment of Shellfisheries in the Estuarine and Mangrove Ecosystems of the Gambia', https://www.crc.uri.edu/download/WSFS2021-The-Gambia-Report-FIN508.pdf. The *Excursions* are not referenced in either.

28 See Tim Ingold, 'Hau Debate. Anthropology Contra Ethnography', *Hau: Journal of Ethnographic Theory* 7, no. 1 (2017): 21–26. The history of anthropology lies outside this chapter, but is

considered from a number of vital angles in Peter Pels and Oscar Salemink, eds., *Colonial Subjects: Essays on the Practical History of Anthropology*. Ann Arbor: University of Michigan Press, 1999, and by Regna Darnell, 'History of Anthropology in Historical Perspective', *American Review of Anthropology* 6 (1977): 399–417.

29 The direct link is made earlier ('Narrative', *EM*, 208), but reported without Sarah mentioning her own knowledge of Arabic: 'For the medicines used by the Mandingoes, and for their vegetable food, I must refer my readers to the Botanical Appendix. Their manner of eating is like that of other blacks, clawing out of the same calabash with their fingers. Most of them profess Mahometanism, and speak Arabic, using the ancient form of salutation, "Peace to thee," now banished among the eastern Arabs.[r] [...] [r] See Burckhardt'.

30 Sarah's account is not devoid of various derogatory terms, such as 'barbarians' (*EM*, 207, note q), clichés about 'idle' blacks (*EM*, 207), and the 'filth and nakedness of children' (*EM*, 208). However, she made no reference to 'savages'.

31 See for example W. Winwood Reade, 'Efforts of Missionaries among Savages', *Journal of the Anthropological Society* 3 (1865): clxiii–clxxxiii (clxv): 'In plain words, I found that every Christian negress was a prostitute, and that every Christian negro was a thief' (clxv).

32 See note 17. Mahoney's *Gambia Studies,* Part 1: 'Mulatto Origins in Senegambia', 12–40 quotes extensively from, and also paraphrases unacknowledged (31–32; 40), selected parts of Sarah's description. See also Mahoney's earlier *Stories of the Gambia*, 38–42, a shorter version of 'Mulatto Origins', which also cites 'Mrs Bowdich' (42).

33 For recent research into the medicinal practices of the Gambia's various ethnic groups, see Madge, 'Collected Food' and 'Therapeutic Landscapes' in note 26. Madge makes no reference to the *Excursions* as a precursor study.

34 See Megan Vaughan, 'Healing and Curing: Issues in the Social History and Anthropology of Medicine in Africa', *The Society for the Social History of Medicine* 7, no. 2 (1994): 283–95, and the work of Jean-Pierre Willem, *L'Ethnomédicine, une alliance entre science et tradition.* Genève-Bernex: éditions Jouvence/Gaillac: Biocontact, 2006.

35 Intriguingly, Sarah's party included another nameless European woman, who may have been Kilham.

36 The note 'u' is the cross-reference Gray and Frederiks above both quote regarding Kilham, although she appears here only as an unnamed 'elderly lady'. The purpose of Sarah's note, triggered by 'cultivation', is to record the Quakers' intention of forming 'a colony in Bakkow'.

37 See Reade, 'Efforts of Missionaries among Savages', note 31. He interestingly argued for a better understanding by Christian missions of West African Muslim faith.

38 For a definition and details of this Parliament, see https://www.newworldencyclopedia.org/entry/Inter-religious_Dialogue.

39 David N. Livingstone's *The Geographical Tradition: Episodes in the History of a Contested Discipline.* Oxford: Blackwell, 1992 spearheaded and encapsulated various debates in the history of geography concerned with repainting its imperialist roots. See Clive Barnett, 'Impure and Worldly Geography: The Africanist Discourse of the Royal Geographical Society, 1831–73', *Transactions of the Institute of British Geographers* NS 23 (1998): 239–51 for postmodern-postcolonial rejections of Livingstone's model of nineteenth-century colonial encounters as 'conversations', and McEwan's 'Gender, Science and Physical Geography' cited in note 19, which claims that women were excluded from 'physical geography in the years before the institutionalization of the discipline' (215).

Chapter Five

A FOREMOST (WOMAN) EXPLORER'S FIRST-HAND 'NOTES': *STORIES OF STRANGE LANDS AND FRAGMENTS FROM THE NOTES OF A TRAVELLER* (1835)

As the introduction outlined and *contra* Donald deB. Beaver,[1] Sarah resourced both her young family and her continuing work in natural history after 1824 by publishing original stories commissioned by Rudolph Ackermann's *Forget Me Not* and Thomas Pringle's *Friendship's Offering* (see Appendix 1), the two most important of Britain's new Gift Books.[2] At first sight her *Stories of Strange Lands and Fragments From the Notes of a Traveller* in 1835 was therefore little more than a money-spinning reprint. Its collection of ten of the 'Stories' and four 'Fragments' in the above-named Gift Books even adhered to the chronology from 1826 to 1833 of their first appearance.[3] Indeed Sarah's prefatory rationale for the 1835 collection specified that '[m]any fruitless enquiries have been made for them, as the first are out of print; and many questions asked, and explanations demanded, by those who have read them; and to satisfy all, I have gathered them together, and added such notes and remarks as would tend to their elucidation' (*SSL*, xiii). Yet several had been reprinted in this period, often in the first volume of newcomer reviews in the burgeoning print market discussed in Chapter 2 (p. 53–54).[4] 'Story VI, Samba', and 'Fragment V, The Voyage Home', are not traceable to them, however. Rather Edward Moxon, a new entrant publisher of major writers and poets including Elizabeth Barrett Browning (1806–1861),[5] saw the reputational gains precisely in 'reissuing' Sarah's already-published 'West African' stories and fragments' for what they 'added' not only quantitatively but also qualitatively. The entirely new West African material of 'Story VI' and 'Fragment V' included their strikingly choric 'Notes to Ditto' also figuring for those preceding them on the contents page placed, unusually, at the end of the volume. In effect, the original Gift Book version in each case saw supplementation by at least a further third through its respective 'Notes'. These also concealed five interleaved plates, to which chapter 7 returns, unmentioned on the inside title page. Their descriptions are again found only at volume end in the table of contents. Moxon's edition of the *Stories of Strange Lands* therefore aligned with the original Gift Book format with selected illustrations, by now including Sarah's own as integral to its new material. As the final words of Sarah's Introduction confirmed: 'The following pages close with the *sketches* belonging to *my first voyage to Africa*; and having been encouraged beyond my best hopes, I again launch my little bark into the broad ocean, myself at the helm; and venture to look forward to a return to port, with a cargo in exchange' (*SSL*, xv, emphasis added).

This chapter focusses on the importance in 1835 of Sarah's signatory 'I', 'my' and 'myself' here as indicative at once of her 'return to port' – the new Fragment V is examined in the second section – and of her 'cargo in exchange' in the first, because also spearheading her future work(s) in natural history-making. If the *Memoirs/Mémoires* had directly addressed her sole authorship under her second married name, 'Mrs Lee', the *Stories* further consolidated her larger independence through their severally artful renegotiations of her original work as 'Mrs T. Edward Bowdich'. Here was a differently articulated 'marriage' of the now six stories and five 'Fragments' deriving from her extensive parts in the fields of natural history and exploration during her 'first voyage to Africa' (with her first husband). Its new telling in 1835 is then no 'rather predictable division of labor between female and male writers' as Mary Louise Pratt contended in her important *Imperial Eyes*, discussed in the introduction (p. 10). Rather, and in line with my first study of 'The Booroom Slave' in *Stories from Strange Lands* for its salient 'Notes',[6] this chapter further unpacks their Humboldtian precedents for comparative scientific observation in the field overseas. After Cuvier's death and on the heels of her *Memoirs/Mémoires*, Sarah's 1835 *Stories* in effect creatively reset her authority in pioneering natural history-making independently of its French Muséum frames. The lights of the (world) expert 'Notes' by Cuvier and Humboldt for the French *Excursions* discussed in Chapters 1 and 4 now also illuminate her own. She could differently combine 'comparative anatomy' details and 'plant geography' purviews for the larger ends of her 'habits of early years' as her Introduction intimates:

> It is impossible to shake off the habits of early years, and my matter-of-fact studies and reflections have remained with me […]; therefore, every story is founded on truth; every description of scenery, manners, and customs, has been taken from the life: as much of the language of the actors have been preserved as is consistent with civilized ears, and the Fragments have not been in the slightest degree embellished. *In these, and in the Notes, my great difficulties have been to repress an exuberance of observation and circumstances*, in the fear that I might become trifling, and to avoid an egotism which would be tiresome. The number of I's that I have scratched out […] to avoid this provoking monosyllable, almost surpass belief; for it is very natural that wish to impress on the reader, that we have been witnesses or actors in unusual scenes, in order to stamp them with veracity […] and we are apt to forget that we ourselves are not as interesting as they are. (*SSL*, xiii–xv, emphasis added)

Crucial to the optics of the *Stories* were the long-standing 'habits' of Sarah's abolitionism, formed and informed by her Unitarian upbringing and education discussed in the introduction (p. 13) as emancipatory pursuits incorporating natural history at home and overseas. But she could now also explore its report in new modes with wider public appeal. The 'note'-worthiness of 'unusual scenes' determining the expanded *Stories* as a collection also more overtly related her five clearly autobiographical *Fragments from the* Notes *of a Traveller* as 'sketches', that is unfinished figured and written impressions from her (private) travel journal. In effect, Sarah's first voyage to West Africa formed the essential preparatory 'drafts' – via her formerly piecemeal Gift Book contributions – for her larger more finished 'pictures' of the 'Africa' question with unusual insider-outsider eye-witness view. As Jill Rappoport has observed, 'slavery makes a regular appearance within the annuals' of the period.[7] If Sarah's 'Boomroom Slave' was indicative, and also innovative as I have shown in its protagonist Inna's specifically female, indigenous observational that is scientific points of view,[8] this chapter contends that the collected *Stories* more concertedly confronted the intersecting issues of the 'Africa' question by means of similarly pioneering intercultural natural history perspectives. Key to the issues of slavery and emancipation was the question of racial difference, especially when used in support of the allegedly innate superiority of white races. According to late eighteenth- and early nineteenth-century Western 'scientific' theories, all human species derived either from one common race (monogenism, and largely Cuvier's position), or emerged variously (polygenism) in different climatic regions of the globe.[9] Of immediate relevance to publication of the collected *Stories* in 1835, and as the opportune moment for Sarah to revisit her long-standing 'habits', was the passing into law on 1 August 1834 in Britain of 'An Act for the Abolition of Slavery throughout the British Colonies; for promoting the Industry of the manumitted Slaves; and for compensating the Persons hitherto entitled to the Services of such Slaves'. The *Stories* thus provided a 'ledger' of British progress since 1807 (the Abolition of the Slave Trade), through its author's pioneering scientific and personal, as well as non-conformist political, social and economic perspectives in 1816 and in 1835, thanks to her newly recollected 'Notes'. These consistently document 'the language of the actors (for) civilized ears.' .

In consequence, the more significant appeal of Sarah's 1835 collection through the Moxon imprint was its overt address to the wealthy, cultured white British lady in her drawing room as the same genteel target audience for Gift Books. To educate and entertain her through the *Stories*' rich West African scientific and intercultural interest was concomitantly in the *Fragments* to hold up Sarah's own scientific and intercultural enfranchisement as a reality, not a fiction. The number of I's that she did not scratch out in her new 'Notes' reveals their authority. Space in the next section permits fuller analysis only of those for 'Samba' (the added story six) and for 'Adumissa' (story one), since they newly correlate in 1835. Appendix six supplies the relevant 'Note' materials. The longest 'Note' for 'Agay the Salt Carrier' (story five) then also offers insight into Sarah's careful self-positioning as pieced together more fully in 'Fragment V' in 1835, discussed at the chapter end. Her first journey in 1816–1817 to 'Ashantee' included the Gabon on the return voyage. The importance of Sarah's noted observations from each country then magnified in 1835 in their comparative West African interest, because this was also informed by her first report from the Gambia in 1823. The discomfiting aspects of the 'Africa' question above that chapter four examined for the Gambia pertained to the wider region. Sarah's larger West African 'intelligence' on them – her pioneering perspectives – therefore derived from her unusual position reiterated in 'Notes' to the Fragments. She was the first white woman that her interlocutor subjects, male and female, had encountered. The British lady reader of the *Stories* would also find herself the subject of their gaze.

'A Cargo in Return': Renewing the 'Continuance' of the Journey

Key to Sarah's 1835 collection is the 'added' Story VI, 'Samba'. It also stands apart by its country of origin as indicative of its original provenance. 'Samba' has a bit-part role in Sarah's 'Narrative' of the Gambia in 1825 (*EM,* 204–206). We cited its contexts in Chapter 4 (p. 98), Sarah's report of Albreda's 'known emporium for slaves and smuggling', and of Salem 'situated to the north of the river's mouth; where the slaves [...] meet their purchasers' (*EM,* 205). Chronologically, therefore, his story antedates and informs her other Gift Book publications on West African subjects from 1826 to 1833. Indeed, the anonymous reviewer of the *Excursions* for *The New Monthly Magazine and Literary Journal* in 1825 called attention to 'Samba' in a résumé of his story specifically to qualify the remark: 'We regret to find from Mrs. B's narrative that the French still continue to carry on the infamous traffic in slaves, notwithstanding all our efforts to prevent it'.[10] Sarah's double collection of Samba's story in 1825 and 1835 is therefore significant as an *exemplum* for the abolition of slavery everywhere, that is in and beyond the 'British Colonies' in 1834. In the 1825 version, 12-year-old Samba's story of escape during his forced march to Salem, and pleading for protection from the Commandant (of Bathurst), whence he is 'sent to take up his abode with other liberated Africans' (*EM,* 205), triggered an extensive explanatory footnote to his original purchase by 'a black woman [n]' (who had sold him at Albreda):

> [n] This woman had long been suspected of slave-dealing, and a poor slave girl belonging to her applied to Captain Findlay in my presence, for protection against the ill-usage of her mistress, and shewed marks of severe blows. She was of course immediately taken care of, but her owner assembled some friends in the evening, and tried to force her from the person to whom she had been temporarily confided by Captain Findlay. This was previous to Samba's escape, which with other instances, amply justified the seizure of the woman, and she was in confinement when I left Bathurst, waiting to be sent to Sierra Leone for trial. (*EM,* 205)

Footnote '[n]' in the *Excursions* therefore sets the precedent for Sarah's Gift Book stories and their 1835 recollection with 'Notes' in its combination of factual, 'ethnographical' report in a neutral third person, supported by first-hand observation to verify Samba's story. These significant double optics also hone Sarah's use in the *Excursions* of the carefully framed story-within-a-story: Samba's direct speech is recorded in his framed narrative, to answer back directly to colonial points of view.

The much-expanded 'Samba' of 1835, however, also offered a reply to the newly enshrined 1834 Act as the most reworked of the *Stories of Strange Lands* collection, aside from its four extensive 'Notes' (*SSL,* 190–197). Significantly, these did not reprise the 1825 footnote '[n]' above, despite the one-page introduction to the 'Notes' for 'Samba' in 1835 verifying 'as no fictions' Sarah's earlier report from the Gambia of continuing clandestine (French) 'slave-dealing' (*SSL,* 190). Rather, the 1835 version retells Samba's story of capture during his extensive search for his wife-to-be Zaina, seized by slave-drivers when she was collecting kola nuts with other (Muslim) girls from her village. The British Governor's searches had failed to locate the cache of slaves containing Zaina, until Samba encountered 'the poor slave girl' of footnote '[n]' above, who pointed him to where he should look: '"Oh, Gubbernor! Missy beat Yahndi, and Yahndi tell all. You no find slaves here, but take one boat, and go to ——," (pointing to a small village some little distance up the river,) "and you will find all. Yahndi run away; for if Missy see her talk to Gubbernor, she beat her too much"' (*SSL*, 187). The cameo part and recorded pidgin of the named Yahndi here – to which Sarah called reader attention in her Introduction – allow an otherwise anonymous female protagonist and literal 'footnote in history' in 1825 to speak out in her own name against cruel and brutal slavery (as evidenced in the marks upon her body in footnote 'n' above). Her small act of moral agency reports resistance here against widespread indigenous, as well as transatlantic, slaving practices. Although the main plaudits in the story ostensibly go to the white English 'Gubbernor' (as the heroic and just rescuer from the middle passage of the hero and heroine, Samba and Zaina), the attempts of the 'Commandant' to attach them to his household more interestingly fuelled their larger resolve, namely the restoration of their rightful independence concluding the story, and holding up its moral:

> They too ardently desired to see their respective parents to accept any offer [...] and they joined a party of gold traders, proceeding to their home, which they reached in safety, to gladden the hearts of fathers who were mourning for their lost children. They never forgot the humane Commandant, who had restored them to life and liberty; and the yearly presents they conveyed to him, by means of traders and travelling Marabouts, proved that their gratitude was not evanescent. (*SSL*, 189)

That Samba and Zaina decline the 'liberty' of the Governor's offer of 'service' is because it represented less than the fuller liberty of return to their own people. Their actions therefore also critique the 'solution' of Abolitionists at the time, who had proposed the status of 'Liberated Africans' for returnees from Caribbean plantations to new settlements in the British colony of Sierra Leone. Readers in 1835 could then also not avoid the comparison between the benevolent, pro-abolitionist, 'Gubbernor' of 'Samba' and the similarly 'enlightened' white mistress of 'The Booroom Slave' (Story III).[11] The upstanding moral actions of both white 'saviour' figures are recognized, but then firmly decentred in the endings of both stories: they are the key *recipients* of larger gifts of gratitude integral to longer West African customary heritages. The non-white givers therefore have the final word, whether according to Samba's (Muslim) Gambian or to Inna's ('fetish') 'Booroom' (eastern Ghanian) customs. Both stories end on the work that these central figures then undertook upon their marriages. Their 'happier thereafter' was an onward crusade among their own peoples against the collective scourges of slavery as upheld by powerful indigenous rulers, and by local customs reinforcing such traditions. The challenge of these endings for the (white lady) reader in 1835 lay in their ultimate questioning of the moral superiority and benevolence of abolitionist 'civilizing missions'.[12]

'Samba' therefore provides an important template for Sarah's reframing of the five Gift Book African stories for their major connecting themes and 'resistance' narratives on several levels: indigenous (and female) points of view, intrinsically *dialogical* form, and the key framing device of the story-within-a-story also guaranteed the larger authority and veracity of the eponymous 'native' protagonists, and their (oral) story 'collector'.[13] Sarah's *Stories* then take the story-within-a-story technique of their Gift Book versions to a second degree: doubled report in the extensive 'Notes'. As 'Samba' also exemplifies, they contain eye-witness, 'insider' knowledge and authority for their facts, whether these derive from field observations of natural history authenticated by Western scientific sources, or from reported anecdotes – to which we return in Chapter 9 – and oral histories illustrating local customs.

Frequently, therefore, a 'Note' takes the form of a framing 'scientific' statement that is further endorsed by indigenous information and Sarah's field-observational (and hence autobiographical) account. The four 'Notes' to 'Samba' are indicative. The first, for the 'Kola nut' that Zaina was collecting when captured, allowed Sarah to describe it for British readers from her botanical and personal knowledge in the field, to include not only its indigenous and scientific names and its uses, but also its taste and local costs (from personal encounter). The second, by contrast, challenges the reader's assumed and potentially erroneous knowledge of Africa's dangerous *fauna*: 'another was eaten by a lion (²)' (*SSL*, 182). The three-page 'Note' is not about lions, but other members of Africa's cat species (SSL, 194) that Sarah encountered during her first journey. 'Anthropological' observations and explanations constitute a third type of 'Note': '*They said that* the yellow people (³) liked to buy slaves' (*SSL*, 185, emphasis added). Because Samba is the speaker here, his intermeshed insider viewpoints on racial difference challenged Western 'scientific' classifications and hierarchies. The three-page explanation of 'yellow' (*SSL*, 194–196) exemplifies 1830s' understandings of, and sensitivities to, descriptions of 'races' (by the colour hierarchy of 'white', 'yellow', ['red'], 'black'[14]):

> (³) By the yellow people are meant Mulattos, who, generally speaking, are the worst of the inhabitants of the colonies. But one degree removed from a black man, they assume an immense portion of consequence and conceit, which renders them much more incapable of improvement than the black man would be. Their conduct to their slaves was always cruel; and after slavery was abolished among the English, they treated their apprentices equally ill, and frequently enticed them [...] to the French colonies, in order to sell them. (*SSL*, 194)

Such generalisations about 'Mulattos' now offend modern sensibilities, but need careful rereading in context, especially when plural in this 'Note'. Without passing judgement directly on the common occurrence of white (British) slave owners siring children by their women slaves, Sarah next explains to her reader the technical classification terms for 'mulatto' offspring sired by a white (or 'yellow') father at each successive union with the mixed-race child (by inference necessarily female) as 'Mulatto', 'Mustee', 'Mustafee': 'after this the children are supposed to be white. The speckled people, of whom I saw a few instances, appear to owe their strange piebald appearance to disease, and are more frightful to look at than can well be imagined' (*SSL*, 195). Long pre-dating Mendel and later nineteenth-century interest in photographing 'piebald' people as freaks, Sarah's 'Note' here clearly distinguished cases of probable vitiligo (and other congenital forms of melanism) from albinism. Her report of them also underscores her no less striking self-awareness in this lengthy 'Note', chiming with her observations of the 'Mulatto' women in the Gambia (discussed in Chapter 4). Sarah continues: 'In pronouncing judgment so severely on the Mulatto people, it is proper that I should mention the females of Cape Coast; who, from Mulattos to Mustafees, are mostly kind and well behaved; several of the Mustee women of rank were very superior in intellect and manner; and I can never think of several of them without gratitude' (*SSL*, 195). There follows Sarah's lengthy personal account – to which Chapter 9 returns – of a head woman's exceptional care of her and her unnamed child (Florence) when both were very sick and, on her departure from Cape Coast, the gift of 'the most precious Aggry bead' (*SSL*, 195). It triggers a full page discussion of the cultural origins of these beads from recent scientific authorities such as Sir Joseph Banks, M. Caillaud and Mr Bowdich. Sarah concludes that '[a]fter my last return from Africa, Mr Phillips analysed some of the Aggry beads for me, and found that the colouring matters of several were manganese and cobalt; the two latter substances, it will be recollected, were rediscovered in Europe long since the Aggry beads have been dug up in Ashantee and Fantee' (*SSL*, 196). Sarah's quietly persistent distinction of women's cultural knowledge and standing (and often their superiority) is therefore demarcated severally in this 'Note' as indicative of her own self-insertions as first-hand authority. If the final 'Note' of 'Samba' is to the travelling 'Marabouts ([4])' (*SSL*, 189) who conveyed annual gifts in gratitude to the 'humane Commandant', such observations about non-Christian faiths and their 'charitable' values further challenged the first premises of 'Christian' white European civilising missions, especially among enlightened Britons.

For the reader of the stories only in their 1835 collection, the name 'Samba' was already familiar, however. It is that also of the 'extremely wealthy' father of the eponymous Adumissa, the first story-heroine (*SSL*, 1). A 'kind master' to his numerous slaves, Samba allows them 'many hours of rest, or labour for their own profit' (*SSL*, 2). More importantly, he also (over)indulges his only daughter to be 'uncontrolled and unconstrained' (*SSL*, 2), whether in her 'love of superiority [and] taste for finery' as she parades in her home town of 'Igwa'[15] with her cortège of slaves, or in her pursuits in the 'wildness of her native forests [that] had imparted a hardihood and courage possessed by few girls of her country; and, having had no associates of her own sex and age, she had been accustomed to trust to her own resources for amusement, and to think and act for herself' (*SSL*, 3). The reputation of Adumissa for her unusual female autonomy is therefore of 'proverbial' proportions, as the first 'Note' puts it, both for Europeans who 'never saw a finer form, nor a sweeter countenance' and for the peoples of Fante: 'Her coldness to, and rejection of, all her suitors, were proverbial; and the whole story is well known throughout Fantee and the neighbouring nations' (*SSL*, 19). Sarah therefore collects the story, to elaborate upon the 'manners and customs' centring upon her that are its shocking core. Clarifying these, John Parker recently stated 'she killed herself in an ornate public ceremony, having already presided over the opening stages of her own funeral', continuing that '[i]t is largely thanks to [Sarah] Bowdich's account that the event can be disentangled from legend and situated in a broader history of suicide in the Akan world'.[16]

Parker's important study of the taboos of suicide 'in the Akan world' focusses on their lack of discussion both in the West African cultures in which they occur and in social anthropology. Despite noting Sarah's unusual female perspectives in 1835, Parker took up instead Edward Bowdich's earlier collection of Adumissa's story in his *Mission to Ashantee* (1819).[17] In so doing, Parker misses Sarah's clearer identification – importantly through her fictional Adumissa's eyes and recorded words – as to why her act of suicide both constituted the only outcome for her and, despite its taboos, was memorialised. Sarah's first 'Note' in fact formulated the rationale for both in the heroine's 'coldness to, and rejection of all her suitors' above. It included the arranged marriage even to Amoisee, the most suitable of her superlative suitors. Moreover, Adumissa's final public taking of her life was the crescendo of other similarly self-destructive acts as statements of protest. First was her refusal to eat as her refusal of marriage in any form. Her consultation of the chief fetish man, Adoo – the higher authority than her father with powers to secure her status as un-marriageable – then involved her perilous journey through a forest and near fatal encounter with a 'panther', avoided thanks only to her timely rescue by the same love-struck Amoisee. Her summons of the latter to thank him for her life was precisely her *anti*-fairytale narrative, voicing why she rejected him most of all:

> I shrink from marriage with you, with a disgust and horror that I cannot overcome. But it is not for you, individually, that I feel this; my disgust is with marriage altogether. I have been an only child; I have been the principal person in my family; every one has obeyed me; and, *free and happy, I have never known what it is to submit.* If I were to marry, I should become nothing; I should lose my rank and my freedom, and be mingled with the multitude. [...] *But neither you nor I can alter the customs of our country.* Look at my mother; she also was an only child and brought my father great wealth. What has she been ever since she married? *Confounded amid a number of other wives*; exposed to their jealousies and intrigues, *and not retaining a shadow of influence.* And yet I am no better than she is. People say I am handsome; but beauty will soon pass away, and then [...] *I should sink into a condition very much resembling that of a slave.* This, then, is my firm resolve—I will never marry any one (*sic*); *and if compulsion be used, I will put an end to my life.* (*SSL*, 11–12, emphasis added)

Adumissa's was no 'distinctly Romantic sensibility, with occasional fantastical and Gothic elements' that might explain her suicide,[18] but rather more shockingly for genteel lady readers in 1835 her articulate critique of the stark reality that women in her culture, as others in the region, endured. Arranged marriage and the custom of polygamy were their female condition, plight and destiny. For those like Adumissa knowing their own minds and self-worth, 'choosing' a condition tantamount to a slave's above takes on a further twist in the 1835 retelling, when seemingly stopped by Amoisee successfully taking his own life on the evening of her summons of him. His suicide first triggers the necessary appeasements by Adumissa's father to Amoisee's family that are the equally 'unalterable' customs of the country: 'You know that blood requires blood, even if a slave be killed; but Amoisee was nephew to a king, and we must have gold also. Our first requisition is one hundred ounces; and our second, any young man of your family who will die for you' (*SSL*, 13–14). Since reprisals by Amoisee's uncle included the kidnap and killing of male relatives in Samba's family, despite the latter's payments of gold ahead of any suitable second requisition being found, Adumissa's larger perspicacity pinpoints the root of the (male) suicide taboo: 'nothing but my death will appease the manes of poor Amoisee; and I deserve to lose my life for having so long disobeyed you' (*SSL*, 16). A long description follows of Adumissa's overseeing of the public preparations for her customary death, including her fearless securing from Amoisee's uncle that he end all further reprisals against her father, before she publicly takes her life (by one fatal gunshot wound to the heart as Amoisee had done before her). Herein was her 'legendary' measure of autonomous female defiance and resistance against the seemingly unchangeable cultural forms protecting male honour (and virility). If her death causes her father, Samba, to die of grief at the same moment, a 'Note' assures the reader that 'I have had the description of this scene from my uncle [Mr Hope Smith], who had been admitted to the solemn ceremony' (*SSL*, 30).[19] This scene of customary death is then amplified by ('anthropological') descriptions of the funeral rituals (lasting six weeks), including consignment of the bodies of Samba and Adumissa 'to the earth, under the floor of the room where they had died ([30])' (*SSL*, 18). Note [30] attests to Sarah's own first-hand account of such practice:

> The cellar of the house in which I lived for many months in Igwa, was the burial-ground of a numerous race; and such a circumstance attaches the people to every spot that has been once inhabited. It is also the custom to expose all dead bodies for three days before interment, dressed as handsomely as the circumstances of the survivors will allow. (*SSL*, 30)

If the intelligent reader already imagines the realities of such practices in a hot country, the remainder of 'Note [30]' spares no details from Sarah's personal knowledge, as well as cultural and moral judgement:

> At one period when small-pox and typhus fever raged in the town, a number of deaths took place daily, and as my house stood higher than the rest, I could never look out from my window without seeing several corpses lying across the doors to keep them cool. The treatment of the poor people seized with small-pox is barbarous in the extreme; a temporary hut is erected in the bush, and the patient taken to it by some one who has had the disorder, and left day and night with nothing but a little boiled corn and water, by way of medicine or provision. The sick person is visited once a day only, by any one who will perform the office, and the visitor throws off his cloth long before he reaches the hut. On his return, he plunges into a large panful of water, put outside, and after thus purifying himself, seeks and resumes his cloth and returns home; but the sufferer is not infrequently devoured by beasts of prey. (*SSL*, 30)

The 'barbarous' Fante customs in exacting 'blood for blood' in retributions above were, in Sarah's view, one with their cruelties regarding 'care' of the sick and dying, and their (drunken and disorderly) excesses in funeral revelries involving further human, and often female, sacrifice.[20] The 'happy ever after' of 'Adumissa' thanks to her legendary intervention publicly with her own life was her securing of more humane new bonds of friendship between otherwise rivalrous families returned to 'their quiet state. At each anniversary of the event, three weeks are devoted to its celebration, and Adumissa will never cease to be quoted by her countrymen as the model of every thing that is lovely and heroic ([31])'. As the finale and moral of the original Gift Book story in 1826, its West

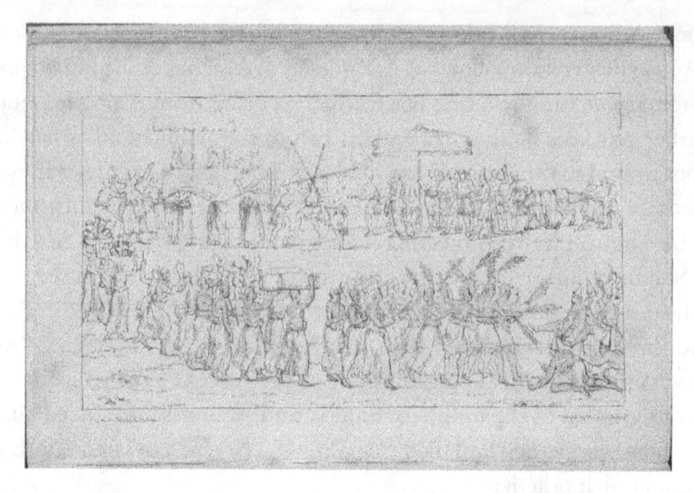

Figure 5.1 'The End of the Funeral Procession, PLATE 1'. Courtesy of the University of St Andrews Libraries and Museums, *Stories of Strange Lands*, sPR4161.B374.

African version of 'Romeo and Juliet' for the reader is more ambiguous in the long final 'Note [31]' in 1835 reproduced in Appendix 6. Sarah's own eye-witness account of the effects of these celebrations includes her interventions to overturn them. She will reprise this episode in Blanche's similar actions in, *Sir Thomas the Cornish Baronet* (1856).

The thirty-one 'Notes', for 'Adumissa' importantly include the interleaved illustration that is Plate 1 (in Figure 5.1) and its 'Description' of the 'victim about to be slaughtered for the evening feast' (*SSL*, 363):

> This Plate represents the end of a funeral procession. At the right hand corner are the executioners, with their pointed caps and large knives, meeting the women with their palm branches. Behind the latter, are men with flat sticks, drums and rattles; and these are followed by women bearing brass pans filled with cloths, twisted into various shapes. Four priestesses, or fetish-women, are dancing at the right hand corner, before the four chief mourners, seated in a palanquin. After the mourners come four captains, dancing, with flags in their hands; then a near relation of the deceased, carrying a string of Aggry beads in his hand, and supported round the waist by a favourite slave: his umbrella is carried by followers, and they precede the fetish-men and victim about to be slaughtered for the evening feast. The whole is taken from a large original drawing, the sketches for which were made in Fantee. (*SSL*, 363)

The 'Notes' therefore vouched for the natural productions concerning indigenous manners that elaborately costumed the difficult truths about Fante customs and 'fetish' belief systems that bound women. As in 'Samba', the four categories of 'Note' and their intercalation of autobiographical participant observations allowed Sarah to stamp their science with personal authority not only deriving from her field but also from 'Muséum' knowledge. Two 'Notes' stand out, therefore, within the contexts of her Paris training *after* she had stayed in actuality at 'the fort of Annamaboo' (*SSL*, 32).[21] In 'Note [4]' despite its succinctness, Sarah makes two different contributions to recent scientific knowledge:

> Naturalists declare that there are no humming birds in the Old World, and in my first journey to Africa I was not aware of this. Still, however, with better knowledge now, I am not inclined to change the appellation, because nothing else will convey an idea of the size of the birds with which the people deck their heads; and, moreover, I have some suspicion, that humming-birds *are* to be found on the western coast of Africa. How the natives prepare them I know not; but, after extracting the inside of the bird, the whole is well dried before it is worn. (*SSL*, 20)

In 1826 and 1835, Sarah had full knowledge both of Cuvier's recent *Ornithology* and of scientific taxidermy, especially the preparation of birds about which she had published in her anonymous *Taxidermy* (1820).[22] In 1835, she also published the still anonymous, updated, fifth edition. Her final remark in this 'Note' therefore signals her informed professional interest in the clearly successful indigenous techniques for drying and preserving bird specimens, including with the feathers, as well as her authentication of their provenance, when 'humming birds' should not exist in this part of Africa according to Western authorities.

The second scientific 'Note' is triggered by a description of Adumissa's makeup, skin preparations and costuming for town: 'her whole skin was polished by the perfumed *shea-tolu*, and her *pagnes* were always of the most costly Chinese silks ([16])' (*SSL*, 3). The three-page 'Note [16]' (see Appendix 6) not only offers a comparative 'anthropology', botany and (Paris Muséum) chemistry but also collates Sarah's various (personal scientific) sources for *shea tolu*, including its double description in T. Edward Bowdich's *Mission from Cape Coast Castle to Ashantee.* The first is in the chapter on 'trade' describing 'Ashantee grease', with an asterisked cross-reference taking the reader to the second in its Chapter XIII 'Sketch of the Gaboon': 'Before I understood them to be distinct trees, I concluded the odica and the butter both to be the produce of the cacao-nut, but the butter answers closely to Mr Park's description of the shea-tolu.'

The reader automatically assumes that the 'I' in the 1819 *Mission* is its author, Edward. Sarah's 'Note [16]' in 1835, however, makes her 'I' unassailable, by revealing retrospectively both her real presence in the Gabon expedition to see the 'fat tree', and her intermediation in translating the 'Booroom' servant's recipe for the butter. Her scientific cross-referencing therefore unmasks Edward's 'I' as the 'we' that included Sarah invisibly in '*we* relished the meat fried in it exceedingly' (in the Chapter XIII reference, Appendix 6). But it is Sarah's drawing for 'Adumissa' in 1835 (Figure 5.1) that further proves her larger presence in Edward's allegedly sole authored *Mission*, since her description of the figure concludes: 'The whole is taken from a large original drawing, the sketches for which were made in Fantee' (*SSL*, 363). Chapter 7 will take up the case of its identification. Sarah's authority in her 'Notes' was therefore not second-hand compilation from the best book authorities, including her husband's, but her first-hand knowledge from the field, before and after her scientific training in Paris. Her 'Notes' in 1835 therefore documented the latest natural and 'anthropological' knowledge and understanding of West Africa, because she had also contributed to that knowledge 'anonymously'.

Sarah's very significant decision to place the narrator alter ego of herself in 'Igwa' in the introductory frame of 'Adumissa', the better to enable its local woman inhabitant to speak from inside its cultural perspectives, constituted a most concerted challenge to (scientific) travel narratives of the period. To portray Adumissa and herself in the many 'I's of her 'Notes' in 1835 as inter-reflecting 'participant observers' enabled Sarah to attest to their autonomy and agency outside Fante and British 'customary laws' for women. Although no proto-feminist, Sarah could then tackle the question of polygamy and *female* immolation in 'Gift Book' disguise in 1826 and then 'scientifically' in 1835 as her brave determination to address the 'shackles' of marriage, whether arranged, forced or unequal as the sole 'end' for women's lives. To report women's insider knowledge and actions, especially when these could break powerful chains – literal and metaphorical – would make way(s) for cultural change. Sarah's own nonconformist experience with Edward in their differently equal scientific partnership to undertake their journeys of exploration to West Africa could only in 1835 reveal her earlier superior expertise to his concerning the 'plant geography' and 'taxidermy' of the region, and her very unusual circumstances that informed its recording.

Let us then recall the wealthy urban British female readerships for Gift Books in the 1820s and 1830s.[23] Such women had no direct political involvements 'at home', let alone in British 'Realpolitik' (although the term was not coined until 1851). British setbacks in Asante were many upon the death in 1824 of its most forward-looking King, Osei Bonsu, the model of Sarah's 'King' in 'Agay the Salt Carrier' (story five). His successor and younger brother Osei Yaw was also dead in 1834, a ruler much less tolerant with regard to his imperial subjects (other Akan tribes and the Fante), and actively anti-British.[24] Sarah's 'Agay' with 'Notes' in 1835 therefore challenged by analogy and inference – in word and image – the ignorance not of the Asante, but of her female readers, thanks to the failures of British society to educate women in matters of state as well as of taste. The differences were indeed almost tangible, in their very hands: the Gift Book as decorative object – gilt-embossed, silk or morocco-bound – was as beautiful but less useful or symbolically valuable as the many silk, gold and leather refinements – for example Adumissa's clothes – in the descriptions of Asante civilisation Sarah recorded. Two Empires, each in the scrutiny of the other, were therefore held up to her British readers for comparison. Sarah's record of her self-positioning as the unusual white woman in their double gaze in the longest 'Note [18]' for 'Agay' – reproduced in Appendix 6 – is then striking given her target readerships.

Agay's rise from being a 'salt-carrier' to his position as fourth linguist and the King's diplomat-interpreter triggers the longest among several autobiographical revelations of Sarah's revised 1835 collection: 'At length, as he was sitting at a solitary mess of yams and ground nuts, an Ocrah ([18]), (a confidential servant,) of the king summoned him to follow in silence to his majesty' (*SSL*, 166). Sarah's 'sketch' in 'Note [18]' concerned her similar audience, in a day in her life at Cape Coast Castle (in 1816). It also offers an extraordinary 'comparative anatomy' of British society 'at home' in its reconstituted Cape Coast drawing and dining-rooms. Her report then plays specifically and deftly on the kinds of social etiquette dilemmas that her lady readers might also have encountered as hostesses in difficult situations, particularly when alone. The description of her encounter with Ocrahnameah, a local native chief, therefore offered a slyly witty reprise of the 'Silver Fork' literary genre in its concomitant deriding of the snobberies of the upper classes.[25] If Sarah then appears in her 'Note' to mock the attempts of the 'natives' to use their knives and forks correctly (by barely concealing her laughter), she nevertheless manages to 'keep [her] countenance', because matters of face (and state) were paramount on both sides of this encounter, not least given the potentially vulnerable situation she found herself in: 'an anxiety which taught me to conciliate the Ashantees as much as laid in my power'. A mis-step on her part could have had further direct impacts on Edward's diplomatic efforts. Indeed, Sarah openly admits her intercultural 'faux pas' in choosing chicken, which is matched by her guest's over-eating and drinking. Unusually, then, in descriptions of 'native' manners, the inappropriate behaviour of Ocrahnameah is not censored through appeal to the higher order of British manners and standards of civilisation. Rather he is censored, and severely punished, by the even higher standards of civilised Asante habits and customs embodied by YokoKroko in 'Note [18]' (*SSL*, 178–79). The larger moral message of Sarah's recorded 'palaver' is clear: there was much that Europeans might learn through better diplomatic engagement and use of 'interpreters' of Agay's stripe in West Africa, especially when the two parties were alert to intercultural sensitivities and differences. Able and gifted 'linguists' – interpreters of culture as well as language – including women such as Adumissa and Sarah herself were vital to recording difficult cultural 'truths' that could speak directly to power in the changing realpolitik that the abolition of the slave trade and of slavery provoked.

In 1831, the first autobiographical novel by a black woman slave from the West Indies, *The History of Mary Prince*, was published through the agency of Thomas Pringle, secretary of the Anti-Slavery Society. Sarah's 'Agay the Salt Carrier. An African Tale' was first published in his *Friendship's Offering: a Literary Album* the same year. Her differently significant 'African Tale' in 1835 also deployed autobiographical disguises and masks, but in the form of scientific 'Notes'. Where Edward sought to measure his diplomacy – by citing Agay's story in 1819 as a model for his own – his superior British cultural 'leadership' point of view in the *Mission* did not account for Sarah's less front-of-house diplomacy that better negotiated new common cultural ground that might also transcend prejudicial assumption and misunderstanding. The retelling of 'Agay' with a woman's-eye view on Edward's *Mission* in 1831 already made her intellectual and scientific emancipation in her own voice distinctive. In 1835, Sarah's 'Note [18]' added important (ethical) 'thou' perspectives to her un-silencing of women as intercultural mediators and 'translators'. Appearance in the eyes of the other is very different from self-presentation, especially in her precious record of (male) Asante responses to herself as the first white woman with whom they had had an audience. In not scratching out the many 'Is' of the 'Notes' to her *Stories*, Sarah better prepared her readers in 1835 to apprise the unusual 'Is' and eyes of *The Fragments.*

Notes of a 'Mad' (Pioneering Woman) Traveller

The five concluding 'Fragments' were no 'sketches' (Sarah's word above). For the first time in her corpus, Sarah overtly meshes her (scientific) authority – the 'I' of the 'Notes', the autobiographical 'I' – into a *singular* account (in five acts) of her first journey to West Africa, without need of a self-effacing preface. Its model is the *scientific* voyage of discovery, exemplified in the Travels (Voyages, and their accounts) of a Cook, a Flinders, a T. Edward Bowdich and a Humboldt, and in 1835 being conducted by 'The Beagle' (with Darwin among others aboard). If Sarah's *Fragments* therefore recount a woman's journey of (scientific) exploration to West Africa – Cape Verde, Goree, the Isles of Los (Fragment II), Sierra Leone (Fragment III), before arrival at Cape Coast, then the Gabon in Fragment IV on her return journey – long before Mary Kingsley's to the region, they also capitalised on the powers of their (self-selecting) 'Note' form. Despite clear adherence to acceptable 'travel diary' format,[26] their unusual *incompleteness* precisely as 'Fragments' was one with their important 'tailpiece' positioning that could present the amalgam of new report for 'science' as hybrid fragmentary forms, including 'Notes' and (travelogue) 'Appendix' matter. Precisely because unfinished, Sarah's *Fragments* also challenged the paradigms of the (male) quest and scientific exploration narrative with its all-conquering I. The eminently self-situational 'myself at the helm' could pay tribute to the unequivocal fact of her major accomplishments as a traveller 'in skirts' as no lesser a 'discoverer'.[27]

'Fragment I' set the tone of personal revelation, to play on (first-time) inexperience: 'I had never been at sea, had never seen the inside of a vessel' (*SSL*, 249). Worse, '[d]azzled by six miles of rough water passed in an open boat' (*SSL* 252) to reach the vessel, Sarah tried to find her cabin only to meet further confusion, including the motley crew, triggering its first 'Note'. Only there does the reader discover that she had not set sail alone,[28] but was travelling with her baby daughter (Florence) who, after inadvertent injury by a fall into the hold, became the ship's 'mascot' (*SSL*, 263–64). The painful 'truth of the captain's remark, that the ship was not fitted up for passengers' (*SSL*, 253) heralded the many (untold) discomforts that Sarah later encountered, endured and overcame. Before the vessel finally sailed, however, Sarah records the 'self-evident' truth of her unusual position. When the captain boarded, he was 'followed by several Liverpool people, who were anxious *to behold the lady passenger mad enough to go to Africa*. Having been stared at for a few hours, I rejoiced when the pilot gave the signal for sailing' (*SSL*, 254, emphasis added). The impolite humiliation and labelling of her as 'mad' by such 'Liverpool people' charted their disbelief in a 'lady passenger' despite their very eyes. In therefore deftly certifying at her own expense her 'correct' observation henceforth, Sarah ensured the veracity and authority of her account that both followed and eschewed the standard topoi of the sea voyage narrative – seasickness, lack of sea legs, appalling conditions on board – by *not* describing them, or through distracting 'anecdotes' instead (*SSL,* 256 and 268). Her 'unmad' eye as the 'participant observer' observed of the ship's crew guaranteed her factual, non-judgemental, account of their actions. Two aboard reappear in 1847 as important protagonists of *The African Wanderers*.

The second 'Fragment' then finally introduced the singular events of Sarah's voyage that distinguished it from its models; the 'frolic' of 'the day we crossed the tropic' (the ritual ducking of crew members by a dressed up 'Neptune', *SSL*, 269), and her early (scientific) observations directly in the first-person during a brief landing at 'the settlement of Goree' of its very 'handsome' mulatto women (*SSL*, 272, discussed in Chapter 4, and above). The positive effects of her encounter with these women – including their delight in encountering her and her (white) daughter as its 'Note [3]' further reveals – counterbalanced the description of the (more alarming) rumblings of mutiny on board her ship, and the (lifetime) event for Sarah on the onwards journey to the Isles of Los, where she hooked a shark (*SSL*, 274) in the spirit of the best male adventure story, remade in her fictional *The African Wanderers* as chapter 6 investigates. When she recounts her landing on the Isles of Los, she does so by directly inviting the reader also to see the 'huge millipedes [...] brilliant lizards [...] birds with the most splendid plumage [...] the beauty of the flowers, and the perfume [...] so new to me, that I almost doubted if it were real' (*SSL*, 276). Her 'excursions' on these Isles (*SSL*, 277–79), to note their natural history and 'ethnography' retrospectively 'preface' her (published) natural history endeavours in Madeira and the Gambia.

But 'Fragment II' more strategically reveals Sarah's larger 'role' among the ship's crew: 'I had taken a positive interest, not only in their characters, but in the successful trading of the vessel, and had helped the first mate to write his log (7), till I fancied myself almost responsible for its correctness' (*SSL*, 238). 'Note 7' unmasks this 'help' as her keeping of the first mate's log of eye-witness events aboard ship 'for three or four weeks; and yielded me considerable amusement' (*SSL*, 292), as the story which readers have in fact been reading. 'Fragment II' also presents Sarah's 'log' as her 'anthropological field notebook' in all but name, as well as the 'cargo in exchange' that will become the 1835 'collection' of stories, fragments and notes bound together. Her 'log' also prepares the abolitionist observations of her *Stories*. Her new brig 'had been taken [...] from the Spaniards, in the act of conveying slaves; and how the hundreds of human beings which she contained were crammed into her, I have already described' (*SSL*, 294). The reader in 1835 was to shudder at the ongoing trade Sarah then met with *at sea* – the brig encounters a Portuguese vessel laden with slaves, and tries in vain to give chase – and at the unspeakable conditions (worse by far than any discomforts Sarah had earlier experienced) of the middle passage, to applaud all efforts for its end.

When 'Fragment IV' then plots Sarah's safe arrival at Cape Coast, the ironies of being a (lone) 'lady passenger' aboard ship are more apparent on shore: 'the then Governor, Mr Dawson, sent his own canoe to take me and mine ashore, where he feasted, and fêted, and indulged me (9), till he could with safety place me in the care of a relative (10)' (*SSL*, 302). Both 'Notes' confirm the crux of Sarah's (unexpected) arrival just as Edward's uncle, Mr Hope Smith, was to take over as the new Governor at Cape Coast. Her loss of independence because her husband was absent lay in the 'romance' of his having set sail for London to be with her (and their child) with her twist. With similar initiative and independence of spirit – 'the lady passenger mad enough to go to Africa' – she had set sail from Liverpool to be with him. Sarah's double vantage points in the *Fragments* of 1835 offered comparative reflections on her *two* journeys of scientific discovery to West Africa which, in each case, only had the company of Edward on one leg of *her* journey. Hindsight also unusually revealed the motivations of Sarah's creative foresight:

> Why is it, that every one who has lived in North-Western Africa [...] should retain so deep an attachment to that barbarous land? It is not, like other tropical countries, a scene of luxury; on the contrary, it is a life of incessant danger and privation. It possesses not the charm of refined and intellectual society; its European inhabitants, with very few exceptions,

professedly try to get money as fast as they can [...] and yet when they do return, there is no place on earth so dear to them as the land they have left. (*SSL*, 309)

The third-person perspective offers a preliminary 'sketch' for *The African Wanderers* (chapter 6) and *Sir Thomas the Cornish Baronet* (chapter 9). But its views are qualified by its later, first-person, companion piece (modestly) withheld to the end of her *Fragments*, to make their larger point as an open account for (female) science endeavour:

And so it is with myself. I have visited other lands, nay, lived in them, and my path has been broken and rugged. Still more thorny was it in Africa, and yet my thoughts and feelings incessantly recur with indescribable affection *to those wild scenes*; every circumstance vividly rushes before me as if it were the occurrence of yesterday, and my very dreams are of that magnificent land, *where Nature has lavished her treasures with such limited profusion.* Perhaps these treasures form one of the secret links of that chain which binds us all to her; and her lofty primitive mountains, her mighty rivers, her impenetrable forests [...] the strange and uncontrolled forms with which her wastes are peopled; [...] her children, rude and disgusting as many of them are, even form objects of compassionate interest, from the very curse under which they seem to labour; and all these perhaps create feelings in residents, which, to those who have always dwelt in civilized nations, are inexplicable. There may yet be another cause, which is, the constant excitement afforded by a life which often presents danger, and constantly requires contrivances for comfort and enjoyment (3). (*SSL*, 310–11, emphasis added)

Here, on the one hand, emerges Sarah's unusually philosophical, Humboldtian, speculations on the geo-bio-anthropological interconnectedness of 'Nature' and its rich untold phenomena. On the other hand are the double dimensions of her self-positioning as a writer of West African natural history, articulated in the final sentence in the impersonal and universal third person. Its anecdotes and sentiments in its 'Note 3', however, reveal the creative 'contrivances' in Sarah's needle (in the absence of supplies of 'comforts', when crates of Bibles arrived instead of clothes) as part of her (pioneering) woman's-eye-view of scientific work overseas outside British colonial life and missions such as the Bible Society. As concomitantly a wife and mother, who has just lost baby Florence of the 'Voyage out', Sarah's particular alertness to recording indigenous 'children' and their mothers distinctively augments (male) scientific travel narratives.[29] Indeed, the final full stop of the *Fragments* – landing and breakfast at Penzance, and onwards journey to 'our friends at Falmouth (14)' (*SSL*, 347) – triggers one of the two longest 'Notes' of the final fragment, providing Sarah with a last opportunity for comparative 'anthropology' of her 'homeland', including self-reflection and self-revelation:

Our appearance was so grotesque as to raise many conjectures in the town whence we came, for the news of an arrival from sea had caused many a peep from between the window curtains as we passed. [...] and to increase our extraordinary costume, Mr Bowdich had on a pair of the yellow boots of Madeira, and I a large and thick African cloth, which had been my bed the preceding night. [...] We had not tasted bread made of yeast, nor fresh butter for years, and, as we then ate it, we thought that *our childish* opinion of the insignificance of bread and butter must have been very erroneous, and we sent the waiter out of the room, that he might not see us take the last slice. On our road to Falmouth the scenery appeared very tame and confined; it is true that the birds and flowers looked like old friends, but we had been spoiled for the sober tints, the dwarfish trees, the gentle hillocks, and studied cultivation, of our native land. (*SSL*, 361, emphasis added)

As a pioneering woman scientific explorer on two occasions to West Africa, always as also a wife and a mother, Sarah's intimate, autobiographical recollections in her *Fragments* and other scientific accounts long precede the no less 'ground-breaking' women's field exploration (by a Jane Goodall or a Dian Fossey) in the later twentieth century. It is then not only in the 'wild scenes' of her work[30] but also in her mother courage that Sarah's scientific West African writing 'with herself at the helm' still gives much pause for thought, especially in the enduring importance of her precious, and expert, scientific 'Notes' from the field. They are, in consequence, no footnotes *in* histories of nineteenth-century exploration, geography and science, including with gender lenses, discussed in the introduction (p. 4–5). The sole pilot of her 'little bark' indeed, Sarah's 'cargo in return' in 1835 affirmed her manifold gains, not major losses (her first child, first husband, her specimen crates and barrels, her access to the Paris Muséum after 1833). Her enlargements of her ventures in natural history-making in 1835 now also co-extensively harnessed scientific, literary and artistic genres and audiences. Chapter 6 discovers their integration, further extending the skills and acumen of her 'trade'.

Notes

1 See deB. Beaver, 'Writing Natural History for Survival', 19–31 and Appendix 1 for Sarah's contributions to *Forget Me Not* from 1836 to 1847.

2 For the genre and its British history see Anne Renier, *'Friendship's Offering': An Essay on the Annuals and Gift Books of the Nineteenth Century*. London: Private Libraries Association, 1964. For Sarah's regular publications in *Forget Me Not*, see Katherine D. Harris, *Forget Me Not: A Hypertextual Archive of Ackermann's Nineteenth-Century Literary Annual, An Edition of the Poetess Archive* (2001), revised 2007 http://www.orgs.miamioh.edu/anthologies/FMN/Site%20Index.htm)/. Harris re-classifies all Sarah's contributions under 'Bowdich' rather than 'Lee' (although a former collator had done the opposite).

3 The list is Story I, 'Adumissa', II, 'Amba, the Witch's Daughter', III, 'The Booroom Slave', IV, 'Eliza Carthago', V, 'Agay, the Salt Carrier; and Fragments I, 'Going to Sea, and the Ship's Crew', II and III, 'The Voyage Out', IV, 'A Visit to Empoöngwa; or a Peep into Negro-Land'. Story VII (about India), 'The Life of a Hero', and Sarah's 'French' stories VIII–X, deriving from her stays in Paris, lie outside this chapter's discussion because only two brief 'Notes' were added to story VII and none to the Paris set.

4 For example, 'Eliza Carthago' (Story IV) was republished in full in *The Gentleman's Magazine and Historical Chronicle* xcviii (July–Dec. 1828): 347–49 as part of a review of the 1828 'Forget Me Not'. 'A Fragment' (heralding the five 'Fragments' with Notes) first appeared as 'A Scene in Negroland', commissioned for the launch volume of the *Court Magazine and La Belle Assemblée* (Aug. 1834): 95. 'Jacqueline' appeared in its September number (117–21), but had already been published as a story by 'Mrs Lee (formerly Mrs Bowditch) (*sic*)', *The English Annual* 1 (1832): 245–56, edited by the Hon. Mrs. Norton.

5 For biographical information on the poet-publisher Moxon see https://www.browningscorrespondence.com/biographical-sketches/?nameId=1284b, the electronic amplification of the life-work of Philip Kelley (Ronald Hudson and Scott Lewis), *The Brownings' Correspondence: A Checklist*. Winfield: Wedgestone, 1984–2023.

6 Orr, 'Amplifying Women's Intelligence'. This chapter builds upon this study.
7 Jill Rappoport, *Giving Women: Alliance and Exchange in Victorian Culture*. New York: Oxford University Press, 2012, 28.
8 See Orr, 'Amplifying Women's Intelligence'.
9 See Hannah Franziska Augstein, ed. and intro., *Race: The Origins of an Idea, 1760*–1850. Bristol: Thoemmes Press, 1996, and Sarga Moussa, ed., *L'idée de "race" dans les sciences humaines et la littérature* (xviiie et xixe siècles). Paris: L'Harmattan, 2003, including for their contextualizations of Cuvier's work within intersecting European theories.
10 *The New Monthly Magazine and Literary Journal* (1 July 1825): 317. Because the name 'Samba' evokes the pejorative racial connotations of 'Sambo', modern critical sensitivities may have short-circuited larger probing of Sarah's account(s) for their particular French and West African contexts.
11 See Orr, 'Amplifying Women's Intelligence'.
12 See Sarah's conclusions to her Introduction to the 'Notes' for 'Samba' for her further questioning of progress against slavery when gullible good 'Commandants', such as Sir Charles McCarthy, had been taken in and killed by 'Sam Brue [...] the most notorious slave-dealer in Fantee' (*SSL*, 190–91).
13 The story-within-a-story device therefore operates in the same manner as Sarah's use of the modesty topos in her prefaces, whereby she can downplay her position as a (mere) woman contributing to science, yet own the successful results of her work.
14 See the works cited in note 9.
15 Sarah chooses the native Fante, not British name for key places in her stories. 'Igwa' is Cape Coast.
16 John Parker, 'The Death of Adumissa: A Suicide at Cape Coast, Ghana', *Africa: The Journal of the International African Institute* 91, no. 2 (Feb. 2021): 205–25 (205). Parker also notes that T. Edward Bowdich had earlier recorded Adumissa's story in the *Mission from Cape Coast Castle to Ashantee* (of 1819), for its important verifications of Asante and Akan laws concerning 'types of self-destruction involving the uttering of an oath' (209).
17 Parker, 'The Death of Adumissa', 208. Tellingly, his article takes *Edward*'s version as the more authoritative, because 'shorn of the literary elaborations of Sarah's tale' (209).
18 Parker, 'The Death of Adumissa', 208.
19 Sarah thus reveals the first-hand provenance of the story for both Edward's (1819) and her own (1826) very different accounts.
20 Space does not permit discussion of story IV, 'Eliza Carthago' (a fort, not a female protagonist), which also treats a different report of 'treacheries' and retaliations, to explore their many causes and effects.
21 Annamaboo was 'ten miles to eastward of Cape Coast and is the great mart on the Gold Coast', in the account of Capt. John Adams, *Remarks on the Country extending from Cape Palmes to the River Congo*. London: G. & W. B. Whittaker, 1823, 8.
22 See Orr, 'The Stuff of Translation'.
23 There has been renewed interest in the medium of the Gift Annual, allowing gifted women writers an outlet for gender and social criticism, despite the saccharine, highly 'chaperoned' nature of their content. See, for example Jill Rappoport, 'Buyer Beware: The Gift Poetics of L.E.L', *Nineteenth-Century Literature* 58, no. 4 (2004): 441–56 and Serena Baiesi, *Letitia Elizabeth Landon and Metrical Romance: The Adventures of a Literary Genius*. Bern: Peter Lang, 2009. Both depend heavily on Renier cited in note 1.
24 See, for example Robert B. Edgerton, *The Fall of the Asante Empire: The Hundred-year War for Africa's Gold Coast*. New York and London: The Free Press, 1995.
25 For a fine bibliographical introduction to the genre, see Tamara Wagner's 'Silver Fork Novel' (Fashionable Novel). https://www.oxfordbibliographies.com/view/document/obo-9780199799558/obo-9780199799558-0136.xml.
26 See Orr, 'Pursuing'. On women's practices of keeping a 'diary' to record observations while abroad on art connoisseurship, see Palmer's '"I will tell Nothing the I did not see"', 248–68.
27 See Mary Russell, *The Blessings of a Good Thick Skirt: Women Travellers and their World*. London: Collins, 1988.
28 Sarah may have travelled with a maidservant, although the *Fragments* do not disclose her presence.

29 See also the ending for her observations of children, *SSL*, 361–62. When Sarah first landed at Cape Coast she logs 'so many strange things presented themselves, that I scarcely knew which to look at first. The shape of the women, however, excited a special curiosity; for I had never heard that anything approaching to the Hottentot form had strayed beyond that country' (*SSL*, 331). Readers of the time would know of Saartjie Baartman 'the Hottentot Venus' (1789–1815), whom Cuvier dissected to ascertain the anatomical veracity of the distinguishing features – extended buttocks and labia – that had famously made her a freak-show exhibition. Sarah's realisation that the women of Fante have this 'strange *contour* [...] by artificial means' (*SSL*, 311) – a long strip of cloth is rolled into a cushion – is interestingly part of her recorded response as a woman who cannot deny subjective feelings as wholly part of 'scientific' observation: 'I know not why, but I felt perfectly relieved on finding this ungraceful appendage was not caused by a sport of nature. The women of Fantee, however, were quite at a loss to comprehend my dislike, for it plays a very important part in their dress' (*SSL*, 311). The mutual discussion the final sentence reports is a precious record of their intercultural exchange. Sarah discovered that the size and costliness of the materials for the 'appendage' correlated directly with the woman's social and childbirth status, and hence (cultural) wealth.

30 In picking out this phrase in the title of the first book-length study of Sarah, Strickrodt's *'Those wild Scenes': Africa in the Travel Writings of Sarah Lee (1791–1856),* did not ground its interests in the French scientific frames and models that inform my work.

Chapter Six

A REFIT FOR LARGER SCIENTIFIC PURPOSE? PIONEERING NATURAL HISTORY FICTION ABROAD AND AT HOME IN *THE AFRICAN WANDERERS* (1847)

The seeming hiatus of 12 years between Sarah's publication of *Stories from Strange Lands* (1835) and her first novel, *The African Wanderers* (1847), raises the stakes for the 'continuance [...] to completion' (*EM*, inside title page) of her book-length ventures in expert natural history. Indeed the leanest period of their production, which Chapter 7 addresses, falls between final numbers for *The Fresh-Water Fishes* in 1838–1839 and *Elements of Natural History* (1844). The Introduction clarified how Sarah's regular publication of Gift Book stories from 1824 supported her family and new work in natural history to book-length result. Yet the marked change in this pattern is 1845 (see Appendix 1), between the *Stories* in 1835 and her first novel in 1847, and despite the former collection clearly and variously inspiring important episodes in the latter, as outlined below. Sarah had no further need of supporting Gift Book stories when, as Silke Strickrodt first contended, *The African Wanderers* was 'one of the first novels in the English language, which is set in West Africa', making its study part of the necessary redress of the 'minimal attention so far paid to Sarah in the literature of English and African Studies'.[1] Since Strickrodt did not elaborate further, this chapter investigates how *The African Wanderers* not only meets, but also exceeds these claims for it through its differently 'equinoctial' Humboldtian inspirations geographically and generically than were either the *Excursions* or the *Stories from Strange Lands* as discussed in Chapters 4 and 5. Of particular salience for Sarah's new departures in *The African Wanderers* in this chapter were the historical contexts that framed the writing of her first novel in 1847 and its immediate reception. Indeed as Chapter 3 disclosed for her *Memoirs*, the indicative lenses of an extensive initial review of *The African Wanderers* in 1848 provide similarly invaluable insights for how to read this novel in its own terms today, and within Sarah's multi-genre scientific corpus in the final decade of its major book-length production(s). Her reputation as a major writer of natural history was not in question. Rather the snapshot review of *The African Wanderers* in 1848 as 'literature' spearheads how this chapter differently argues for the work's 'refit' of natural history-making for Sarah's larger purposes.

In February 1848, the anonymous reviewer for the *Dublin University Magazine* identified the distinctiveness of *The African Wanderers*. Judging it alongside canonical works, such as 'Sacred Scriptures' and 'Bunyan's "Pilgrim's Progress"', here was an unusually worthy remake of Daniel Defoe's *Robinson Crusoe*:

> Robinson Crusoes, old and new, have often appeared, but fell so far short of the original, as scarcely to be readable [...]. We have, however, now before us a charming little book [...], and which, while it is perfectly free of affectation, or attempt at imitation, possesses in its narrative – in the simplicity of its style – in its graphical description of scenery – in the amount of information which it affords, and in the moral lesson which it teaches, all the fascinating power of the Selkirk story.[2]

Significantly, *The African Wanderers* had appeared simultaneously in the United States under the title *The African Crusoes.*[3] However, the reviewer could expatiate on its many merits only after dismissing potential disqualification of its writer because

> *not well known to the novel readers and crochet-manual students of the middle of the nineteenth century,* [yet] has long been known, and *her works estimated as they deserve,* by the learned of Europe. In the museums of natural history in the cabinets of the Jardin des Plantes, in the bibliotheks (*sic*) of Germany, and the boudoirs of Russian naturalists, the efforts of Mrs Lee's pen and pencil are to be found.[4]

The reviewer's unusually knowledgeable and lengthy bio-bibliographical portrait of Sarah confirmed her status as a '*scientific* traveller', a contributor to 'knowledge of comparative anatomy and zoology, which very few ladies have ever acquired', to underscore her distinctive and intimate acquaintance '[d]uring her sojourn in Africa [...] with the customs, habits of life, and general natural history of the swarthy tenants of that burning region; and with this knowledge, and with those recollections, she brought to the task of this present work *qualifications which few could be found to combine.*'[5] Before offering a résumé of the main storylines and notable episodes, including among its 'curiosities' a 'white negro', the reviewer also set out why *The African Wanderers* was unusual for the novel genre, including by inference Robinsonades:

> *Her design was,* by drawing attention to the condition of the aborigines of Africa, the genuine negro population, and by a fair representation of their natural qualities, their many talents and capacities for improvement, their comparative superiority in benevolent feeling over all other known savages, their ready reception of Christian and missionary instruction, *to move the heart of Christian England in their behalf, and urge on the promotion of the very movement for their civilization and conversion from idolatry.* [...].[6]

In highlighting the multiple scientific importance of Sarah's *African Wanderers* for the first time, this chapter now responds directly both to the 1848 reviewer's alleged 'design' here for it, and to Strickrodt's double claims in 1998 for its place 'in the literature of English and African Studies' concerning West Africa. The immediate historical contexts of its writing in the next section more crucially inform understanding of how *The African Wanderers* already offered a critique of civilising missions of all stripes, whether religious (the 1848 reviewer's above) or imperial. For example, Strickrodt asserted that it pitted 'bad savages' against the 'good natives [...] in contact with Europeans', whereas the novel's moral, racial and other 'camps' are not so clearly cut.[7] If *The African Wanderers* is therefore at least an unusual variant of 'Literature of Empire' because penned by a woman,[8] its adaptation of classic (factual) colonial travel writing and adventure fiction counterparts of the period – by a Captain Marryat or a G. A. Henty for example – or of Robinsonades (that the 1848 reviewer underscored, but did not name[9]) is more significant. Sarah's 'refit for larger scientific purpose' of *The African Wanderers* in this chapter's running head emerges precisely when serious critical attention is paid to the Humboldtian purviews of its inside title page (Figure 6.1), namely 'its graphical description of scenery – in the amount of information which it affords, and in the moral lesson which it teaches' noted in 1848 above.

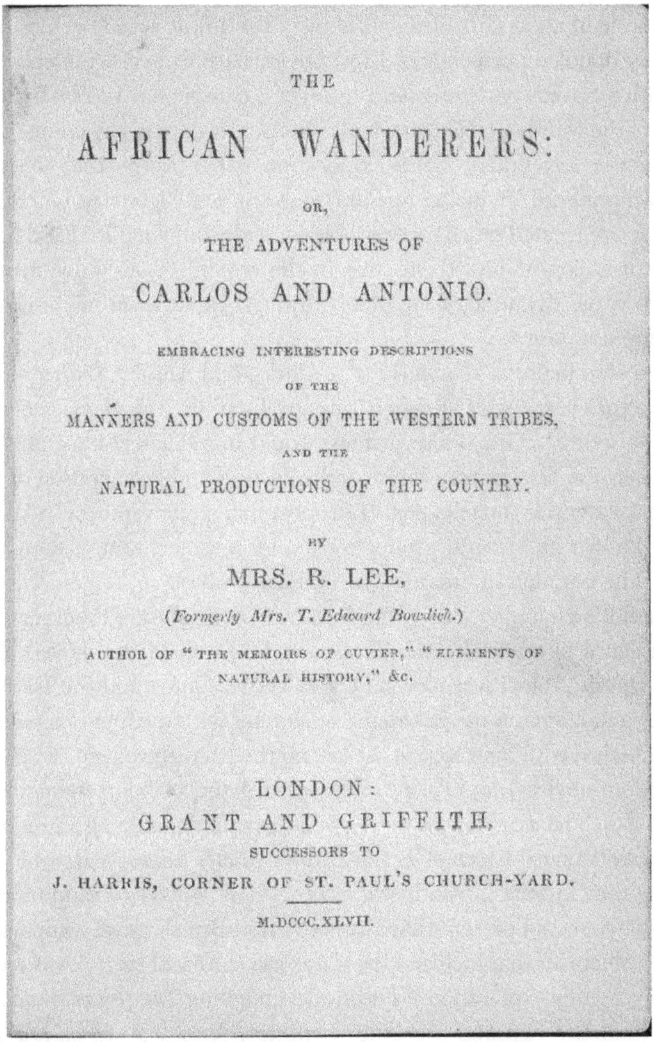

THE
AFRICAN WANDERERS:
OR,
THE ADVENTURES OF
CARLOS AND ANTONIO.
EMBRACING INTERESTING DESCRIPTIONS
OF THE
MANNERS AND CUSTOMS OF THE WESTERN TRIBES,
AND THE
NATURAL PRODUCTIONS OF THE COUNTRY.
BY
MRS. R. LEE,
(*Formerly Mrs. T. Edward Bowdich.*)
AUTHOR OF "THE MEMOIRS OF CUVIER," "ELEMENTS OF NATURAL HISTORY," &c.

LONDON:
GRANT AND GRIFFITH,
SUCCESSORS TO
J. HARRIS, CORNER OF ST. PAUL'S CHURCH-YARD.
M.DCCC.XLVII.

Figure 6.1 'Inside Cover Page of *The African Wanderers* (1847)'. Courtesy of the University of St Andrews Libraries and Museums, *The African Wanderers*, rPR4879.L44A64E47.

The Generic and Ideological Refit of *The African Wanderers*: A First Natural Science Fiction

Despite core Robinsonade elements – shipwreck, marooning on an island, heroic survival, eventual rescue – Sarah's novel unmakes and refits each one. Most significantly, there is no 'Robinson' and no 'Man Friday'. Rather, mutinous white crewmates maroon the two white European protagonists, Carlos a Spaniard and Antonio a Venetian, on an island. Their immediate escape in their small boat then precedes their eventual shipwreck in chapter six as the 'launch-point' for the main plot: they land somewhere on continental West African shores, to encounter the region in a fictional journey of its exploration through necessary survival. By thus eschewing the 'civilizing' narrative of the Robinsonade in all its core elements above, *The African Wanderers* also capitalises on its consistently doubled perspectives. Most obvious are its two 'wanderer' protagonists and their native counterparts met with in several continental West African lands. But the purpose of the wanderers' journey is also unusually doubled by counterpoint: it is no imperial overseas 'adventure', that is exploration as bio-prospecting conquest in other guise,[10] nor an imperial 'civilizing mission' to promote enlightened Christianizing values since these are treated much more problematically only in initial and final chapters. Instead, the protagonists find themselves in the region – the double meaning of this reflexive verb is the dynamic of the plot – through exploration of countries that they had never set out to visit.

The unforeseen fictional 'wanderings' of Carlos and Antonio therefore more closely connect to factual precedents. Readers already familiar with Sarah's *Stories of Strange Lands* (and their earlier Gift Book versions) would immediately have identified in *The African Wanderers* the further translation, (re)collection and re-narration of Sarah's (and the Bowdichs') extensive, independent (European) scientific ventures in West Africa, to make better known its 'natural productions' and 'manners and customs' on the title page above. The parallels are many. The fictional 'wanderer', Antonio, is drawn from the larger-than-life character enlivening Sarah's 'Voyage Out' ('Fragment I'). Her eyewitness experience of Sierra Leone ('Fragment III') informs the novel's fifth chapter. Carlos's sea voyage from Liverpool via Cape Verde, Goree and the Isles of Los (*AW*, chapters 3–4) amalgamates the Bowdichs' encounter with the first two on their journey in 1823 to Madeira, with their stop at the last on their return voyage of 1817 (from Cape Coast to London), after a delay of some seven weeks in the 'Gaboon' (treated in 'Fragment IV' of the *Stories*). The experiences there of Carlos and Antonio also re-narrate the scientific detail of chapter thirteen of T. Edward Bowdich's *Mission to Ashanti* (1819). By not incorporating the Gambia in its itinerary, *The African Wanderers* could more freely refit the ideological scope and persons that are its two non-British travel companion 'wanderers'. Its larger blueprint that included the *Excursions* (*EM* and *EMFr*) was recent French-international scientific exploration.[11] The model impelling *The African Wanderers* was none other than Humboldt's ground-breaking scientific *Journey to the Equinoctial Regions of the New Continent* undertaken with Bonpland, newly reapplied to comparable West African 'equinoctial' destinations where Humboldt (and Bonpland) never set foot. Moreover, the harnessing of Humboldt's scientific travels and exploration of global plant geographies brought coherence of plot for similar 'wanderings' in fictional guise. Sarah's equal 'attention to the condition of the aborigines of Africa' (the 1848 review) in *The African Wanderers* then further enriched its comparative scientific interest. In current parlance, Sarah's first novel is as striking for its indigenous 'ethnic diversity' as for its 'biodiversity'. Humboldt's global attention to the latter left little space in the *Journey* for the former. However, his avowedly anti-slavery stances return in fictional guise in the Hispanic 'wanderer' in *The African Wanderers*, Carlos.[12] To argue for Sarah's fictional translocation of Humboldtian principles to include, but not causally explain West Africa's various humankinds, as we will see in more detailed analysis of key 'graphical descriptions' below, is therefore to promote the much larger 'refit' of her otherwise merely 'recycled' material from the semi-fictional *Stories*. If this reading appears too modern, it is vital to recontextualise both the 'Manners and Customs of the Western Tribes' (the novel's 'ethnic diversity') in pertinent and controversial 'African' debates in 1846–1847 and the 'natural productions of the country' in relevant travel accounts of the period.

The life-long abolitionist, Thomas Clarkson, died in 1846. As one of the twelve founding members of the 'Society for Effecting the Abolition of the Slave Trade' in 1787, and a key player in passing the 1807 Slave Trade Act into law, he was the invited opening speaker at the first World Anti-Slavery Convention at Exeter Hall (London) in 1840, because the work of abolishing the Atlantic Trade was still not won in Latin America. Sarah's former suitor, Thomas Hodgkin,[13] was among convention delegates, alongside a small group of prominent women not permitted to speak such as US delegate Elizabeth Cady Stanton, promoter of rights for women.[14] In 1845, the 'Aberdeen Act' permitted the British Navy to intercept slave ships bound for Brazil as it had earlier intercepted Portuguese, French and other slavers in West Africa (as witnessed by Sarah on the River Gambia in 1823–1824 in Chapter 4). That the abolition of slavery and the slave trade was also clearly not won at its source or destination markets constitutes the recurrent major theme of *The African Wanderers*: it depicts different indigenous, as well as transatlantic practices and heritages respectively of the 'Kaylees' (*AW*, chapter eight), and 'Yellow Gaston' (*AW*, chapter ten). As a spoiler alert, more shocking for readers (in 1847 and today) is the capture of Carlos and Antonio by native 'Filatahs' (*AW*, chapter fifteen), and their selling as white slaves at a market for the purpose (*AW*, chapter sixteen) to a wealthy 'Moor' as representative of intra- and transcontinental African trading networks.

The question in 1847 of final abolition of world slavery spearheaded in consequence the related issues of liberation and emancipation, both for former slaves – the experiment of Freetown in Sierra Leone to rehome former plantation slaves of the American South and the Caribbean – and closer to home for women on both sides of the Atlantic. But (non-)equality of status regarding civil liberties pertained also to 'white' denominations in Britain, long disenfranchised by their religion from attending Oxford and Cambridge Universities and holding public office, including memberships of scientific Institutions and learned societies. The 1846 'Religious Opinions Relief Act' removed most final bars to 'equal' participation in public life for (men who were) Jews, Dissenters and Roman Catholics. As Chapter 5 discussed, Sarah's *Stories of Strange Lands* (with scientific 'Notes') had already enabled her to speak in 1835 against the horrors of the indigenous African slave trade, and misplaced colonial and missionary interventions. Her *African Wanderers* was a more audacious parallel venture in shaping opinion. Completed in 'Nov. 1846' as dated by her preface to its first edition, Sarah's first full-length novel was clearly bio-(geo)graphical:

> 'Where we have suffered much, we love much,' is a saying verified by the undying interest which the Author of the following story takes in the western coast of Africa. To call the attention of the wise and good to a part of the continent (the river Gaboon) but little known, has been one of the chief objects of her undertaking; […] One kindly impulse, one mite added to the exertions made on behalf of this magnificent land, will indemnify the Author for her labour, and the anxiety which necessarily attends publication. (*AW*, i–ii)

In 1846–1847 Sarah could enjoy not only greater freedoms of scientific but also religious and political expression in *The African Wanderers*, to set her interconnected abolitionist, emancipatory and broad-church scientific-educational interests more centrally within its fictional expeditionary mission, 'call[ing] the attention of the wise and good' to the 'western coast of Africa'. In line with her factual report of the Gambia also 'but little known' (see Chapter 4), Sarah's 'design' for the fictional *African Wanderers* also renewed her commitment to 'this magnificent land' through her overt use above of the language of (Biblical) parable, the widow's 'mite' (Mark 12:41–44; Luke 21:1–4). Redolent of her own status (as 'formerly Mrs T. Edward Bowdich'), her 'small' giving of all her means to extending natural knowledge of West Africa had larger private and public intent. Although she remained disenfranchised in Britain and France in 1847 from public scientific life as a woman, her new recourse to third-person *autofiction* (the 'wanderer' guise of Carlos and Antonio),[15] could refit her first-person encounters with Sierra Leone, 'Ashanti' (Ghana), the 'Gaboon' in her 1835 *Stories of Strange Lands* in further double 'report'. The new fictional guise and genre of *The African Wanderers* reincorporating them also pragmatically included selected Gift Book stories published from 1835 to 1844 that were her regular means to support her family and larger work from 1844 in natural history-making. Significantly, therefore, Sarah's first novel in 1847 spearheaded strategic successors such as *Adventures in Australia* (1851; revised 1853) that replaced her further need annually to publish a Gift Book story. Herein lies a key to her 'prolific' final decade that has remained unexplained in deB. Beaver's DNB entry. But her first novel concomitantly reprised Edward's earlier works on North-West Africa as third-person accounts that also benchmarked more recent 'progress' in the region (see Appendix 7), regarding both the (non-)abolition in 1847 of African-World slavery, and advocacy for (more radical) scientific and religious 'mission' than well-meaning Abolitionists in Britain were preaching in pulpits, conventions, parliamentary and Learned Society chambers, but without ever stepping foot beyond them.[16] Indeed, the more preposterous ideas of these 'armchair' well-meaning are among the butts of the novel (*AW*, 232–33).

For readers in 1847, *The African Wanderers* had additional prescience at home, however, through its parallel thematic resonances in mass forced movement and expatriation of British citizens. The Highland Clearances, collapse of rural 'crofting' economies and expansions of an industrial urban workforce saw men, women and children undertake back-breaking drudgery in Britain's low-paid mining, factory, building, textile and other industries. When work failed, punishment for starving petty thieves and prostitutes was the (white British) 'middle passage' to Australian penal colonies.[17] But destitution and mass exodus had been greatly exacerbated by the Great Famine in 1845, devastating the Scottish Highlands as well as Ireland; the event epitomised the fatal outcomes of crop disease when plant 'biodiversity' is supplanted by human over-reliance on 'monoculture' for staple foods. The British Relief Association was founded only in 1847 to address the Famine's agricultural and human impacts. Of note, therefore, in *The African Wanderers* are its representative secondary protagonist-foils aboard the fated, ironically-named vessel 'Hero' on which Carlos sets sail from Liverpool. Its surgeon, Scotsman (Sandy) Fraser, plays key roles (*AW*, Chapters 3–6) before his death of fever in 'mentoring' Carlos's (and the reader's) knowledge both of (tropical) medicine – treatments of tropical fevers, dysentery and alcohol-related inflictions – and of marine natural history. Fraser identifies the various fish and marine mammal species the 'Hero' encounters, each generating a footnote to provide the scientific (Latin) name. Crewman Johnstone from Ireland is the 'Hero''s maker of merriment, sharing character traits with Antonio, whom he tricks as a 'Venetian' to look out for 'Portuguese Men of War' not in other ships, but in the vernacular for the species of stinging jellyfish (*AW*, 40). But the 'Hero''s precarious, all-male micro-society and hierarchy also refracts the pivotal 'domestic' engagement and work at home of aristocratic and bourgeois ladies in Britain in their philanthropic missions to alleviate the sufferings of the poor 'overseas' and at home. The *Dublin University Magazine* reviewer's very particular designation of 'Mrs R. Lee' would then have had especial resonance. She is a 'lady' authoress for readers of her sex, despite being 'not well known to [its] novel readers and crochet-manual students of the middle of the nineteenth century' (above). The 1840s saw major promotion of crochet-work among the leisured classes, with various manuals written by ladies for ladies on the subject, with patterns for philanthropic projects adding to their 'useful knowledge' and charitable accomplishments.[18]

Sarah's first novel thus endorsed and newly articulated for wider audiences her foremost commitments and life's work as a 'lady' *scientific* traveller and science author in continuity with, yet clearly distinct from her late husband's, and her own sole-authored contributions to making West Africa better known. No differently to her (non-fictional) scientific works, the Preface to *The African Wanderers* left its readers in no doubt concerning the (unusual) first-hand scientific veracity of what they are to encounter in its fictional pages. The paragraph omitted in the quotation from it above highlights both her extensive personal authority and her explicit use of others' recent 'graphic descriptions' of West Africa:

> With the exception of the field of battle, the finding of the two boys, the journey to Santander and Liverpool, the history of Carlos has been invented as a vehicle [...] Antonio is only an imaginary personage after he lands in Africa. *Every production is true; and most of the circumstances are drawn from the personal experience of the Author or her friends.* Among the latter she has to thank Captain William Allen R.N. for allowing her to extract materials from

> his graphic description of Fandah; also Mr. Freeman, whose conversation and reports (the latter supplied by the kindness of Mr. Beecham) have afforded her useful statements. (*AW*, 1–2, emphasis added)

A further point of fact is that *The African Wanderers* follows closely in the footsteps of major British and French exploration and travel accounts published on the region since 1799 (see Appendix 7), because they variously engaged in what by 1830 was the major geographical debate and object of (British) West African exploration. As Charles W. J. Withers clarifies: 'The Niger problem was one of the most important geographical questions discussed in late Enlightenment Britain [concerning] the dual question of the course and direction of the Niger River and its terminal point'.[19] In her Preface, Sarah's scientific accuracy in the 'wanderings' of her novel derived from specialist geographical sources – see Appendix 7 – to include her own as 'Bowdich' among them. She was therefore alert to the major and ultimately fatal Niger expeditions of Lieutenant Hugh Clapperton (1788–1827), whom she cites as we see below, and the success of Richard Lander (1804–1834). His second expedition in 1830 (with his brother John) to explore the lower Niger confirmed its previously unidentified (contrarian) course: its many tributaries flowed indeed into the Bight of Benin. The work by 'Captain William Allen R.N.' (1792–1864) then provided more than the 'graphic description of Fandah'. Its final chapter elucidates his staunchly abolitionist views, further ramified in the 'reports' of 'Mr Freeman'. Freeman's first 'Journey' not only footnoted the earlier account of 'Bowdich' (1819), making Sarah's prefatory acknowledgement poignant. Freeman's name and ancestry – the son of an African father and an English mother[20] – also made his 'black life matter' in current parlance in her 1847 novel. In its times and until too recently, missionary activity in the region was assumed to be (imperial) white. Freeman's exception and (missionary) journal accounts from the field (Ghana) adopted the 'standard' (white) travel account in their chapter synopses and running heads. Sarah's larger familiarity with writing in these genres herself allowed her to amalgamate into her *African Wanderers* in its latter part a fictional re-adaptation of William Allen's 'manifesto' combined with Freeman's 'true' reports of West African 'missions'. When Carlos and Antonio reach 'Koomassee', they importantly encounter the fictional Wesleyan minister, 'Mr F', clearly modelled on Freeman in the particularity of giveaway personal detail: 'The officiating minister was standing on a raised platform, and his complexion showed that he had a slight claim of affinity with the dark inhabitants of sunny climes; but he was gentlemanly, pious, and intelligent' (*AW*, 311). Black cameo parts constituted central scientific realities in Sarah's fictional report.

To make West Africa, and more importantly contemporary indigenous knowledge of it, better and differently known to the juvenile (and adult) target market for *The African Wanderers* in Britain and its Empire in 1847 was therefore to follow the *lead* of Sarah's pioneering female scientific footsteps – her exceptional 'widow's mite' – for the novel's singular autonomy, independence and prefigurative modernity. Here indeed was 'one of the first novels in the English language, which is set in West Africa' as Strickrodt contended, but with more singular critical design. As crystallised in her prefatory acknowledgements and portrayal of 'Mr F', Sarah's pioneering natural history fiction refitted the contrarian geographical course of the Niger in the novel's wandering 'design' for a larger scientific moral purpose, that natural history knowledge is unraced. Sarah's remarkable scientific double narrative informed by French and British natural history knowledge and first-hand experience is encapsulated in the following passage reconfiguring 'Clapperton' among her sources:

> and when they came near the trees, Carlos exclaimed, 'These must be the mighty baobabs of *which I have so often read, many of which, according to botanists, have stood since the creation of the world, and which Captain Clapperton described* as having a number of green velvet purses hanging from them. Here they are. Look up, Antonio, at the fruit: it is full of a delicious, slightly acid, and farinaceous substance, in which the seeds are imbedded, and of which the monkeys are said to be so fond, that the tree is often called the monkey-bread. It is, however, named by *botanists after the great French naturalist, Adanson, who was the first to describe it properly, after having seen it in Senegal, where he lived for some time. I have read a beautiful eulogium on him, written by the immortal Baron Cuvier.* But now for the tree. (*AW*, 211, emphasis added)

For such non-conformist interventions to strike home (the novel's British readerships) with traceable narrative authority (as here), Sarah first sets up her 'contrarian' and 'unitarian' perspectives in the no less striking double narrative frame of the novel's first two chapters. These directly target narrow (British imperial) assumptions and cliched 'African' knowledge – prejudicial attitudes concerning white superior race(s) and 'Africa' as a hot, exotic land inhabited by dangerous wild animals including human savages and cannibals – fuelling the country's double paradox as the 'white man's grave' and the 'dark continent'.[21] Our exploration now looks explicitly in Sarah's first novel at its unusual combination (doubling) of non-conformist scientific content and 'unitarian' (Humboldtian) views of nature and humankind.

A Refit with Scientific Purpose (1): Explorations in (Contrary) Black-and-White Framing

Unusually for 'British' fiction, the novel opens in a battlefield in Spain, during the (recent) First Carlist War (1833–1840), with unequivocal direction of the reader's gaze:

> [...] the eye rested on none but the images of peace, and the bounty of the Almighty; but on looking immediately around him, the spectator might receive a painful proof of how much man does to deface this fair creation. His strife and contention had been there, and the plain was strewn with the dead and the dying; for it was the evening after a fearful battle, which had taken place between the two struggling parties, called the Carlists and the Christinos. On the occasion to which these pages refer, the Carlists had remained masters of the field, and the Christinos had fled in all haste, leaving their dead upon the ground. (*AW*, 1)

The first false 'fact' to fall, upon which all others in the novel depend, is that civilised Western society in 1840 is all 'peace'. The clearly (white) 'savage', uncivilised, irrational partisanships of this scene also 'deface this fair creation', formulating what is an unusual 'environmental' observation for the period. Savagery and salvation cannot therefore be neatly polarised or 'coloured', as is underscored by the 'Good Samaritans' amid this white-on-white conflict between Monarchists (Carlists) and Liberals (Christinos):

> A fatigue-party, commanded by an English officer who had entered the service of Don Carlos, and formed chiefly of men of his own nation, was searching amidst the slain for those who might, though wounded, be still alive, and was followed by carts, in which straw was laid to receive the sufferers. Strangers to the vengeance which men of the same country feel towards each other in civil war, the soldiers indiscriminately sought for both friends and foes, and, carefully filling the vehicles, were about to return, when a low, wailing cry met their ear. 'Surely that was the cry of a child,' said Captain Lacy to his sergeant; 'how can it have come here? Look for it, Brown.' (*AW*, 1–2)

The aptly named 'Lacy' and 'Brown' (combining the 'crochet-manual' threads, and non-black and white colours), return to rescue and adopt the two orphaned boys, Carlos the main wanderer protagonist (named after Don Carlos of the Carlists), and his brother Henriquez. 'Sides' therefore do not hold before the greater humanity that is no respecter of persons here, a mentality that also has its negative face:

> The country was in such a state that no party was safe; the Carlists robbed Christinos, the Christinos plundered the Carlists, and the banditti stripped them both. For greater security, those who were obliged to travel generally assembled together in numbers, and hired waggons (*sic*), [...] and in this manner [Lacy, Brown and the two boys] reached Santander. (*AW*, 7–8)

This 'feudal' white Spanish world then proves but a different shade of white colonial, industrial and modern British 'civilisation' in Chapter 2. It is set in affluent mercantile Liverpool-Wavertree that is Lacy's home, to puncture its 'opportunity' for the adopted boys in the novel's representation of industrial progress:

> The air of Liverpool, impregnated by the smoke of many chimneys and factories, did not agree with Colonel Lacy's health, accustomed as he had been for years to live almost entirely in the field; he therefore took a house at Wavertree, within a walk of the town, and close to a day-school for boys, where his proteges commenced their education. (*AW*, 11)

This early cameo in 1847 of 'air pollution' and its adverse health effects even in the airier Wavertree may again shock the twenty-first century reader in its already prescient diagnosis in 'environmental medicine'. But the greater shock and counter-cliché lies within the anti-idyll of Lacy's Wavertree home. The 'unchristian' Mrs Lacy is determinedly inimical to housing her husband's adopted Spanish children with their own, especially the 'impetuous and noisy' Carlos causing 'the destruction of her nerves' (*AW*, 16). The character differences between the 'industrious' Henriquez and rambunctious Carlos further add to the text's setting up and knocking down of false dichotomies in Carlos's idleness – viewed as negative – and Henriquez's industriousness (*AW*, 12) as positive, to determine the plot. Despite their guardian's benevolence, Carlos's motives for setting sail to 'seek his fortune' highlight the real mediocrity and bureaucratic tedium of industrial Liverpool when Henriquez inquires of his brother's state:

> 'Do you not see?' replied Carlos; 'do you not feel the glorious spring, the sun high in the heavens, the birds soaring in the sky, the leaves and the flowers expanding in the glittering rays, and everything inviting us to freedom? I hate the counting-house, I hate the desk, I hate the narrow dirty streets – I hate them all, as I hate Mrs. Lacy,' he added.' (*AW*, 25)

In the bidding farewell of Carlos as he sets sail from Liverpool at the end of Chapter 2, its final words clinch the novel's inversions of 'white hero' stereotypes in epic adventure-travel that depend upon benevolent lady-mother 'Penelopes' waiting at home:

> The parting scenes had better be passed over without description; they served to convince all parties, except one, of the depth of their mutual affection, and she wondered why the Colonel should be in such low spirits about a willful boy, *who was nothing to him but an object of charity*: this, however, she wisely kept to herself. (*AW*, 25; emphasis added)

'White' in its many negatives left behind in the opening double frame of 'Europe' is in further contention aboard the ship, 'Hero', in the microcosm of its hierarchy with white Captain (and lesser command), and black cook. Aboard, Carlos not only meets Antonio, his wanderer companion, and Fraser, the Scottish surgeon. His white British antagonist Gray (clearly mixing black and white) is also the novel's chief 'villain' (blackguard), determining the 'course' of the novel's plot to the lands of the Niger ('Black' Africa). Gray's revenge – because Carlos has seen his otherwise concealed scars from floggings, when Fraser with the Spaniard's help had removed his shirt to treat his fever – is deliberately to leave Carlos and Antonio stranded on the island chosen as the burial ground for the 'Hero''s principal fever-victims (its Captain and surgeon), to sail off in usurped possession of the 'Hero' (in actions prefigured by the Spanish 'banditti' on land above). Structurally as well as representationally, the differently wrong-footing 'white' perspectives of the frame, and of the 'Hero', thus shake readers' superior assumptions and narrow prejudices, to prepare a differently orientated 'Niger' as the novel's main interest from Chapter 8. Its main contents are thus geographically and figuratively emblematic of Sarah's 'contrarian' pioneering scientific perspectives on West Africa's indigenous biodiversity and ethnic diversity. Through the eyes of Carlos and Antonio as they 're-acculturate' (their no less white beliefs), Sarah's 'fictional' travel narrative consistently non-conforms to 'adventure' and 'Robinsonade' genres by making West African others (non-human and human) important protagonist-'heroes' in the wanderers' story.

A Refit with Scientific Purpose (2): Exploration of Equatorial 'Plant Geographies'

A scientific travelogue is nothing without its 'interesting descriptions' (of the inside title page). The long descriptive passage is also a staple of nineteenth-century realist, naturalist and adventure fiction, slowing the main action and its pacing either to increase cliff-hanger tension or to present didactic content. The impatient reader often skips such descriptions as dispensable and irritating disruptions of the main plot. Four main description chapters of the twenty in *The African Wanderers* – Chapters 7, 12, 13 and the first part of Chapter 18 – are then salient for their contents and their *placing* as extended 'plant geographies'. They each follow main action and distinguish 'native' *flora* and *fauna* specific to the particular next 'habitats' (equatorial regions of the African continent) that Carlos and Antonio encounter, understand and hence survive. Their and the reader's acquiring of new botanical knowledge is therefore quintessential to the novel in terms of their plot, because plant identifications and descriptions also constitute (death-defying) 'cliff-hangers'. In *The African Wanderers,* Carlos and Antonio depend at different points for their very life on recognising 'useful' plants and trees, whether signalling water or for food, shelter, medicines and essential implements, such as the 'calabash' for precious drinking water (*AW,* 105), although beautiful species are also appreciated. In Chapter 7, for example,

> they saw some broad green leaves floating gracefully over the water's edge, and hanging from them were bright scarlet flowers, and large bunches of oblong fruit. They proved to be plantains and bananas. 'This is indeed providential,' said Carlos; 'the plantains are not ripe, and will supply us with bread, which is the staff of life, and keep for months when roasted'. (*AW,* 102)

In consequence, the two wanderers vitally – in all senses – orientate and better equip themselves in each successive new alien 'plant geographical' landscape for better onwards survival, by building on what is familiar – previous book knowledge in Carlos's case (see the 'Clapperton' example above); in Antonio's from foods enjoyed in previous journeying (the 'sweet' flesh of the 'bonita' caught 'between Sierra Leone and Cape Mount', *AW,* 108) – and/or comparable in kind to fruits, grains, leafy plants, trees in temperate (European) climes. The altogether Humboldtian point of the *flora* and *fauna* chapters is therefore not to exoticize *different* nature in its African kinds of (scientifically known) plants, birds, animals, insects, reptiles, but to familiarise the reader with 'local' habitats as the only means by which the white 'heroes' can be sustained.

Chapter 7 is therefore standout for its plant geography, and as a 'transition zone' not only in terms of the wanderers' literal and physical but also cultural-geographical revictualing: they encounter no other human beings, not because they are 'masters of all they survey' in 'empty' overseas worlds of 'discovery' narrative, but rather to focus on the precariousness of their onwards survival through necessary quick learning and adapting to 'nature' around them, and as if 'natives'.[22] For example, in identifying 'coconuts', Carlos can reach them only because he 'watched the negroes, both at Cape Coast and Sierra Leone, as they climbed the palms; and I dare say that I could do the same' (*AW,* 100). If Carlos and Antonio lose their small boat to its disintegration in this chapter, forcing them to journey henceforth on foot, their 'discovery' of *fauna* (in lizards, birds, a 'Manatus or Sea-Cow', *AW,* 105) before encountering human species challenges reader perception of 'acclimatisation' on two counts. The first is a passage that would seem to endorse the 'colonial gaze' in the cliché of the 'indolent native':

> The bush, as the people of Africa call the 'wild wooded' country, is not such a bad place after all,' said Antonio ; 'everything you want is found there ready-made.' 'Yes,' rejoined Carlos [...] Those who have only to stretch out their hands to gather their food will never work for it; and then they want so little clothing. Thus, with many of these nations, wants must be created among them before they will become thoroughly active and industrious. *That is also one of the reasons why they persist in the slave-trade, for it gives them much less trouble than tilling the ground, or carrying on manufactures.*' (*AW*, 103, emphasis added)

The second chapter, however, already set up the *false* opposition between Carlos's 'idleness' and Henriquez's 'industriousness'. The important qualification here is rather in the second feature of 'indolence': Carlos sees it as an explanation for the persistence of the *indigenous* slave trade. Both wanderers will, however, revise such general views (the uniform 'those', 'they') upon further encounters with different 'nations', to which we return shortly. More interestingly for contemporary debates about human species in 1847 that chapter 5 considered (p. 114) is Antonio's cameo portrait concluding chapter seven of the wanderers themselves, seen as if through 'native' eyes:

> I'm thinking that if we meet with natives, they will take me for a monkey of some sort. Your hair, Lacy, is hanging over your shoulders, and your mustachios are very respectable for your age; but your beard does not trouble you much; whereas mine reaches to my waist, and, joined to my hair, whiskers, and mustachios, must produce an effect which may be serviceable, for I am sure it would frighten the fiercest enemy. (*AW*, 112)

Indeed, Antonio later frightens a slave girl less by his hirsuteness than by his white face (*AW*, 175). To compare himself comically (because of his character) to a 'monkey of some sort' here, indeed almost as an Orang-utan ('wild man of the woods') is, however, an unusual end to the chapter: the swarthy *white* is 'monkey-man' (not the cliché 'negro' of the period). If this reading again appears too modern, the pivotal 'biodiversity' chapters (12 and 13) only confirm the novel's unusual non-conformism to accepted animal-racial ideas and hierarchies of the period, including pre-Darwinian ideas about African 'natives' (as the least developed and 'civilised').

The all-important, mid-novel, 'rainy season' description (Chapters 12 and 13) not only puts its 'nature' uppermost because it is synonymous with equatorial 'equinoctial' Africa, but also demonstrates its agency in these regions as the supreme arbiter of *common* animal and human life. Shelter, not travel, is therefore paramount, with the stocking of suitable provisions essential for humans if they are also to overcome the related illnesses of the season. Chapter 12 opens with Carlos's positive evaluation of how far the wanderers have 'travelled' in terms of acclimatisation and acculturation, but his enumeration of 'dry season' lessons learned will prove woefully incomplete:

> we are better off than we were before, for we have more means of defence (*sic*) with us; we are more expert in using those means; we know much better how to manage about provisions; we are better acquainted with the natural productions around us; we are more used to walking; and altogether know better how to provide for our safety and comfort, than when

> we were first deserted; so, as there is now light enough, let us read a prayer, seek the blessing of God, and then bravely set forward for the huts. (*AW*, 182)

The wanderers' now different equatorial 'plant geography' is not another, but the same place in (climatically) new conditions. The static intensity of its nature description is no (didactic) digression to be skipped, but rather experienced for its heightening drama of 'plot': (how) can the wanderers survive 'tornados' and, worse, the concomitant rainy season tropical fever?

The careful comparative construction of Chapter 12 – it mirrors chapter seven in its lessons of necessary adaptation and in its absence of local peoples – represents its contents with an all-important change of perspective from the wanderers' habits on the move – of climbing trees to sleep in at night for example – to sedentary survival. Suddenly, trees are no longer safe in storm and lightning. Parallel encounters then become the chapter's primary comparative focus concerning species affinity and difference. While in their initial 'night' tree, Carlos and Antonio espy 'a little old man striving to mount [it]'; and beat 'him off in case they would kill a fellow human' (*AW*, 187). Readers probably guess before Antonio reveals the next morning that the 'old man' is a chimpanzee with 'very little hair in the front of its body, which gave it a more disgusting appearance. They presented a piece of plantain to it, which it ate with avidity, and then looked at them with a vacant, stupefied air' (*AW*, 188–89). Its apparent hairlessness and its vacancy are because, like the 'wanderers', it was 'drenched to the skin' (*AW*, 189) 'and thoroughly chilled' (*AW*, 187). Both contribute to the creature's speedy end as easy prey to a leopard, thus prefiguring similar potential outcomes for its human doubles. As in Chapter 7, the wanderers' urgent construction of a suitable hut of 'bamboo rafters, and […] palm leaves' (*AW*, 190) to withstand adverse weather and marauding creatures also includes its stocking with 'provisions which would not require cooking' (*AW*, 191) with honey, limes, roasted plantains and cassava roots, kola nuts and sugar cane, corn, roasted birds (*AW*, 191–92). Such 'Western' industry, however, fails to account for catching rainwater, and for the inevitable physiological consequences of rainy-season drenching. In equal detail in the latter part of the chapter, Carlos manifests the various symptoms and stages of tropical fever (as the chimpanzee) before Antonio also succumbs, both 'wanderers' very narrowly escaping death in their slow recovery (*AW*, 194–99) from sickbed immobility: 'The ague still tormented them, and as soon as they gained a little strength seemed to pull them back again; and their pale, sallow faces and wasted limbs, told but too truly how nearly they had fallen victims to this deadly disease' (*AW*, 199), by implication because 'Europeans'. The point is less to endorse their narrow escape from the 'white man's grave', and more to debunk their allegedly superior knowledge. Having taken up residence in their hut rather smug in their cleverness in its making upon also discovering similar nest constructions by weaver birds 'in one common roof, each [with] its separate abode' (*AW*, 193–94), Antonio's (and the reader's) surprise on his recovery from fever is to discover near chapter end another hut at the back of their own, 'built while we were ill, for it was not there when we came; it is not as large nor as good as ours, but I saw some one moving in it' (*AW*, 200–201). Expecting its dwellers to be 'some natives' clearly with less sophisticated skills, they discover 'a female […] with her young one, which she held *with one hand*, while *with the other* she dexterously mounted to the top of her hut, where she sat, seemingly to enjoy the pleasure of nursing her *infant* in the open air' (*AW*, 200, emphasis added). The words 'hand' and 'infant' delay the identification by Antonio of 'a wild man, or rather wild woman of the woods', corrected by Carlos's citing of their Naängo (Gabon) friend, Wondo's local name and natural science knowledge of it:

> I think it may be the Inghena [...] These animals imitate men and women as closely as possible, but are worse companions than they would be, for they are extremely fierce, and one blow of their *paw* will kill a man. They build a hut and then live on the roof, as you see that one does now; they carry things about on their shoulders till they drop with fatigue; and none of them have ever been taken alive, their strength is so great. There must be another not far off, *for they live in pairs*; and the best thing we can do is to retreat as soon as we can, pack up our things, and go; for they are very cunning and treacherous. (*AW*, 200–201, emphasis added)

The graphic illustration of this description in the novel (AW, 201–202, see Figure 6.2) heralds the 'treacherous' presence of the male Inghena (Gorilla) as the twin danger with imminent fogs that cause the wanderers' immediate departure. But this scene is complicated by its unusual focus on his female mate as the main interest of the (monogamous) 'pair'; she is wounded after killing the leopard attacking their (now dead) 'infant'. Herein is the rationale for the male gorilla's increased aggression and attack on the wanderers' hut, so that they have to kill him in self-defence. The scientific importance of this elusive West African top primate for recent London scientific knowledge and collection will be discussed in Chapter 8. The 'whole body of natives pouring down upon them' that Carlos and Antonio again expected, but turns out to be the wounded female gorilla's cry when protecting her young, thus casts a different comparative light on accepted animal: human classificatory, and hence behavioural, differences allegedly explaining the 'higher' human, including 'pairing' of the gorilla couple for life. But it is Sarah's scientific cameo of the female gorilla here that is doubly striking scientifically and for fictional representation. First, female animal behaviour was barely reported in explorers' accounts, with female specimens more rarely collected, traded and then mounted in museum collections. The males of species (like birds) commanded the greater public (and expert scientific) interest for their 'plumage', including in this case the gorilla's sheer size and fearsomeness.[23] Second, Sarah's female gorilla is most 'human': she is the builder of her hut (home) and in her cradling of her 'infant' in the image almost a primate 'Madonna with child'. Readers cannot but contrast this animal mother nurturing and protecting her young from all attack with the 'un-nurturing' and unprotective Mrs Lacy. But the scene neither anthropomorphises nor racialises the (male) gorilla as will be increasingly the case in later paintings and sculptures depicting alpha male gorillas carrying off (black) women.[24] By suggestively setting the huts of the 'Europeans' and still mysterious Inghena-Gorillas back to back, Sarah leaves moot in the culmination of this chapter the differences or family resemblances between 'bimania' – the description is clear in the female's 'hands' (and the male's 'paw') – and human 'wanderers', whether West African or European peoples. Sarah's descriptions of African 'biodiversity' and 'ethnic diversity' together and separately thus question (European

Figure 6.2 'The Ingena with her Infant' (1847). Courtesy of the University of St Andrews Libraries and Museums, *The African Wanderers*, rPR4879.L44A64E47.

scientific) human 'superiority'. The novel's various parallel scenes speak more largely by inference rather than by peddling recent 'scientific' race theory or Christian religious conviction concerning orders of Creation (in Genesis for example) and of procreation in monogamous wedlock. *The African Wanderers* thus unusually structures its various West African 'plant geography' chapters to preface, contextualise, but not causally explain their human kinds. Such encounters also provide a mirror in which to look afresh at the European 'wanderers' as (hairy) counter specimens.

A Refit with Scientific Purpose (3): Exploration of Indigenous ('Aboriginal') Ethnic Diversity

The first 'natives' that Carlos and Antonio experience (in Chapter 8) are therefore significant: 'the body of a negro, who appeared to have recently died, and which Carlos proposed to bury; but Antonio dragged him away, saying, "We had better attend to ourselves, and he is past any good that we can do for him"' (*AW*, 119). In not replicating the actions of Lacy and Brown in Chapter 1 (burial of the children's dead parents), Carlos here identifies the different work of a recent slave kaffle and their need to escape it. Antonio's questioning in '"are we not white men?" will be taken up in a later mirror scene that will also endorse Carlos's reply here to Antonio's question: '"Certainly," replied Carlos; "but such inhuman brutes as slave-dealers would not hesitate ill-treating two defenseless men."' (*AW*, 119). Whether the victim is European or indigenous, the bestial nature of 'slave-dealers' of any stripe is the point. Its further evidence is the second 'native' of unspecified ethnic group that they encounter: '"Another body?" asked Antonio. "It is a child," said Carlos, stooping to raise it. "He is not dead, and perhaps we may save him. He must have been abandoned by the unfeeling wretches, because he could not keep up with them. Give me some water"' (*AW*, 120). Inadvertently replicating the action of Lacy that secured Carlos's own life, the 'wanderers' become 'Good Samaritans' and 'maternal' carers in one – reinforced by the lithograph of the scene – with the comic relief that this child reacts to their 'monkey-like' hirsute (male) appearance entirely as Antonio had predicted at the end of Chapter 7. The child, however, also reciprocally 'nurtures' them by his superior bushcraft, showing the wanderers where to find the tubers of a yam and hog yams (*AW*, 121–22). Their return of the child to his home village then distinguishes his mother from its other inhabitants with 'filed teeth' as 'savages' (*AW*, 125), because the wanderers discover their slaves incarcerated in a shack as intertribal war booty, to then have their guns wrested from them as prisoners also. In the 'filthy excess' (*AW*, 129) of victory celebrations, however, the child and his mother differently act by enabling the wanderers' escape by canoe through riparian 'plant geographies' that reveal mangrove swamps, with brilliant fishes and insects, parrots and monkeys as well as oysters (Chapter 11), to land in the first place in the novel (and near its centre) with a proper name, 'Naängo' (*AW*, 136). The source Sarah used as her authority is none other than T. Edward Bowdich's 1824 *An Essay on the Geography of North-West Africa*, with report of its people as 'courteous, hospitable and intelligent'.[25] The 'true hospitality' (*AW*, 138) that Carlos and Antonio receive for the first time fictionalises the 1824 authority importantly through Sarah's adept double-mirroring of intercultural perspectives. At the feast in the wanderers' honour (in counter-point to the 'filthy excess' above), the governor (brother to the king, father to Wondo) requests their 'history', but does not comment except through clear body language until Carlos finishes. As a West African, his comment from insider perspectives (rather than Bowdich's informed white outsider account) on interrelated subjects of slavery and cannibalism is significant:

> Those people whose child you saved were Kaylees. Lucky for you, you took that child, for without that they would have killed you. They have had war lately, and taken those prisoners whom you saw in the shed: one part will be sold as slaves, one part you saw eaten, and the rest will be eaten soon. Very bad people those; they don't care for goats, pigs, or fowls when they can get man. You would have been saved to the last, and then eaten too. They eat father, mother, children, who die in sickness – anything to get man's flesh. That lad who brought you back belongs to some people below here, and was trading up the river when the Kaylees came to the place where he was, and took him prisoner with the others; your

> coming away saved his life. Now you must not go from here till English ship comes to take you. English sometimes trade with us, and they will carry you home. Mostly French people seen here. (*AW*, 139–40)

To view 'natives' through 'native' eyes in this major chapter of *The African Wanderers* is therefore also to review the important description of the 'white negro' (the 'curiosity' picked out by the 1848 reviewer), belonging to the speaker above. His son, Wondo, glosses the non-verbal sound effects of his unusual 'song story' performance (*AW*, 143–45, reproduced in Appendix 7). This 'white' (albino) figure had already appeared in Sarah's *Stories of Strange Lands* (see Chapter 5). Here, Antonio's important interruption of Wondo's 'translation' underscores the fragile ethnic demarcations between slave-taking and non-slave-taking groups enacted in the song: 'So then you do not deal in slaves?' said Antonio.

> 'No,' replied Wondo, 'neither I nor my father will have anything to do with that. But my uncle and the Queen, as you call her, will always trade in them; and those long, low houses behind theirs are the barracoons, where they keep the slaves to wait for the ships. Some will come before long; we buy as many as we want for servants, but we never sell them again'. (*AW*, 151)

The wanderers then encounter the trader, 'Yellow Gaston' (whose father was Portuguese, *AW*, 173), come for the human captives in these 'barracoons'. Only then do they understand too late the above 'native' injunction, to await an 'English ship'. Because the trader takes them to be abolitionist spies, the wanderers have to flee Naäga for fear of his reprisals (*AW*, 178–79).

Nothing, however, prepares the reader for the further 'injunction' that is contained in the song of the 'white negro' in how Sarah turns the wanderers' onward plot. After themselves being captured in a 'kaffle' on one of the Filatahs' 'predatory excursions' in Chapter 15,[26] and sold as slaves in Chapter 16 to the wealthy Moor Sidi Baba, Carlos and Antonio themselves become the 'white negro' and his 'inspiration'. Like him, they use their talents in harmonised singing and music-making to define their different status among other slaves. That Sarah should put her wanderers into the shoes of 'black' slaves, including this 'white negro' freeman in the household of Wondo's father, was clearly strategic. It turns black- and white-raced European assumptions inside out and, more importantly, highlights their blind spots. If this reading appears too modern, its larger point was already made in Edward Bowdich's similar sentiments, recorded two pages after his exposition on 'Naängo': 'Naturalists have too often overlooked their own species in their travels, and the insects of a remote country are sometimes better known than the people.'[27] Sarah's pioneering perspectives in *The African Wanderers* included also turning the lens of 'North-West Africa' back on the 'white civilizing mission', to encompass the region's more complex intracontinental Arab and Muslim 'ethnic diversity'.

The final stage of the 'wanderings' then puts Carlos and Antonio in the shoes of Allen's report of 'Fandah' and in canoes on the Niger (*AW*, 239, 246, 259), but also 'passing as Arabs' due to their 'bronzed' skins (*AW*, 239). Description of Fandah's mosque, and an encounter with a Muslim cleric Ibrahim who admires the 'holy book' that Carlos carries in a bag round his neck, do not quite prepare readers' shock when the wanderers are then paraded at the slave market (see the final passage in Appendix 7) and bought with cowries by the Moor 'of consequence' (Sidi Baba).[28] The wanderers' 'Thank God!', neatly encapsulates the religious and racial hierarchies in tension here, including the Moor's scorn and definition of them as 'Nassareen' (Christians or early Jewish supporters of Christ), confirmed in their immediate treatment. Despite Carlos's struggle to keep the bag containing his prayer-book, they are stripped of their clothes and possessions, made to 'beat corn incessantly' (while tied to two wooden stakes) and whipped if they relaxed (*AW*, 270). Worse lies in the greater test of their ('civilised Christian') forbearance after a month's hard labour in Sidi Baba's unsuccessful attempts to 'convert' them to Islam (a further mirror of Christian missionary practices). These involve the shaving of their ('monkey') heads, their wearing of turbans and the further 'degradation' of being 'fastened to black slaves, and sent to cut and carry wood and water from the neighbouring forest' (*AW*, 270–71). If the wanderers' 'firmness' of inner and outer resolve could immediately be explained by their 'Christian' values, the novel's larger examination of *all* coercive dehumanising practices here interestingly proposes intercultural resistance as the surer means to overcoming. Carlos and Antonio rapidly learn the Arabic of their captors and, on Antonio's suggestion, sing 'a buffo song' (*AW*, 273), which immediately changes their status (like that of the 'white negro') within Sidi Baba's circles.[29] Through song and then their wood-working (again like the crafting of the harp by the 'white negro') enabled by the good auspices of the slave overseer, Adoossee, the reputation of Carlos and Antonio as Sidi Baba's rather different 'white slaves' increases, but to their larger detriment despite better treatment. They are now indispensable to the Moor's status among his superiors, including the King (*AW*, 273–76).

The 'escape' of Carlos and Antonio then confounds any neatly raced or 'credal' plot solution in its new 'wandering' with multiple outcomes: 'They were bound for Yahndi, the capital of Dagwumba,[30] where the Moor had promised to go and organize the affairs of the Mohammedans who had settled there; he being a high priest and lawgiver of the tribe of Ali' (*AW*, 276–77). This 'holy' journey provides acceptable refusal by Sidi Baba of the King's persistent requests to buy his two white slaves, and the wanderers' passage in the Moor's cavalcade towards Cape Coast (and a ship for England). Escape, as previously, combines opportunity to bring larger human good by also overstepping (ethnic, creedal) 'rules'; in other words, the repeating topos of the novel is the intervention of 'good Samaritans'. En route, Carlos exchanges places with a sick Arab guard, Yusuf, and shoots one of two marauding lions on his watch before Yusuf returns to his post. Resulting reward by Sidi Baba of Yusuf for 'his' bravery is promotion to captain of his army. In Yusuf's acknowledgement of Carlos's act and resulting silence in 'Christian, you are my brother', with Carlos's reply that 'all men are brothers among Christians' (*AW*, 279), the novel is not interested in interfaith exchanges, or larger 'reward' for Carlos's similar promise of silence to Gray (about his stripes). The more strategic outcome is Yusuf's facilitation of the wanderers' escape plan (on horseback), because his position as captain can arrange it.

The novel's 'happy ending' may then seem more problematic generically and ideologically for the modern reader of nineteenth-century (imperial) travel fiction and adventure stories. The classic safe return to Liverpool delivers the hirsute and robed 'wanderers' to a *double* recognition scene in Colonel Lacy's Wavertree home of Carlos by Henriquez, and of them both by their long-lost relations. But Liverpool's place in the Spanish brothers' onwards 'commerce' together potentially unsettles the gains of Carlos's self-understanding as a 'white man' through inhabiting the 'plant geographies' of his 'wanderings':

> Commerce with the western coast of Africa was a principal feature in the transactions of Carlos, *in the hope of benefiting a country in which he took an undying interest; and when he reflected on the immense riches of that beautiful land, and the universal spirit of traffic which pervades its inhabitants, he hoped that sooner or later its natural productions would wholly supersede the degrading and inhuman slave-trade, which stamps it with the seal of barbarity*. The increasing consequence of the brothers, and Carlos' experience in one quarter of the world, gave weight to their opinions with men high in the service of their adopted country; and their suggestions being acted upon, were most important to the welfare of their sable brethren: [...] Convinced that the enormous quantities of calomel, colocynth, quinine, etc., with occasionally profuse bleedings, in cases of fever, only increased the danger, Carlos suggested that other systems of medicine should be tried, till the baneful effects of the climate were ameliorated, and white men found that they could live in Western Africa. (*AW*, 356–57, emphasis added)

The lenses of the 1848 reviewer return readers of Sarah's first novel then and today to 'its graphical description of scenery – in the amount of information which it affords, and in the moral lesson which it teaches'. In the context of 1847, its more pioneering perspectives here identify 'world slavery' concerns, not only (still) extant in 'Africa' but also hidden in plain sight in the competing world 'trade' of industrial, 'abolitionist', Britain. The different enlightenment of Carlos's self-discovery in 'graphical description' of West African 'scenery' is not so much his scientific report as his means to elicit 'plant geographical' response on 'new systems of medicine' in the 'opinions with men high in the service of their adopted country'. The 'moral lesson' that Sarah's first novel teaches, which her *Excursions* could not, is a larger respect in 1847 for 'the immense riches of that beautiful land, and the universal spirit of traffic which pervades its inhabitants'. In 'Carlos' is therefore a fictional embodiment of a Humboldt as 'scientific explorer' and as the extender of his 'equinoctial' plant geographical 'views' as lessons in *reciprocal* human civilising endeavours – cultural, scientific, religious and medical – that take world natural history resources fully into human account.

But behind her *African Wanderers* as her first 'natural science' fiction in these regards, Sarah could also put herself in cameo in Carlos's 'adopted' shoes, both to decry the 'baneful' effects of slavery and of 'malaria', and the similarly unsatisfactory treatment and recognition of the *scientific* traveller as clearly only the male of the sex in her novel:

> To the solicitations of learned bodies who sought for his contributions, Carlos was glad to answer by communicating all the knowledge he possessed; but he found it very difficult to check the accumulated entreaties of the fairer sex that he would give them his autograph, or make a sketch for their scrap-books. At length, however, some new wonder came, and the collectors left the African traveller in peace. (*AW*, 359)

Sarah's pioneering first novel 'of West Africa' like her first 'plant geography' of the Gambia, demonstrate how such reports in different genres need to be much better known in studies of travel literature(s). If women continue to have minor roles in *The African Wanderers*, the larger lesson for the British lady reader is not to emulate its 'Mrs Lacy', or these unnamed members of the 'fairer sex' entreating autographs, but rather to demand more vociferously an education in natural history that would prepare women to become (lady) scientific travellers and authors themselves. If the final section of this book takes up the larger question of how better to educate interested newcomers of both sexes in expert natural history-making, Sarah had no further need after 1844 to write Gift Book stories to support it. In the wrappers of her further 'natural science fiction' – *Adventures in Australia* (1851 and 1853), *Playing at Settlers* (1855), *Sir Thomas* (1856)[31] – from its generic invention in *The African Wanderers* of 1847 are her literary-scientific supports to advance the final decade of her natural history-making work(s).

Notes

1 Strickrodt, *'Those Wild Scenes': Africa in the Travel Writings of Sarah Lee (1791–1856)*, 54 and 103, respectively.
2 Anon., 'The African Wanderers', *Dublin University Magazine* XXXI, no. CLXXXII (Feb. 1848): 252–57 (252). This review is unknown to Strickrodt.
3 Mrs R. Lee, *The African Crusoes*. New York: Dick and Fitzgerald, 1847, with later imprints in Philadelphia: Lindsay and Blakiston, 1854 and 1860 and in Boston: Lee and Shepard, 1873 and 1876.
4 Anon., 'The African Wanderers', 252 (emphasis added).
5 Anon., 'The African Wanderers', respectively 252, 253 and 254 (all emphasis in the original).
6 Anon., 'The African Wanderers', 254 (emphasis added).
7 See Strickrodt, *'Those Wild Scenes'*, 85 and 94, respectively.
8 Greenstein, 'Sarah Lee: The Woman Traveller and the Literature of Empire', 133–37. See also John Mackenzie, ed., *Imperialism and Popular Culture*. Manchester: Manchester University Press, 1986, and his *Imperialism and the Natural World*. Manchester: Manchester University Press, 1990 (although they do not treat Sarah's work).
9 See, for example Ian Kinane ed. and intro., *Didactics and the Modern Robinsonade*. Liverpool: Liverpool University Press, 2019 (also covering Robinsonades by women or with women protagonists of the period, such as Agnes Strickland's *The Rival Crusoes* of 1826 and Barbara Hofland's *The Young Crusoe* of 1828).
10 See, for example Shane Mulligan and Peter Stoett, 'A Global Bioprospecting Regime: Partnership or Piracy?', *International Journal* 55, no. 2 (Spring 2000): 224–46.
11 See James E. McClellan III and François Regourd, 'The Colonial Machine: French Science and Colonization in the Ancient Regime', *Osiris* (2001): 31–50, especially 49.
12 See Orr, 'New Observations'.
13 See Orr, 'Women Peers'.
14 See Douglas H. Maynard, 'The World's Anti-Slavery Convention of 1840', *The Mississippi Valley Historical Review* 47, no. 3 (Dec. 1960): 452–71.
15 Serge Doubrovsky's allegedly seminal articulation of the genre 'autofiction' in 1977 is clearly and differently challenged by Sarah's unusual novel already in 1847. Space does not permit our investigation here of the many 'autofictional' passages in *AW*.
16 See Rosamaria Alibrandi, 'British Parliamentary Abolitionists: Sir Thomas Fowell Buxton (1786–1845) and the Political and Cultural Debate on Abolitionism in the Nineteenth Century', *Parliaments, Estates and Representation* 40, no. 1 (2020): 21–34 online at https://doi.org/10.1080/02606755.2019.1704403.
17 Sarah's *Adventures in Australia* (1851) is a differently critical, natural science fictional response to such policies. See Orr, '*Adventures in Australia* (1851) by Mrs R. Lee'.
18 See, for example Mrs Jane Gaugain, *The Lady's Assistant for Executing Useful and Fancy Designs on Knitting, Netting and Crochet*. Edinburgh and London: Ackermann Co., 1841, Miss Francis Lambert, *My Crochet Sampler*. London: John Murray, 1844, Mrs Eliza Warren, *The Court*

Crochet Collar and Cuff Book. London: Ackermann Co., 1847, and the prolific Mlle Eléonore Riego de la Blanchardière's *Knitting, Crochet and Netting*. London, 1846, the alleged originator of 'Irish Crochet' as a mode of famine relief for Irish workers – men, women and children – in penury due to the Great Famine.

19 Charles W. J. Withers, 'Mapping the Niger, 1798-1832: Trust, Testimony and "Ocular Demonstration" in the Late Enlightenment', *Imago Mundi* 56, no. 2 (2004): 170–93 (170).

20 See John Flint, 'Freeman, Thomas Birch (1809-1890)', https://doi.org/10.1093/ref:odnb/47629.

21 See respectively Philip D. Curtin, '"The White Man's Grave:" Image and Reality, 1780–1850', *Journal of British Studies* 1, no. 1 (Nov. 1961): 94–110; Lucy Jarosz, 'Constructing the Dark Continent: Metaphor as Geographic Representation of Africa', *Geografiska Annaler: Series B, Human Geography* 74, no. 2 (1992): 105–15, and Stuart L. Pimm, 'Africa: Still the "Dark Continent"', *Conservation Biology* 21, no. 3 (June 2007): 567–69.

22 See Carlos's self-reflective comment in chapter 13 when they find a cooking pot in an abandoned native village: 'As you say, we will stew our flesh and vegetables together; but I really am afraid, if ever we should reach England, that we shall never lose the habit of thinking a great deal of eating and drinking, to which we must now plead guilty.' 'Show me the man in the wilderness,' exclaimed Antonio, 'that is not' (*AW*, 221–22).

23 See Isidore Geoffroy Saint-Hilaire, *Description des mammifères nouveaux ou imparfaitement connus de la collection du Muséum d'Histoire Naturelle et remarques sur la classification et les caractères des mammifères*. vol. 10. Paris: Archives du Muséum d'Histoire Naturelle, Gide Libraire, 1858–61, reprinting his three *mémoires* of 1843 (pertinent here), 1845 and 1852.

24 See, for example Marek Zgórniak, Marta Kapera and Mark Singer, 'Fremiet's Gorillas: Why Do They Carry off Women?' *Artibus et Historiae* 27, no. 54 (2006): 219–37.

25 Page 76. The passage also references the longer chapter 13 in the '*Mission to Ashantee*'.

26 Allen, vol. 1, 377–93 (in Appendix 7:1).

27 Bowdich, *An Essay on the Geography of N-W Africa,* 78 (in Appendix 7:1).

28 See Jan Hogendorn and Marion Johnson, *The Shell Money of the Slave Trade*. Cambridge: Cambridge University Press, 1986, and Green, *A Fistful of Shells.*

29 The wanderers' songs also have the same power and force as 'negro' spirituals. See Thomas P. Barker, 'Spatial Dialectics: Intimations of Freedom in Antebellum Slave Song', *Journal of Black Studies* 46, no. 4 (2015): 363–83; Frederick Douglass's *Narrative of the Life of Frederick Douglass, An American Slave*. Boston: Published at The Anti-Slavery Office, 1845.

30 See Bowdich, 1819, 177–78: 'seven days from Sallagha NE according to the Moors through the Inta town of Zongoo is Yahndi the capital of Dagwumba [...] to be beyond comparison larger than Coomassie.'

31 See Orr, 'Rethinking the Pioneering Text' and '*Adventures in Australia* (1851) by Mrs R. Lee' for the first studies of both novels. Chapter 9 addresses *Sir Thomas*.

Part Three

OPENING ACCESS TO EXPERT NATURAL HISTORY

Chapter Seven

SCIENTIFIC ILLUSTRATION SECOND TO NONE: DOUBLY EXPERT PEN AND INK, AND THE FOREMOST USES OF (WATER) COLOUR

In 'canvassing Cuvier' and 'harnessing Humboldt', Sarah's remarkable book-length publications from 1825 have clearly demonstrated in chapter studies thus far how she creatively developed their different (world-leading) expertise and parallel scientific priorities – in the Muséum; in the intercontinental field respectively – for optimal contribution to leading natural history endeavour outside its many bars to women on both sides of the Channel. The leanest period for her scientific contributions after the last numbers of the *Fresh-Water Fishes* (1838–1839) and publication of *Elements of Natural History* (1844) then starkly stands out. It coincided with the final illness and death of Sarah's mother as Donald deB. Beaver has clarified, but his DNB entry cannot then explain how she 'entered on her most prolific period of writing [...] (both fiction and non-fiction), for both children and adults'.[1] Sarah's natural history expedition 'in forty-four British Fishes' (Chapter 2) had no (commissioned) field project at home to replace it. How was her new work in natural history after 1839 inspired and fostered by experts and leading peer communities as the constant benchmark for her contributions since 1825?

Chapters in this final part address both the 'gap' that is 1839–1844 and its pivotal importance for the 'most prolific' last decade of Sarah's work(s) in natural history. The enduring significance of both Cuvier and Humboldt as her 'French' mentors and models lay also in their common cause for world natural history-making, namely to enhance its understanding in two interconnecting ways. First was their unwavering commitment to science pedagogy and wider public dissemination of new scientific knowledge, in order to inspire newcomers besides informing experts. Among Cuvier's many functions as the Chair in Comparative Anatomy at the Paris Muséum were his public lectures and demonstrations, with women in these audiences. His famed reconstructions of extinct fossil creatures from a single tooth or bone that demonstrated his new classifications of vertebrate life forms were one with his curation of the Galleries of Comparative Anatomy and Palaeontology for expert and general public understanding. In the second part of her *Memoirs*, Sarah had accounted for his 'pedagogical' publications (also listed in its bibliography of Cuvier's many works) as summed up in the 'Tableau Elémentaire, and the two editions of the Règne Animal, as different stages of the same work, and, with the Fossil Remains, and Natural History of Fishes, as the results of his discoveries in comparative anatomy. The collection of M. Cuvier's lectures on this subject is preceded by an introductory letter, addressed to

M. Mertrud'.[2] Humboldt's later works, such as his *Ansichten der Natur* (*Views of Nature*) and *Kosmos* (*Cosmos*) – appearing also in French, and translated into other European languages – gathered together and re-disseminated his prodigious scientific knowledge in accessible, non-technical forms, to set out the many 'wonders' of the natural world, and of the interconnecting worlds of 'Nature'. To neither applied the status of the (later nineteenth-century) 'populariser' in Britain, or its translation equivalent, 'vulgarisateur', as also a cadre in France of professionally-trained science disseminator.

The second priority and necessity that Cuvier and Humboldt both embodied and upheld for all 'experts' in natural history was the ability to make accurate figures – drawings of specimens, geological cross-sections, maps and diagrams taken 'from the life' or from major collections and findings – to explain the accurate (technical) details and configurations of scientific description in immediate 'view'. Indeed, training of the eye and hand for Cuvier and Humboldt made for better (expert) observations in both the museum and the field, just as text and image together enhanced new scientific description, understanding, communication and broader public perception.

These coextensive legacies of Cuvier and Humboldt for *expert public* understanding of natural history inform both the 'gap' and the 'prolific' final decade of Sarah's contributions to natural history through their art(s). She had already honed them before 1844, but they came more prominently to view thereafter as the remaining chapters differently demonstrate. In now examining for the first time the different illustrative media in which Sarah excelled as the (woman) peer of her 'French' mentors, the following chapters reveal how she also pioneered by extending the range of their models. Her creative scientific formats and uses of mixed media, especially in her seemingly derivative last publication, differently show that Sarah was a seriously important expert and forerunner in promotional, next-generational natural history endeavour for her times, and for today. In her own words from the Gambia in 1824, her hope was 'looking to my own support, through *S* and *M* to get employed in different works in natural history, setting up as an artist in that line' (Chapter 2). How she reconfigured the powers of scientific illustration to present 'true pictures' in natural history illuminates her pioneering perspectives on 'scientific' art forms that could visibly include critical counter-narratives.

'Illustrated by Sections, Views, Costumes, and Zoological Figures'

Although no known portrait of Sarah exists, the prodigious number of her scientific illustrations in print – clearly signed '*Sarah Bowdich del*' – total around one thousand black and white drawings (see the bold type in Appendix 1). Part of the scientific significance as well as enormity of Sarah's *The Fresh-Water Fishes of Great Britain* (see Chapter 2) was her superlative illustrations 'from the life' for it. Over the ten years of its publication, she hand-produced the exquisite watercolour plates with gold and silver foils for the some forty-four Fishes she described for the fifty subscription copies (2,200 illustrations in all). The lack of notice by art historians of the quantity and quality of Sarah's work as a foremost scientific illustrator (irrespective of sex) of natural history in the period[3] matches the critical silence by historians of (women's) science on her remarkable scientific publication record from 1825. Hers was the pen and pencil of both the text and the illustrations.

Hers was then clearly very much more than the 'secondary' role that has been ascribed to women in science of the period who were the expert illustrators of scientific works by their husbands or other close relatives (that Appendix 1A might imply).[4] Rather, the glaring lack of notice regarding Sarah's large corpus of signed scientific drawings derives from their conformity with the representational conventions and requirements for science illustration. Her scientific accuracy and acumen, alongside her technical expertise in ready reproduction of images through use of the latest techniques such as lithography, make them indistinguishable in quality from (black and white) images in science publications of the time produced by men. To address Sarah's major unheralded reputation for her doubly scientific pen and ink (in image and text) in this chapter is therefore also to showcase for the first time her even larger portfolio of scientific illustrations in black and white and in colour than are the tallies above. In so doing, the chapter also reopens to view the 'French' and 'British' scientific standards and models for illustration and visualisation of new science in various formats, as integral to and surpassed by Sarah's own.

The wording in the subheading above quotes the inside cover page of the *Excursions* (Figure 0.1 in the Introduction) to remake its point. Invisible behind the impersonal passive is the indisputable fact of Sarah's central importance as the principal scientific illustrator (and writer) of the *Excursions* in not one, but all of the pictorial genres listed here. More than forty figures for the plates in the *Excursions* bear her signature.[5] Sarah's expert illustrations therefore cannot be gendered by the 'technicality' or subject-matter of their scientific illustration sub-genres, or by secondary status as adjuncts to Edward's work, since she was the author(ity) of the scientific *text* describing (almost) all the images. To investigate Sarah's scientific drawings as more unusually integral to the *Excursions* in consequence – whether as carefully interleaved (in the one-volume *EM*) or as a separate Album (*EMFr*) – is then to reconsider them potentially as the *primary* medium for her display of her pioneering natural history work in its polymathic range. In other words, if Sarah's illustrations offered her a better showcase for her (scientific) authority than her words, because the former transcended what her text could (not) convey to her readers as unusually a woman author of science, how does larger notice of her figures also shape reinterpretation of her text? Given the relative paucity due to printing costs of colour illustration in scientific travel and natural history works in the 1820s,[6] the commissioning and inclusion of Sarah's several watercolour plates in the *Excursions,* alongside the majority in black and white, made it notable in Whittaker's (and Fournier's) lists. Does colour illustration augment, or more radically transform, the status and range of Sarah's pioneering scientific perspectives and expertise in the *Excursions*?

A direct response to these questions lies in the last colour plate that the reader encounters: Plate IX 'Costume of the Gambia' (Cover Figure). The most remarkable of the volume, it illustrates our book cover and the 'portrait' response opening the introduction. Among notable rarities in this plate was its production 'taken from the life' in the Gambia by a white woman *in situ* in the 1820s. Since *illustrative* educational and imperial purposes were central to nineteenth-century 'exotic' travel genres and geographies 'of strange lands' for their publishers, the fact that this plate was in colour is not merely decorative.[7] Rather, colour also made a difference to what Sarah could illustrate to further effect for her target readerships, precisely because this portrait

Cover Figure 'Costume of the Gambia'. Courtesy of the University of St Andrews Libraries and Museums, *Excursions in Madeira and Porto Santo*, rP702.M16B7.

shows her accompanying account of the Gambia (discussed in Chapter 4) in a more telling format. This Plate is bound in to face the opening page of the 'Narrative of the English Settlement on the River Gambia' (*EM*, 173), demarcated clearly as Sarah's sole-authored work. The reader in 1825 and today therefore encounters it strategically some thirty-five pages before the long textual description it depicts (*EM*, 208–209), because they have to learn about, and visualise for themselves in the intervening pages its River Gambia (plant) geography and larger cultural contexts.

But the caption for the Plate is also a model of minimal impersonal scientific fact, to identify the individual figures in an order from left to right – via the telling pointing finger of the first listed – that also potentially challenges that order: Plate IX: 'Costume of the Gambia. The figure dressed in blue is an Alcade, or Governor of a town, the woman with a parasol is a Senhara, or Mulatto. The figures passing at the back are, a travelling Moor with his bow and quiver, and his wife and child'. How should we view Sarah's schema, fields of vision and ideological perspectives in this image in 1820s and 2020s view? Her education of her readers concerning the Gambia could not but include contemporaneous questions, theories and hierarchies of race categorised as 'white', 'yellow', 'red' and 'black', as discussed in Chapter 5 (p. 114). Did her illustration with clear description (discussed in Chapter 4, p. 99) align with (scientific) recording and reproduction, or realign and even correct (Western scientific) colour bias, stereotyping and projections thanks to the eye-witness viewer-artist? Did 'Costume of the Gambia' provide Sarah with scientific artistic licence proactively to questioning her readers' cultural assumptions, that is 'coloured' points of view? Can colour plates of this kind therefore continue to speak today to 'colour' assumptions (and blindness) concerning women's contributions to early nineteenth-century 'anthropology' and 'ethnography' before these academic disciplines were founded? For example, the pointing finger, the gaze of the central figure and the focus on the 'travelling Moor' suggest alternative, parallel views – Arabic is read from right to left – that return the reader's gaze to the latter's wife (and child) as equally in the line of the pointing finger.

The execution of this plate in terms of line, perspective, detail and texture – not only the colours of the costumes relative to their wearers' ethnicities but also the colour palette and materiality of the environment in which they stand – also calls attention to the decentred/recentred status and perspectives of the people and place pictured here, as well as to the arrangement of the one in relation to the other. Sarah witnessed first- rather than second-hand the colours, light, sandy topography and accompanying *seasonal* vegetation represented here (see Chapter 4). This was is no schematic or imaginary landscape or, indeed, a reconstruction based on a sketch made first in the field (by her husband), which she then later coloured using colour keys (used by illustrators preparing technical drawings or lithographs).[8] While clearly outside the frame, the (white woman) artist's position participatively *in situ* but decentred, also makes Sarah's sparing use of whites intriguing and potentially strategic.

This colour Plate also has an interpretational companion in black and white (discussed below). Both confirm that Sarah regularly encountered the subjects of her study here, in Bathurst's (Banjul's) market, streets and new customs building, yet the 'town' (in the caption) and its colonial-commercial setting is markedly absent from Plate IX. It therefore concentrates on *in situ* (frieze-frame) documentation, reconstructing for the (white) viewer the *indigenous* (multi-ethnic) Gambian economy instead, to highlight the scientific accuracy of its variously interconnected geographical, botanical and ethnographical elements. For example, the backdrop wash is not only a foil-effect for the greater interest of the representative human subjects in the middle- and foreground but also depicts the quality of hazier morning light in the Gambia before the sun is strongest, making colours/shadows much more starkly defined. In the foreground, the shadows are among many touches of both scientific accuracy and artistic effect. The shadow of the principal 'governor' figure in blue thus falls on the central woman figured in yellow, the main subject of his pointing finger, and particularly her face by attention equally and comparatively with the face of the 'higher class' Gambian woman in the background. The latter is interestingly in our direct line of focus, by necessary positioning of the artist's easel. Is this background woman deliberately 'over-shadowed' as her own shadow merges with other longer shadows in the background? Or is she the necessary foil in form, skin colour and physiognomy to highlight the contrastingly paler coloration of the face and excessively clothed body of her Gambian counterpart? The green parasol, so like and unlike the palm fronds which also provide shelter from the sun for women such as this Moorish 'wife' and mother in the background, would 'explain' the paler complexion of the woman in yellow as more nearly the counterpart of white (British and French) female readers of the *Excursions* in 1825/1826. A racial 'hierarchy' is therefore not only suggested but also questioned. But the striking yellows and reds are complicated by the compositional balance of figures occupying higher and lower status – the representations of two women at the centre of the image – and of the two men in the left and right fields and frames of view. Indeed, the directive male and female gaze and stance of both the 'Alcade' in blue and the 'Senhara' in yellow look back to the presumed white reader-viewer,[9] as located 'behind' the position of the white woman artist drawing them. The viewer outside the frame is then the most ignored by the 'Moor' and his family, travelling left to right on their own business (beyond 1820s British colonial controls). The 'sticks' held by the two native women (in the centre of the frame) are then also suggestive of the representing 'sticks' – the brush/pencil of their woman artist – as items of greater potential might than the proverbial sword, or bow and arrows in this case carried by the male figures in the foreground, yet off right and left field. Plate IX is therefore concomitantly descriptive, illustrative, representative, educative and 'anthropological' for its day, and for modern reappraisal. Its positioning (in the text) also requires the reader to return to it and study it more closely again, upon reading its textual description. We have already quoted key paragraphs from it in the description of 'The mulatto women, who are mostly Joloffs from Goree' in Chapter 4 (p. 99). They are prefaced by

> [t]he superior classes of the Mandingoes, and the travelling Moors of the interior, frequently assume a turban, and this, added to their full and graceful pagnes, their red sandals, their elegantly shaped scimitars, and their light bows and arrows, gives them a very picturesque appearance. The older Alcades wear a large, pointed, grass hat, looking like a portion from the thatched roof of their huts while the younger chiefs have a white cap, beautifully embroidered with coloured cottons, in diamonds, stars, and other devices. The higher classes of the women generally wear a short shift, and two pagnes of equal size; their gold ornaments are numerous and massy, their ear-rings especially, which are often of such a weight, as to require a string passing over the head to support them, as they would otherwise tear the ears. Natives of all shades, and both countries, assume very dark blue for mourning, and lay aside their ornaments. (*EM*, 208–209)

It is therefore not the (white British woman) artist, but the perspectives of the 'Alcade' in blue (as representative of the 'superior classes of the Mandingoes'), the 'Moor' and the 'Senhara' in larger and longer River Gambian society that establish the *punctum* in this image – nicely represented in the 'place' of the clearly pointing finger – where its focus and 'colour' message lie. The viewer's particular attention is drawn literally and figuratively to the central women figures in this 'native economy' (discussed in Chapter 4), particularly the pale-complexioned Senhara in yellow. Her commanding gaze, ease and place compositionally and (inter-)culturally denote her position (message). Hers were the larger equalities and freedoms, including of movement, than her 'civilised' white British/French female viewer counterparts in the 1820s, because her self-determined economic status also defined her. The personal *wealth* of the woman in yellow-gold at the centre of this image is on full display in her clothing, its ornamentation and accoutrements (her jewellery, shoes and parasol). Thanks to the business of the 'higher class' Arab/'Moorish' family behind her, we also understand her wealth and position at the centre of trading interests along various routes to west and east of this frame. The female figure most in the shadows, and in the background, then provides wider lessons of this Plate. Despite sharing the same codes of respectable ethnic and Islamic 'civility' as the other figures portrayed here, the Gambian mother carrying her new-born child and her bundle of goods inhabits a more precarious and dangerous (marginal) place in Arab-Gambian trade networks due to frequent slave raids in the interior. Sarah (re)collects them in her Gift Book and 1835 *Stories* (discussed in Chapter 5). But the profile view of the Gambian mother, like the three-quarter view of the 'over-clothed' Senhara also permitted the white woman artist in their company her 'view'. Sarah's portrayals of them are her clear counter-models to illustrations by white male traveller artists of 'native women' depicted full frontal and semi-naked, so as to highlight (sexualised) parts of the female body for the (white) reader's gaze.[10] By framing instead only Gambian appraisals of its women, wives and mothers in the Alcade's pointed gaze, Sarah's remarkable depictions in 'Costume of the Gambia' encapsulate her pioneering ethnographical portraiture of 'native' women that critiques allegedly 'scientific' (European white male) drawings and 'figures'.

The art-historical and historical significance of Sarah's colour Plate IX is however also indicative of her wider contributions to scientific illustration in (British and French) history of early nineteenth-science. These contributions also emerge only when illustration rather than text is made the primary reference point for reinterpretation. Sarah's strategic uses of colour and black and white, and her composition of 'figures' (human and non-human in her Fishes in Chapters 1 and 2) also demonstrate her mastery of (French and British) scientific illustration conventions as part of her remarkable accomplishments in her learning and communication of natural history. The remainder of the chapter now presses harder on why contextualising Plate IX and Sarah's scientific illustration art-historically makes of her work an important benchmark for her times, and for twenty-first century criticism seeking out alternative, counter- or resistance narratives already present within imperial, colonial and *scientific* records. Sarah's rarity and contributions in these fields cannot be appraised by assuming a standard, impersonal, universal model for them, or by designating a gender, race or class to the paint brush or painter behind the science illustration here.

The pencil and paint brush of Plate XI are therefore more interesting and potentially unorthodox, when examined through the wider corpus and expertise of Sarah's palette and outputs, which strikingly not only visualise her extensive scientific but also technical learning. The *Excursions* was the first text for which she published signed watercolour drawings. Hitherto, she was the sole artist of the more than one thousand specialist black-and-white drawings illustrating the various compendia (in Appendix 1 section A) on the latest developments in French natural sciences for 'use by students and travellers' that the Bowdichs produced to fund their scientific expedition to Sierra Leone. The images in them were therefore as strategic for scientific identification purposes as the textual descriptions, despite the 'hierarchy' of the clearly numbered drawings as 'appendices' to these texts. The increasing scientific quantity of Sarah's signed drawings in these works chronologically is significant for measuring her output, and hence co-authorship. She made them in 1819–1823 (and during her regular pregnancies) as *identification* copies of key specimens in all classes – mammals, birds, reptiles and molluscs – from the Paris Menagerie if alive and, if dead, from the Muséum galleries, collections and laboratories. Sarah therefore executed them according to the scientific standards of the expert Chair in question, and under the mentorships of excellent and exacting scientific draftsmen in Cuvier, Humboldt and Lamarck for the many conchology figures noted in Appendix 1A.

But Sarah's black-and-white drawings were more than an exercise in accurate natural history illustration, recording both the latest scientific classification and figuration-reproduction standards set by the Paris Muséum laboratories for new specimens described in (international-)French scientific print. Her drawings exemplify the exercise of essential scientific field identification practice, in preparation for the Bowdichs' scientific mission to Sierra Leone. In hot climates, speed is of the essence to capture an accurate picture – the key identificatory characteristics – of a creature new to Western science, ahead of making more detailed written observation notes and preserving the actual specimen (in alcohol, prepared as a skin) for later shipment and inclusion in national museum collections. In some cases therefore, as in the anonymous *Taxidermy* (1820), Sarah also copied key illustrations from earlier authorities on how to prepare and stuff a bird for scientific collection. The original (French) labelling is in fact moved into the image – see Figure 7.1 – so that the 'how to' instructions were instantly 'readable' for Anglophone readerships. Moreover, this image testifies in its signature – no 'S' qualifies '*Bowdich del.*' – to the fact that conventions of accuracy governing scientific illustrations and anatomical drawings should make it impossible to evaluate a technical drawing by the name (or gender) of the artist, and hence reattribute secondary quality were 'draftswoman-ship' to be discovered. It was not until the expanded sixth reprint of *Taxidermy* in 1843 with a preface for the first time that the name and hence gender of the author and illustrator was revealed.[11] But the corollary in the case of such 'anonymous' works for their readers today is not automatically to assume that the holder of the scientific paintbrush and pen in this period was a (white, scientifically educated) man. Initials and partial signatures ('Mlle' in French) on plates and illustrations also reveal hitherto unacknowledged female, and non-European scientific illustrators.[12]

By 1825, Sarah's demonstrable expertise in technical scientific drawing in black-and-white charcoals and inks encompassed not only objects of natural history and

Figure 7.1 'Model Taxidermy'. Courtesy of the University of St Andrews Libraries and Museums, *Taxidermy*, sQL63.B6.

anatomy, but also panoramic 'views' in the chapter subheading above. In the *Excursions*, she figured the new British settlement of Bathurst in the triple foldout Plate X – see Figure 7.2 – and 'Bakkow', Plate XI. The group in Plate X – in the same front left-of-centre position facing her reader-viewers as Plate IX – not only guides their gaze towards the 'geographies' of the town. The group are equally 'flagged' from behind, by the literal colony harbour flag and vessel sails, to foreground this long-vibrant trading port, despite its changing colonial resettlement. Once more the 'black and whiter' face of the 'Senhara' amidst her entourage features the most important woman of this group. The Plate thus accurately portrays the longer-standing *indigenous* citizens of this River Gambia settlement, despite its 'Bathurst' name and panorama of new British colonial buildings. British Government House (still the residence of the President of the Gambia) lies at the end of the second, unpeopled, track on the left; the new British military barracks and hospital in 1823 is in the right foreground, but not one British white figure, male or female, can be found in this Plate. Sarah's bird's-eye view (from a balcony or higher ground?) only makes her illustration more strikingly 'women-aware', but without self-revelation of her unusual place in the 'picture'. The human figures not only lend a strongly visual counter-hierarchy to the colonial architecture that seemingly dominates the town, but also lend this Plate a significant longer historical, and ethnic scale. Although apparently off-centre, their feet picture its heart, the markets and trading area of the port, from which the party are walking.

The invisible, but orchestrating, white British woman illustrator of this scene recording its 'slice' of contemporary Gambian history here (see Chapter 4) did so knowingly, even politically, in her choice of subject(s) and perspectives on them in this 'snapshot'. Their visual point in plain view could not more strongly come *home* to (enlightened) British readerships that unreflecting white civilising missions were off-centred (unless) reconfigured in larger and longer multi-ethnic scale. To hold serious abolitionist views

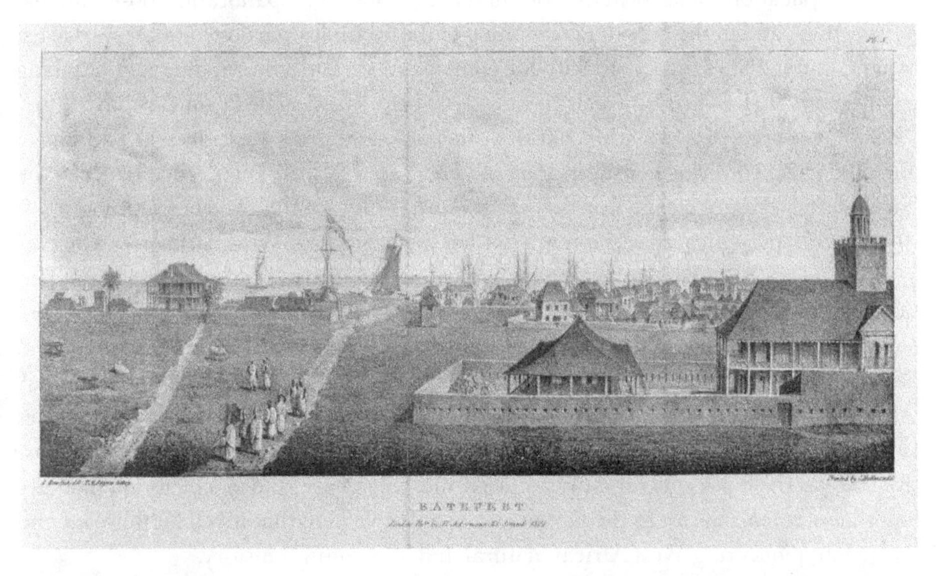

Figure 7.2 'Bathurst'. Courtesy of Hartley Library Special Collections, University of Southampton.

in Britain in 1825 might then also proactively require longer-term colonial (self-)erasure from the picture.[13] If such a reading seems too twenty-first century, this importantly tripartite, foldout image unfolds Sarah's expertly (c)overt intermediations in the Plate's illustrative message in her reader's *hands*: the two right-hand sections specifically depicting the new British colonial buildings 'fold over' onto the settlement's 'enduring' economy and women within it. Even the most liberal-minded of her British readers at home and overseas needed re-educating with regard to West Africa's interconnected, multi-ethnic and Muslim histories (as Chapter 4 discussed).

Sarah's scientific attention to social groups in the black-and-white 'view' of Bathurst and in her colour Plate IX makes her work in early 'anthropological' drawing groundbreaking in its 'portraits' and 'landscapes' of West African societies, because made *in situ* by an unusually non-conformist (white woman) artist. Her attention to native costume and setting was already practised, however, so that her Plates of the Gambia made in 1823–1824 especially in watercolour also require new comparative view and reassessment. The 'foldout' format and perspectives of Plate X immediately reminded me of the widely-acclaimed panorama depiction of 'The First Day of the Yam Custom', signed 'T. Edward Bowdich' as potentially revelatory of its now highly probable *double* paintbrush. His magnificent triple fold-out colour plate (see Figure 7.3) in the *Mission from Cape Coast Castle* offered in 1819 a precious illustration of the foremost vibrant civilisation that was 'Ashantee' (Ghana) in the region. If the image is unforgettable, it also remains widely cited and reproduced in scholarship on Asante. It pays to look and look again at it.

First is the nature and composition of the trees in the middle of the image, echoing the canopies and flags grouping and demarcating the different representative figures. Second is the palette of colours. Third are the costumes. Any one, let alone all three, is suggestive of an authenticating and direct match with Sarah's work in the colour Plate IX above. Further corroboration as the 'double' paintbrush in the 'Yam Custom' lies in Sarah's replications of its formats, both in the triple fold-out, 'panoramic' observational perspectives, and in the outsider positioning of the invisible, 'participant observer' artist of her Gambia Plates, in accord with the complex social and cultural interactions taking place before 'T. Edward Bowdich' as now (also) before *her* for the 'Yam Custom'.

But cross-comparison within Sarah's other sole-authored works in text and image after 1825 also confirms her illustration of 'The Yam Festival'. Interleaved between the 'Stories' and 'Fragments' in *Stories of Strange Lands* (1835) are the 'sketches' (ethnographical scientific drawings) that she made on her *first* voyage to West Africa (see Chapter 5). None of the six plates in the *Stories* indicates their artist, only their zincographer, 'A. Pickan', and printer, 'W. Day'. However, Sarah's textual 'Description of the Plates' concluding the volume (*SSL*, 363–66) correlates in style exactly with her caption and description of 'Costume of the Gambia' above. The final words describing the 'sketches', 'All these plates were taken from drawings, or memoranda, made in the different places to which they refer' (*SSL*, 366), are indeed conclusive evidence of her pencil. The passive voice as the mask concomitantly of scientific 'objectivity' and of a pictorial 'modesty topos' now reveals Sarah as the deft holder of a double pen, that inked her in word and image into pioneering West African natural and intercultural history.

Figure 7.3 'The First Day of the Yam Custom'. Courtesy of Hartley Library Special Collections, University of Southampton.

It is then to the canopies and parasols of Sarah's West African drawings that modern viewers now need to look again, to discover her signature pen. Plate I (reproduced in Figure 5.1, Chapter 5), tucked into the 'Notes' to 'Adumissa' in the 1835 *Stories of Strange Lands*, describes 'the end of a funeral procession'. But Sarah flagged that the 'whole is taken from a large original drawing, the sketches for which were made in Fantee.' Plate I bears the most striking similarities to the 'The First Day of the Yam Custom', particularly to the circular grouping (with respective canopies) in the centre to right panel of its panorama. Sarah's own words in print in 1835 retrospectively confirm that 'S Bowdich *del*' should rightly be restored to both the iconic 'Yam Custom' Plate in the *Mission to Ashantee*, and to all six Plates for the *Stories* in 1835.

Figure 7.4 'Plate V'. Courtesy of the University of St Andrews Libraries and Museums, *Stories of Strange Lands*, sPR4161.B374.

The 'Ashantee' Plates do even more, however. They further confirm Sarah's astute and non-conformist representations in 'Costume of the Gambia' of counter-colonial points of view by dint of her removal of 'white' including 'scientific' reference points. Plate V (*SSL*, 290) in 1835 – see Figure 7.4 – is then also strikingly *déjà vu* by comparison with 'Costume of the Gambia' in 1825 or, more importantly, is pre-figurative of it. Figure 7.4 'replicates' the same group, but with two strategically placed Fantee additions that 'Costume of the Gambia' then removed because outside its multi-ethnic frames. Most obvious is not only the central, male, plumed figure but also the 'Fantee mulatto woman' on the very left:

> This group is intended to represent some of the costumes spoken of in the work. To the left is a *Fantee mulatto woman*; to her left is an Alcade, or Governor of a town in Mandingoe. *An Ashantee captain stands in the centre, with his blunderbuss, sword, bow, and knives; his jacket covered with charmed scraps of the Koran sewn up in silk cases; his spear, his eagles plumes, and gilded ram's horns, his charmed cows, horses, and leopard's tails, and his large boots, with bells around the legs.* To the right of this figure is a Goree mulatto woman, and a travelling Moor, accompanied by his wife and child in the distance. (*SSL*, 366, emphasis added)

Sarah's Plate V in black-and-white first made in 1817 also demonstrates her artful reconfiguration of it in watercolour for 'Costume of the Gambia' in 1824 as entirely consistent with the other-cultural 'scientific' attention, observational accuracy and distinguishing detail of her unusual pencil and pen. The 'Senhara' as newly drawn and contextualised in 1824 within her specific economic, social and geographical networks in the Gambia now brings to her Plate in the *Excursions* above understanding of her much larger trans-regional Fantee economic positioning and perspectives. As a 'Goree mulatto woman', she is not to be confused with her counterpart, the 'Fantee mulatto woman' (also in Plate V, first left) without place in Gambian society, but recorded nonetheless for history in Sarah's 'Notes' because of her clearly large 'appendage' (*SSL*, 331) that we noted in Sarah's 'Note' on it in Chapter 5 (p. 127). Sarah's 'views' of indigenous women in the *Excursions* were no accidental or 'proto-feminist' portrayals, but consistent instead with her comparative 'ethnographical' observations enjoyed thanks to two 'Voyages to Africa'. Her eye-witness drawings of more than she might tell could record the respectively important places of women within the economies and cultures of different West African 'scenes'. But Sarah's drawings also testify to her own positioning as an unusual white woman in their midst as an artist with more accurate scientific view. By comparison with established illustration *practices* and conventions for 'Travels' to Africa, hers – in the pointing finger of her 'Alcade' – offered no racialised or sexualised distorting mirrors of the Western (male) scientific gaze.

'Illustrated by [...] Zoological Figures'

Sarah's exceptional watercolour plates in 1825 of human and 'Humboldtian' geological subjects in the *Excursions*,[14] pale before her extensive corpus of even rarer watercolour paintings. The 44 plates in her magisterial *The Fresh-Water Fishes of Great Britain* (1828–1838) remain 'second to none' as discussed in Chapter 2. To produce them, her work in the heat and different colour palette of West Africa amply prepared her task in Britain, to execute specimen drawings with keen precision, skill and acute eye for distinguishing detail, colour and form. The enormity of Sarah's ten-year undertaking therefore marked the watershed in 1839 of her energies for natural history work thereafter, concentrating on textual, not visual productions of her pen. Indeed, her established reputation with publishers as a writer of natural history meant that after 1840 they paid other well-known artists, such as John Skinner Prout (1805–1876), to illustrate her work. The lessons behind Sarah Bowdich's 'Costume of the Gambia' as a significant scientific illustration for itself, and as indicative of her much larger palette and portfolio of black and white 'sketches' to prepare it, then give pause for further scrutiny of her 'leanest' period of her published corpus between 1835 and 1844. It has now disclosed her signature on an additional six (black and white) drawings hitherto not attributed to her. The impersonal formulation in the subtitles above for non-attribution of the artist (as in fact also Sarah) therefore invites sharper awareness and scrutiny of 'anonymous' illustration of her works after 1840 that may also have been produced by her pen.

The lessons of Sarah's *Elements of Natural History* in 1844 heralded on its inside title page for the 'Use of Schools and Young Persons' are the subject of Chapter 8. More important for this chapter is information on the inside title page that Longman commissioned accompanying figures for this first edition, 'illustrated with engravings on wood'. The 55 drawings constituted substantially more than the three or four that Longman commissioned respectively for Sarah's *African Wanderers* (1847) and *Adventures in Australia* (1851). Although none in the *Elements* is a full-page illustration (unlike those for her later texts), several are half-page drawings: two 'Quadrumana' (*ENH*, 28 and 30), a Vampire (*ENH*, 49), a Barn Owl (*ENH*, 168) and a Capercaillie (*ENH*, 231). The opening illustration (*ENH*, 17) is for the description of the (three) 'Races of Mankind', figured by their different skull types. There is only one other anatomical drawing, of the named parts of the head of bony Fishes (*ENH*, 331) – see Figure 7.5 – although the Elephant (*ENH*, 123) is illustrated by the differential teeth ('grinders') of the African and Indian species. There are 21 drawings of different mammals and 16 for birds, although the 'Caprimulgus' is figured for the detail of its 'large eyes' (*ENH*, 186), and the Weaver (*ENH*, 195) for its nest. Of reptiles, there is a Lizard (*ENH*, 292), a Rattlesnake (*ENH*, 312), a Sea-Turtle (*ENH*, 319) and a Frog (*ENH*, 324). There are seven Fishes from different parts of the globe, and one Selacian (*ENH*, 461). For ease of user reference, all 55 illustrations appear in the relevant paragraph of their textual description, often framed wholly or in part by it. No signature indicates the provenance of any of the illustrations, but those that represent an animal in its entirety share a similar natural history drawing 'style', although arguably two: those without and those with sketched in background details of habitat accord with the main scientific illustration conventions for natural history of the period.

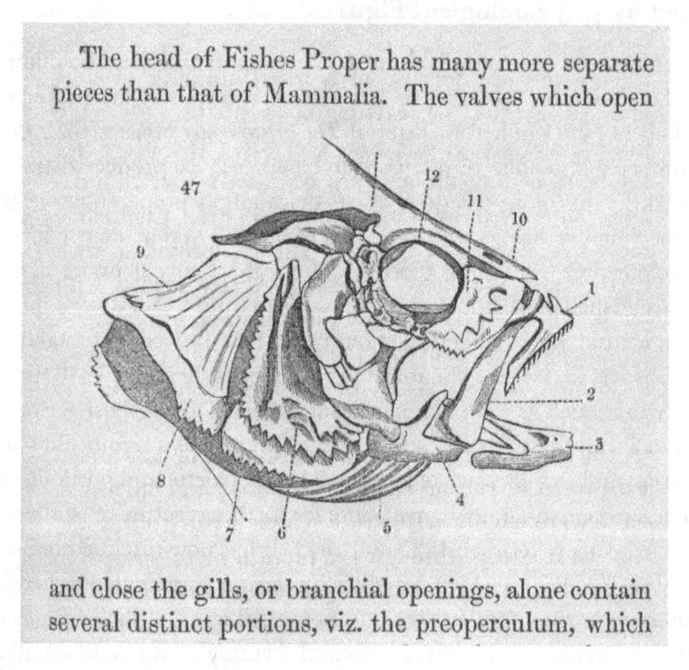
The head of Fishes Proper has many more separate pieces than that of Mammalia. The valves which open

and close the gills, or branchial openings, alone contain several distinct portions, viz. the preoperculum, which

Figure 7.5 'The Perch'. Courtesy of the University of St Andrews Libraries and Museums, *Elements of Natural History*, sQL48.B6.

For readers familiar with French natural history and its major reference books and illustrations of the period, including Cuvier's classifications, some of the drawings are instantly recognisable as deriving from Buffon – such as the rhinoceros (*ENH*, 127) – or the image (by Huët) of the arrival in 1825 of the Giraffe at the Jardin des Plantes (*ENH*, 138), which the accompanying text in the *Elements* recounts. The opening human classification is a copy of a drawing from Cuvier's *Règne animal*, as are the elephants' teeth. The anatomy drawing of the bony Fish (the Perch) is also Cuvier's, opening the second volume of his *Histoire naturelle des poissons.* The one illustrator with direct access to all these originals, who also had the proven expertise in making expert copies of science illustrations (as the taxidermy plate above), was of course Sarah herself.

The further evidence that she was not only the named author, but also the unnamed illustrator of *The Elements of Natural History* is found in her many black-and-white illustrations for *An Analysis of the Natural Classifications of Mammalia. For the Use of Students and Travellers* and *An Introduction to the Ornithology of Cuvier. For the Use of Students and Travellers*, both published in 1821 under T. Edward Bowdich's name. Sarah, however, completed all the accompanying drawings taken from Paris Muséum specimens, the many plates in both volumes clearly signed 'S. Bowdich Lithog'. She had *already* figured the (three) 'Races of Mankind' (*ENH*, 17) for 'Plate 1' in the *Analysis of the Natural Classifications of Mammalia* as well as the 'grinders' of both the Asiatic and African Elephant in its 'Plate IX'. What is notable about both her reiterations is that Sarah does not reproduce an *exact copy* from the original plate of 1821 for the illustrations in *The Elements.* The crania profiles are inverted left to right: the grinders turned from horizontal (Plate IX) to vertical (possibly to save space in the smaller figure in the *Elements*). Several other drawings are more clearly 'reproductions' of images from the original plates of 1821. For example, the 'Aye-Aye' illustrated only by its front half, to show its 'long, slender, middle toe [by which] it conveys food to its mouth' (*ENH*, 104), is figured in exact mirror opposite in Plate XIV. The same Plate also figures the 'Ornithorynchus' (Duck-Billed Platypus) in unchanged view in 1844 (*ENH*, 118), except that it has newly acquired a schematic habitat, see Figure 7.6. Other drawings endorse the certain provenance, yet differentiating representation of the two versions. The half-page drawing of the Vampire (*ENH*, 49) showing the body from the front with outstretched legs and wings with distinctive head/ears in profile, was first classified by Étienne Geoffroy Saint-Hilaire in 1810 as the *Phyllostoma rotundum*. The same representation, but of the insectivorous *Molossus ater* (The Black Mastiff Bat), is found in Plate VII, but Plate VIII then provides much more detailed drawings of the distinctive heads and ears of various species of bat, including Fig. 3 [Head] of a *Phyllostoma.* (Ph. Elongatum, *Geoff.*). The qualification as 'Elongatum' makes this the Lesser Spear-Nosed Bat not the Vampire, but Sarah's knowledge of Geoffroy Saint-Hilaire's classifications – these are all South-American bat species – enabled her to reproduce a scientifically accurate image in 1844 of the Vampire.

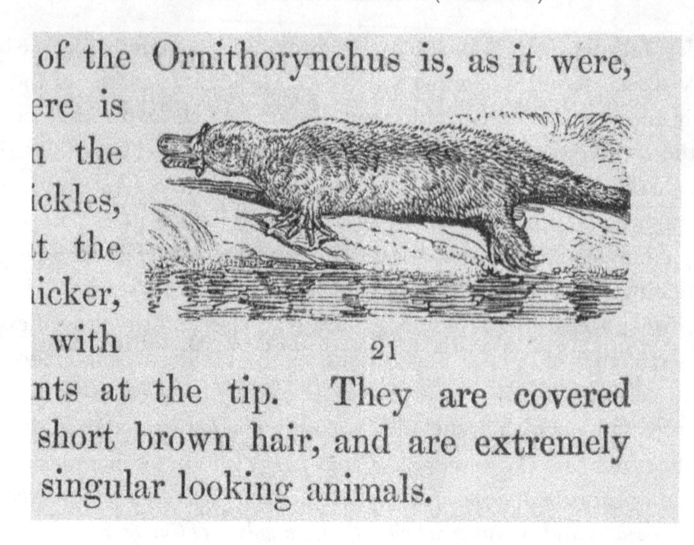

of the Ornithorynchus is, as it were,
ere is
n the
ickles,
t the
icker,
with
21
nts at the tip. They are covered
short brown hair, and are extremely
singular looking animals.

Figure 7.6 'Ornithoryncus'. Courtesy of the University of St Andrews Libraries and Museums, *Elements of Natural History*, sQL48.B6.

The Bowdichs' *An Introduction to the Ornithology of Cuvier* (1821) was much more technical in being a portable compendium for identification purposes of 'Birds of the World' in some eighty pages. The very short descriptions in many cases therefore included an additional numbered reference, where a distinctive detail in the text was further illustrated by a figure. The 261 illustrations in numerical order that Sarah prepared are all 'close-ups' of beaks/heads or feet/claws, not illustrations of the whole bird. In the *Elements* the opposite is the case, with the exception of the profile head and claws of the 'Caprimulgus' (Goat-Sucker). However, Sarah had already drawn this head/beak in the *Ornithology* (Fig. 77) in mirror-image profile, but not the feet. All the other birds she chose to illustrate in complete form in 1844 turn out to be those that she had already figured in 1821 (among many others) by beak, claw and sometimes tail feathers, for example for the 'Lyre Tail' Fig. 109 (*ENH*, 181). The closest 'match' is the distinctive pointed beak, crested head and long neck of the Heron (*IOC*, Fig. 195/*ENH*, 246) making it difficult not to attribute all the bird drawings in the 1821 and 1844 texts to the same illustrator. Here was Sarah's further massive illustration undertaking immediately upon finishing her *Fresh-Water Fishes*, to prepare all 55 woodcut illustrations for the *Elements* alongside writing the text. She could do so in the time because she already had her own published technical studies of 1821 to work from, including the all-important scientific details and distinctions for the species covered. Her new audiences in 1844 therefore view the more familiar, accessible and accepted form of the whole bird as 'taken from the life' in the respective illustration. Although they may not have known it, they additionally beheld a set of scientific drawings taken from specimens in the rich bird collections at the Paris Muséum and its Menagerie, as well as from other specialist works of ornithology. Much more research therefore needs to be done to check the many sources that Sarah and Edward used to compile their *Ornithology* (referenced in its two-page bibliography), as well as to draw out further correlations between Sarah's drawings there and in the *Elements* regarding her specialist knowledge in this branch of natural history. If we see it from the author's eye-view as a very knowledgeable bird illustrator, the *Elements* potentially provided Sarah with the outlet that she was denied in 1821 (for reasons of its specific 'compendium' format) to depict each bird specimen complete (as if 'from the life'). Indeed, Sarah may already have made drawings of the entire birds for the Bowdich *Ornithology* alongside the 'anatomical' studies of their beaks/heads and feet. When their artist emerges from anonymity in the *Elements*, however, what is clear is that its new images were not copies made by a technical illustrator with no knowledge of his (or her) scientific subjects, nor mere reproductions of figures that Sarah had made earlier. Rather, their informed production accorded with the highest scientific standards of both the Muséum and the overseas field as the 'Cuvierian' and 'Humboldtian' benchmarks for works that would inform and mould the scientific education of next-generation museum and field natural sciences. The *Elements* in text, but more importantly image, offered representations of observable natural history (in the wild, in zoos, in museum galleries) with scientific clarity, precision, interest and understanding. If these were fully accessible to 'Schools and Young Persons', they eminently informed general publics that included young girls, women and mothers.

The selection, yet breadth of appraisal of natural history in the *Elements* therefore sat firmly upon its authoritative illustrations and textual descriptions. Because Sarah's new illustrations as 'Mrs R Lee' remade, restored and revealed those she had undertaken as 'Mrs T. Edward Bowdich', however, they also provided her with new outlets and possibilities in the 1840s for her own independent and ongoing scientific learning. New species had constantly swelled the natural history knowledge encapsulated in the Bowdichs' works of the 1820s, especially from the interiors of continents such as Africa and Australia. By 1844, therefore, Sarah had the opportunity to expand her coverage and selection for illustration, to include those unknown to the Paris Muséum collections in 1821. For example, *An Analysis of the Natural Classifications of Mammalia* covered various creatures only found in Australia such as the Kangaroo Rat (48), the Kangaroo (49), the Koala (49) and the Wombat (49), known from the Voyages of Cook and 'Peron' (*sic*), and the work of their artist illustrators. Sarah not only added a very short description but also a new drawing of the 'Jerboa' in 1844 (*ENH*, 109) that bore striking resemblances to the similar images she would have seen in Paris in 1821 and when consulting works in the 1840s to prepare her *Adventures in Australia* (1851).

The illustrations of Fishes in Sarah's *Elements of Natural History* are then of especial significance, given her particular expertise and publications in ichthyology that Chapters 1 and 2 of this book foreground. Significantly, she chose no illustration from either the *Excursions* or her *Fresh-Water Fishes of Great Britain*. The woodcut drawings, numbers 48–55 in the *Elements*, therefore offered her a new medium for her proven expertise in scientific fish drawings in different modes of representation. All her watercolours had been to scale or life-size for example, whereas the very much smaller format for each of her fish illustrations in 1844 presented the challenge of scale details in both meanings of the term; the proportions and indicative markings that can only be represented in careful shadings of black and white. The Flying Fish (number 49) is alone framed against the sea and shown with others 'flying' out of the water, to match the accompanying description:

> It looks very pretty as it issues from the water in multitudes; a rushing noise is heard, and numbers of blue and silver creatures are seen in the air, which dip again into the sea, only to come forth afresh, and but too often fall a prey to winged destroyers. They are attracted by light, and lanthorns (*sic*) were hung at night on the rigging of a vessel in which the author once sailed, and by this means a supply was often secured for breakfast. (*ENH*, 437)

The perspective of personal testimony here was taken 'from the life' ('Fragment II, The Voyage Out', 1835). It thereby further corroborates the illustrator of the *Elements* and her long qualifications as an expert on Fishes thanks to her collaborations on Cuvier's *Histoire naturelle des poissons*, which Valenciennes was still completing in 1844. Indeed, in her description of the 'Silurus' (Catfish) which she illustrated, readers were regaled with its more folkloric reputation: 'M. Valenciennes was told, when travelling with Baron Humboldt in Prussia, that a whole infant had been found in its stomach' (*ENH*, 415).

By comparison with the 'foreign' and exotic New Holland 'Hippocampi', with its 'number of leaf-like appendages on different parts of the body' (*ENH*, 452) which Sarah figured in number 51, almost more intriguing is the 'Plaice' preceding it. As among the best-known fish for her British readers it would not warrant illustration, especially since the description merely adds: 'Plaice are best at the end of May, are northern fishes, and sometimes found in incredible numbers' (*ENH*, 442). The beautifully drawn spiny dorsal fins, spots, scales, lateral line and flatfish eye positioning provide much more than expert *illustrative* detail in figure 50, however, and by comparison with any of the other seven fishes represented. What is a commonplace (flat)fish lies in the discovery of its uncommon 'Plaice' for the keen observer of this image, as Figure 7.7 can now reveal and magnify for the first time. To the left of the pelvic fins below the widest part of the operculum are the tiniest letters, 'S.L'. There can now be no shade of doubt. The

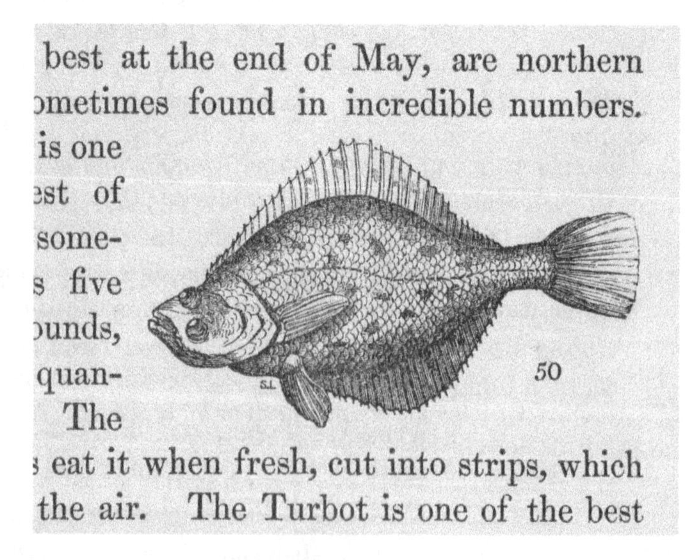
best at the end of May, are northern
ometimes found in incredible numbers.
is one
est of
some-
s five
ounds,
quan-
The
eat it when fresh, cut into strips, which
the air. The Turbot is one of the best

Figure 7.7 'Number 50, The Plaice'. Courtesy of the University of St Andrews Libraries and Museums, *Elements of Natural History*, sQL48.B6.

anonymous artist deftly inserted her signature, placed in her 'Plaice', to indicate the same provenance and place of its illustrator for the 54 other 'illustrations on wood.'

Unlike her bird illustrations for the *Elements*, visualised as a whole from her earlier drawings of their beaks or feet, 'SL' is without a portrait to view. This chapter, however, has made a first and fuller representation of Sarah behind her known artworks by adding another 61 black and white, and one further major colour plate to the list of her expert productions by 1844. By foregrounding her powers of scientific illustration as integral to her illustrative powers in natural history, this chapter has also explored how she could *show* more than she could tell. 'SL' tucked under a fin hidden in plain sight on a scientific plate therefore epitomises Sarah's awareness and negotiations of the problematic authority and status of an expert woman illustrator-writer of natural history in 1844. But this chapter also probed her deft representation and illustration of 'natural productions' as more expert than they first appeared. Not only do her West African drawings look much less 'black and white'. Those that she also coloured recorded what she saw before her more inclusively, thanks to her unusual eye pioneering 'fold-out' perspectives in the field. This chapter can now unequivocally disclose Sarah's larger place and authority as a foremost scientific illustrator (irrespective of sex) for the period as also measured by the iconic 'Yam Custom'. Her work then offers the largest lesson of all for twenty-first century re-readings of scientific travel and natural history of the first half of the nineteenth century. 'Anonymous' work of high scientific quality should not automatically be assumed to be by a man. Sarah's double pen also proved in this chapter that scientific art can convey pioneering and exemplary counter-cultural perspectives and narratives hidden in plain sight in (non)conformist technical practices.

Notes

1 deB. Beaver, 'Lee [*née* Wallis]'.

2 Lee, *Memoirs of Baron Cuvier*, 60.

3 See Orr, 'Fish' and Chapter 2 above as the exception. Women's expertise in botanical drawing lies outside the discussion in this chapter, but its assumed (genteel) occupation is among various critical blinkers for study of the extensive contributions by women illustrators to scientific knowledge. See variously and indicatively Natalie Rowe, 'Sweet Ordering, Arrangement and Decision: The Domestic Nature of Science Illustration by Women in the Eighteenth and Nineteenth Centuries', *Journal of Illustration* 1, no. 2 (Oct. 2014): 211–31; Horst Bredekamp, Vera Dünkel and Birgit Schneider, eds., *The Technical Image: A History of Styles in Scientific Imagery*. Chicago: University of Chicago Press, 2015; Ann B. Shteir and Bernard Lightman, eds., *Figuring it Out: Science, Gender and Visual Culture*. Hanover and London: University Press of New England, 2006; Brian J. Ford, *Images of Science: A History of Science Illustration*. London: British Library Publications, 1992 and Barbara Maria Stafford, *Voyage into Substance: Art, Science, Nature and the Illustrated Travel Account*. Cambridge, MA and London: MIT, 1994.

4 The work of Elizabeth Gould (1804–1841) in John Gould's *Birds of Australia* is an illustrative case. See Melissa Ashley, 'Elizabeth Gould, Zoological Artist 1840-1848: Unsettling Critical Depictions of John Gould's "Laborious Assistant" and "Devoted Wife"', *Hecate* 39, nos. 1–2 (2014): 101–22. See also the work of Eliza Gordon Cumming in Louis Agassiz's *Monographie des poissons fossiles du vieux grès rouge ou système dévonien* as discussed by Orr, 'Collecting Women'.

5 Only two black-and-white drawings concerning the geology of Porto Santo are attributable to Edward. See Orr, 'New Observations'.

6 For the related field of mineralogy imprints see Brian Dolan, 'Pedagogy through Print: James Sowerby, John Mawe and the Problem of Colour in Early Nineteenth-Century Natural History Illustration', *The British Journal for the History of Science* 31, no. 3 (Sep. 1998): 275–304.
7 See T. Edward Bowdich's *Mission from Cape Coast Castle* (1819) and William Hutton's *A Voyage to Africa: Including a Narrative of an Embassy to One of the Interior Kingdoms, in the Year 1820* … London: Longman, Hurst, Rees, Orme and Brown, 1821 as its continuance and 'reply' (citing 'Bowdich' for example 324). Both works were also illustrated by their authors.
8 Comparison with the very different style (and use of shading) in the two plates by T. Edward Bowdich, and with the opening colour plate of the volume, 'Plate I, Costume of Madeira' also clearly by '*S. Bowdich del.*' instantly eliminate both possibilities.
9 The terms 'Alcade' and 'Senhara' are Portuguese creole words indicating previous colonial occupation and its longer legacies in the region.
10 See for example the plates in William Hutton's *A Voyage to Africa*, but especially 'A Mulatto Woman of the Gold Coast' (facing 113) and 'Mode of Travelling in Africa' (facing 177). Sarah's Plate neither exoticised nor sexualised her subjects. Hers was also rarer in depicting children.
11 See Orr, 'The Stuff of Translation'.
12 The Paris Muséum had workrooms and laboratories dedicated to scientific drawing and to training scientific explorers and *aides naturalistes* in these arts, where a small number of French women also worked, on the illustrations for Cuvier's ornithology for example.
13 These include Quaker and Wesleyan Mission in the Gambia discussed in Chapter 4, and as represented by Hannah Kilham and John Morgan in 1824.
14 See Orr, 'New Observations' for my discussion of Sarah's geological drawings according to 'Humboldian' models in her use of a human figure to serve as a measure of the scale of the different strata.

Chapter Eight

TEXTBOOK NATURAL HISTORY: *ELEMENTS OF NATURAL HISTORY* (1844; 1850) AND NEW PARADIGMS FOR SCIENCE PEDAGOGY

As noted in the introduction (p.15), Sarah produced her sole-authored publications in natural history from 1824 at several Camden addresses in close proximity. Her consistent geographical and intellectual vicinity also to London natural science collections and societies was among the ironies of women's lack of presence within them. As explored in the earlier chapters, the scientific authority, resourcing and substance of Sarah's work was manifestly empowered by her French mentors and Paris Muséum connections. Yet to situate the story of her publications from the 1840s onwards always outside the activity of London science creates a history 'from below', or from 'outside' that occludes where and how Sarah's work maintained and developed its pioneering edge concomitantly within British and London scientific contexts. Her *Memoirs of Baron Cuvier* (1833) was indicative of her acknowledged place between its bars. Her *Fresh Water Fishes of Great Britain* also acknowledged in William Yarrell (1784–1856) an expert British, London-based contemporary in natural history similarly working outside 'gentleman science'. This chapter therefore turns to the multiple evidence for Sarah's expert knowledge of, and contributions to, specifically London-based British natural history as the locus of her more interestingly concentric command of its fields.

Sarah's single-sentence dedication of the first edition of her *Elements of Natural History* to 'Richard Owen Esq. F. R. S. Professor of Anatomy at the Royal College of Surgeons' is therefore key to resituating her larger place, and hitherto invisible contributions, at the epicentres of British and London natural science pursuit. By 1844, Owen's reputation and publications as the 'English Cuvier' were clearly established through his major publication on 'the Pearly Nautilus' (1832), and his 1837 Hunterian annual lecture series for informed general publics on comparative anatomy at the Royal College of Surgeons.[1] In consequence, Sarah's concise dedication calls for a major rereading:

> To be allowed to send forth the following work into the world with the name of Professor Owen at its head, *I feel to be the highest sanction which my labours can receive*; but I desire that a dedication of it to him, should be further considered as *a public acknowledgement of the kindness* which I have received at his hands, *and of the respect and friendship which I bear towards him in his private, as well as his scientific life*. S. L. (emphasis added)

The public and private 'Professor Owen' here strikingly match the public and private 'Baron Cuvier' of Sarah's *Memoirs* (discussed in Chapter 3), clearly cited on the inside title page of her *Elements* (*ENH*) in Figure 8.1. Her scientific respect and friendship, reciprocated in the 'kindness' professionally and personally by Cuvier and now Owen for her, were without distinction. In one deferential and referential stroke, Sarah's dedication therefore also offers rare insight into Owen's inner, as well as public scientific circle(s). Given their extensive overlapping interests in Cuvierian comparative anatomy, Sarah's likely attendance at Owen's 1837 Royal College of Surgeons' public lectures indicates major conduits for exchange of the latest (London) scientific ideas, information

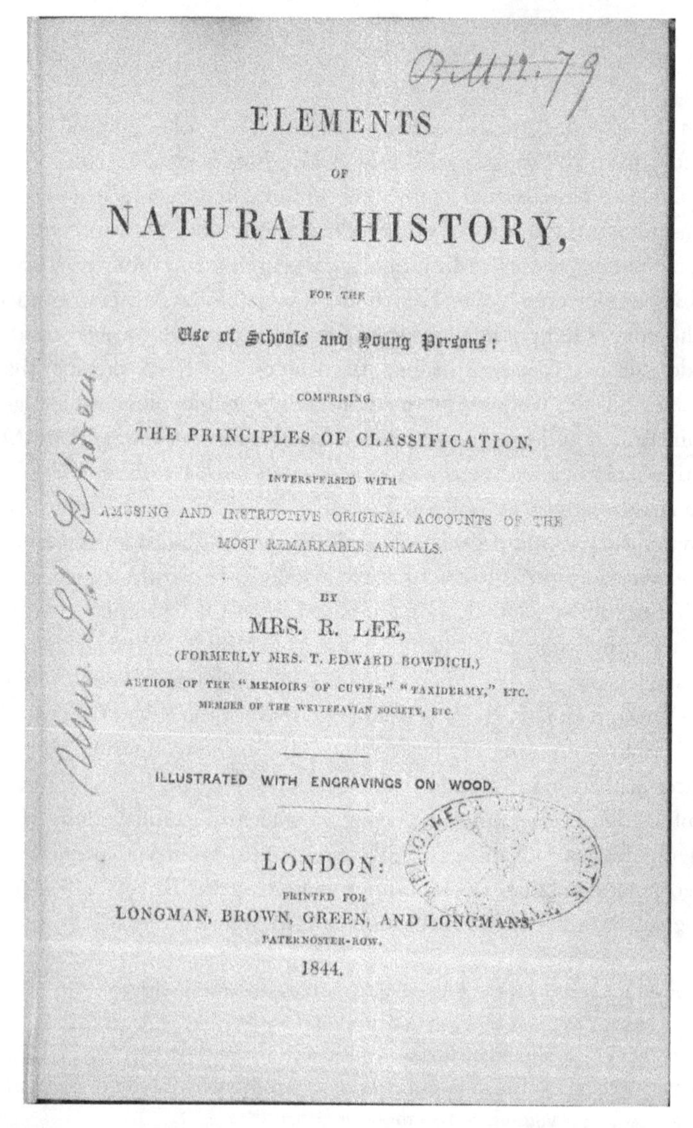

ELEMENTS

OF

NATURAL HISTORY,

FOR THE

Use of Schools and Young Persons:

COMPRISING

THE PRINCIPLES OF CLASSIFICATION,

INTERSPERSED WITH

AMUSING AND INSTRUCTIVE ORIGINAL ACCOUNTS OF THE MOST REMARKABLE ANIMALS.

BY

MRS. R. LEE,

(FORMERLY MRS. T. EDWARD BOWDICH,)

AUTHOR OF THE "MEMOIRS OF CUVIER," "TAXIDERMY," ETC.

MEMBER OF THE WETTERAVIAN SOCIETY, ETC.

ILLUSTRATED WITH ENGRAVINGS ON WOOD.

LONDON:

PRINTED FOR

LONGMAN, BROWN, GREEN, AND LONGMANS,

PATERNOSTER-ROW.

1844.

Figure 8.1 'Inside Title Page, *Elements of Natural History* 1844'. Courtesy of the University of St Andrews Libraries and Museums, *Elements of Natural History*, sQL48.B6.

and publications. But Owen's scientific agency in private would, like Cuvier's,[2] encompass regular exchange with Sarah directly of books, specimens and latest intelligence, as indicative of their longer-standing mutual scientific respect. Indeed, she may have been instrumental in Owen's rise and increased reputation in the early 1830s. Cuvier visited London in 1830, accompanied by Sophie his (English-speaking) stepdaughter and one of Sarah's closest friends. During their visit, Sarah 'brokered' the Cuviers' private meetings with London's scientific community, both established figures like William Buckland (1784–1856) and John Herschel (1792–1871),[3] and more recent newcomers like Owen. The latter's subsequent visit to work with Cuvier at the Paris Muséum in July–August 1831 would immediately have been facilitated by Sarah's introductions. She had long-standing personal connections with its other Chairs (as her *Memoirs* attest).[4] Owen's scientific respect and debt to Sarah may then explain her otherwise intriguing acknowledgement in this dedication: the Owen 'in private' is given precedence over his acclaimed public persona.

British history of science and Anglophone scientific biography rather differently write of Owen in this pivotal period of his scientific life, and after his death, as the envious purloiner and antagonist of others' better scientific work for his own career advancement.[5] The clear evidence to the contrary here of his 'kindness' cannot be explained by Sarah's status as a woman in science, and hence without threat or rivalry provoking Owen's animus. Rather Sarah's pioneering contributions in the *Elements* (*ENH*) to recent natural history knowledge, dissemination and pedagogy shed light on her many points of personal and 'professional' scientific contact with Owen, including through others such as Yarrell in their mutual circles.[6] Indeed, women played proactive parts in London's intersecting and concentric communities of natural science endeavour. Take Caroline Amelia Clift (1801–1873), the scientific illustrator and only daughter of William Clift, the conservator at the Royal College of Surgeons, and Owen's employer. When Owen then married Caroline in 1835, his larger gain was access through the Clift family to their many scientific connections. The untimely accidental death of William then made his conservator position Owen's as already his deputy, before being his son-in-law. But the marriage of Caroline and Richard, residents of the Borough of Camden was celebrated in New St Pancras Church. Since Sarah also worshipped there, she was very probably at the wedding ceremony.[7]

Sarah's dedication above therefore facilitates her seamless move from the 'English Cuvier' to the unsurpassable French original: his were the 'Principles of Classification' on the inside title page. Cuvier also epitomised the intersection of 'public' and 'family drawing-room' science ('Saturday salon'[8]) of both men as its remit. His systems of (Muséum) taxonomy are therefore newly 'interspersed with amusing and instructive original accounts of the most remarkable animals', as the title page clarifies. Sarah's short Preface – reformatted in one view in Figure 8.2 – unusually goes further, to highlight the pioneering perspectives of her work. Expressly designed 'for the Use of Schools and Young Persons' on the inside title page above, it adapted Cuvier's (four-volume) *Règne Animal* to meet the needs of 'the lover of animals' (Sarah's formulation in English of *l'amateur d'histoire naturelle*), whether the 'beginner' or the (budding) expert. The likely English translation (as the most important of several) to which Sarah refers

PREFACE

ALTHOUGH the *Règne Animal* of Baron Cuvier, which now forms the basis of all systematic writings on Natural History, has been ably translated into English, there has, as yet, been no work on this subject adapted to the capacity of the young student. An attempt has been made in the following pages to lay the first four classes of animals before the beginner, in such a manner as shall convey their most important characters, their classification, and economy. At the same time, it is hoped that this book will enable the reader to proceed to the many admirable monographs now published, and the more detailed productions of those whose deep science renders a stepping-stone necessary, to arrive at a full comprehension of their labours.

It was the Author's intention to have included the fossil vertebrata in her pages; but the narrow limits to which an elementary treatise must necessarily be confined, have entirely banished such notices. As it is, much valuable matter has been suppressed by adhering to the size best calculated for general circulation, — a difficult task where anecdotes crowd upon each other, and all are interesting to the lover of animals. The selection that has been made is, as much as possible, confined to those facts which have come under the immediate observation of the Author or her friends. (vii–viii)

Figure 8.2 'Preface to *Elements of Natural History* (1844)'.

is the sixteen-volume *The Animal Kingdom Arranged in Conformity with its Organization. By Baron Cuvier … With Additional Descriptions of all the Species hitherto Named, and of Many not before Noticed*, by Edward Griffith among others.[9] While William Fordyce Mavor had before Sarah published the similarly titled *Elements of Natural History in the Animal Kingdom. Chiefly Intended for the Use of Schools and Young Persons* in 1799, and newly revised in 1833, this work followed Linnean classification as also adopted by Buffon. By contrast, the new system of Cuvier in its four main 'kingdoms' is by Sarah's definition 'the last great revolution' (*ENH*, 2). Her formulation long precedes Thomas Kuhn's similar understanding in 1962 of epistemic breaks in scientific progress.[10]

Sarah's assertion regarding the specific pedagogical contributions of her *Elements* – as the first (in English) to fill the major gap that was a (readily accessible single-volume) 'work on this subject adapted to the capacity of the young student' – was no unfounded or immodest presumption.[11] As the first woman on both sides of the Channel and the Atlantic to undertake a natural history textbook project, her reputation required every fact to be faultless. Her preface therefore draws fine distinctions regarding 'capacity'; her own and that of the 'young student'. While downplaying what she terms her '*attempt* […] to lay the first four classes of animals *before the beginner*' (my emphasis), Sarah's 'elementary treatise' was also designed as the 'stepping stone' to 'a full comprehension' of 'deep science'. Her *Elements* (*ENH*; *ENH2*) was therefore no primer, basic introduction or beginner-level schoolbook.[12] Her 'student' (sex non-specific) of at least upper-school age was already interested in and informed about animals, with clear 'capacity' for further specialist study. The second paragraph of Sarah's preface then underscores the challenging rationale for 'adapting' Cuvier's *Règne Animal*. Her first consideration of comprehensiveness – the possible inclusion of (new) 'fossil vertebrata' – was rejected, but not for want of her own knowledge.[13] Rather, the work's representativeness – 'the size best calculated for general circulation' – was less the reduction of Cuvier's four volumes (and the English translation's sixteen) into one, and more the indicative best coverage (from informed scientific points of view), so as to inspire the 'lover of animals' to greater depth and breadth of specialist knowledge. Sarah therefore brought her own unusually extensive learning (in Paris, in the field) and authority to the scientific selection and substance of her entries. They encompass formal scientific description, 'amusing and instructive original accounts' – henceforth 'accounts' – or 'the immediate observation of the Author or her friends'. Her dedication had already clearly inserted Owen among them.

The *Elements* (*ENH*) therefore filled the very knowledge gap that Sarah and Edward Bowdich had first encountered in Paris in 1819, and how they had resolved it: their compilation works 'for the Use of Students and Travellers' applied the latest classification of Cuvier, Lamarck and others in (French) natural history to their subjects (see Appendix 1). Chapter 7 situated Sarah's foremost roles as their acknowledged scientific 'illustrator', and thereby discovered the name of the 'anonymous' illustrator of the *Elements* (*ENH*). But this new venture far superseded the scope of the Bowdichs' compendia by also adapting Cuvier's second revised edition of the *Règne animal* (of 1829–1830) as part of the deft enormity, yet economy of her task (and why she then excluded Cuvier's fossil species). Only a scientific author of authority had the expert 'capacity' to adjudicate on important contents, and on their *form*, to meet the nascent 'capacities' in her 'student'.

By 1844, Sarah additionally brought very particular expertise and authority to the pitch, production and presentation of her *Elements* (*ENH*). Hers was also the successful support, upbringing and education of the Bowdichs' three surviving children, all now at the age of majority. Sarah had therefore seen firsthand the dearth of authoritative and readable 'textbooks' on natural history (for children of both sexes), inspiring its further study. If her own family therefore model 'the young student' of the *Elements*, she as its author was no 'maternal popularizer'.[14] Rather, the *Elements* crystallised the wealth and insight of Sarah's first-hand French and British scientific knowledge and authority, and how to convey them arrestingly, as this chapter now investigates for the first time.

The Scientific Paradigms of the *Elements*: Classifications of New Natural History

The necessary introductory frame for 'student' readers of the *Elements* (*ENH*) explaining Cuvier's new classification (1817, 1829–1830) as the 'system' that her text adopted was not anodyne paraphrase. In 1844, the controversial question for science and for religion on both sides of the Channel was how the single genus of humans – set apart in the class *Bimania* (two-handed) – fitted with related classifications, whether creatures in the class *Quadrumania* (four-handed, such as great apes), or (four) distinct 'races' that *Bimania* encompassed, as set out in Cuvier's (problematic) race theory. To read this introductory section at face value today as a 'translation' or, worse, endorsement by its woman 'populariser-copyist' would seriously misread Sarah's careful renegotiation of human difference(s) here,[15] also treated in Chapters 4 to 6. She strategically inserted alternative (published) findings into this introduction that implicitly questioned and scientifically distanced her work from Cuvier's – the question of the monogenesis or polygenesis of human races – as well as from Lamarckian (and proto-Darwinian) ideas about the 'transformisme' (evolution) of all animal species. The *Elements* (*ENH*) therefore adopted Cuvier's four widely accepted (non-human) vertebrate orders, by more overtly grafting in 'T. Edward' Bowdich's parallel work (1819, 1821, 1824) in Asante with different evidence on human 'race-relations', in order to take these one step further. Her introduction to Cuvier's system concludes with peoples he did not consider, the further 'mysteries' of the 'Malays and Papoos' (*ENH*, 21–22). These culturally significant migratory (non-white and non-black) peoples demonstrated the arts and sciences of exploration of the Antipodes long preceding those by recent Western voyages of scientific 'discovery'. Sarah's small 'stepping stone' alternatives within Cuvier's classifications thus held open the necessity of further new scientific evidence and interpretation, as provided by future 'students' (modelled on the Bowdichs as 'students' of Cuvier). In consequence, Sarah's careful insertions of alternative evidence here and elsewhere in the *Elements* endorsed rather than contradicted her scientific authority. In then affirming Malay as the source language for the word 'Orangutan' (*ENH*, 25), Sarah neatly reopened to further science the latest discoveries of other important *Quadrumania*. In particular, were the (African) Chimpanzee (*Simia Troglodytes*), as described by 'Mr Bowdich the African traveller' (*ENH*, 26) alongside the 'Ingheena':

> Mr Bowdich was the first to report the existence of a large species which exists in Tropical Africa, and which is called Ingheena by the people of those countries. It was long thought that he had been imposed on by the natives; but the arrival of a male and female at the College of Surgeons, about two years back, verified every description of its size and strength. A person in Bristol wished to procure living examples of this formidable creature, and commissioned the master of a trading vessel to bring them for him. Finding it impossible to get them alive, (and the natives say they may never yet were taken while living,) the captain brought them home dead: the male he put into a cask of rum, and the female into the strongest possible brine. The person who gave the commission, however, refused to purchase them in this state, and Professor Owen secured them for the College. The male (*fig.* 4) is in good preservation but the flesh dropped from the bones of the female [...]. (*ENH*, 27-28)

The very careful structuring of this passage is again as 'stepping stones'. First, Sarah connects the first 'discoveries' by 'Mr Bowdich' to the Bowdichs' similar work on Asante culture: naming of this hitherto unknown animal (the Gorilla) was by adoption of its local, long-known designation, also deployed in *The African Wanderers* (as Chapter 6 discovered), pre-existing Western scientific naming and classification. Second, Sarah confronts earlier (erroneous) scepticism – about 'native' eye-witness report as scientific fact – now proved to be correct in Edward's *Mission*, in order to pre-empt similar scepticism regarding her scientific authority in the *Elements* in 1844 (as penned by a mere woman). These steps prepare the third. It was Owen no less who 'verified' the status and importance for science of the 'Ingheena' specimens, as procured by the Royal College of Surgeons. The provenance of their account and its reporter in the last sentence quoted above is then significant. Without direct access to the specimens in the RCS collections, as authorised by Owen their current Conservator, she could not have seen them, or drawn their figure. She was also in an unprecedented position to judge their preservation as 'good' or otherwise as a published authority on taxidermy and museum mounting. All three 'steps' therefore pre-date and challenge the accepted 'fact' that Paul Du Chaillu was the first discoverer of the gorilla in 1861, as authoritatively asserted in 1992 by Mary Louise Pratt.[16]

The text of the *Elements* (*ENH*) therefore provides clear evidence for the reciprocity of scientific respect between Sarah and her Dedicatee. The evidence of Owen's new specimens at the College of Surgeons then allowed Sarah also to present, and supplement, the comparative anatomy and 'classifications' of the French and 'English' Cuvier. Significant are her disclosures of otherwise discounted females of key species, including, by inference human: she figures a *female* chimpanzee in the *Elements*; hers is then the drawing of the female 'Ingheena' also for *The African Wanderers* (discussed in Chapter 6). Sarah's adept and concerted reinsertions overtly displaying allegedly 'secondary' and less well 'preserved' (females in) natural history, thereby proposed *complementary*, and therefore pioneering, scientific perspectives on 'official' natural history that point up its major blind spots in every scientific class (and human race).

The one further direct reference to Owen in the *Elements* is then no less significant:

> The Monotremae are the most singular animals in creation. Although they have no pockets like the Marsupialia, they yet have the bones which support these appendages. It was for years doubted whether the Ornithorynchus had teats, and, in consequence of some eggs having been found in their holes, they were supposed to be oviparous. Recent observations, however, and chiefly those of the learned Professor Owen, have ascertained that they possess these organs, and consequently bring forth living young. (*ENH*, 117–18)

Owen's place as the final adjudicator on controversial and problematic scientific classification and understanding in the past is again central to Sarah's account here, which proceeds to list other anatomical peculiarities of the duckbilled platypus. It has a second collar bone, males have a spur on each side of the hind foot, it lacks external ears, its tongue is double. Together with thick short brown hair these are 'extremely ugly, as well as singular looking animals' (*ENH*, 118). Sarah's repetition of 'singular' to frame her description endorses her opening superlative that placed this paradoxical animal (*Ornithorhynchus paradoxus*) as viviparous or oviparous at the centre of key debates on its classification on both sides of the Channel. But Sarah's naming of Owen's arbitrations, and their timing, also reveals her indirect part in his '[r]ecent observations'. Owen consulted Cuvier and others in 1831 at the Paris Muséum, when he examined its platypus specimens that Sarah had also seen and drawn there (see Figure 7.7 in Chapter 7).[17] Owen's subsequent papers between 1832 and 1834, and not Cuvier's *Règne animal* (in first or second edition), were therefore the source of Sarah's latest information in the *Elements*, because Owen newly treated the reproductive organisation of platypus anatomy that defined its classification as viviparous.[18] His contributions in 1834 therefore superseded Cuvier's 'definitive' knowledge in 1829 (which Owen cited directly). But in 1834 Owen also acknowledged that this classification was confirmed only thanks to the expertise of field-based 'observers' and specimen collectors in New South Wales, by naming 'Bennett', among the major names collated in Appendix 8. The 'Owen platypus' – in scientific text, image and real thing – thus hallmarked the quality of Sarah's report of recent natural history in the *Elements*, including its 'anonymous' figures that proved in Chapter 7 to be her own. Her work in 1844 accorded clearly to the latest (French and British) scientific models *together*, because it continued her earlier 'Paris' publications and African fieldwork as '*Mrs* Bowdich the scientific traveller'. The 'Owen platypus' in the *Elements* therefore also exemplifies Sarah's pioneering perspectives. To take female comparative anatomy in text, image and real object seriously was to adjudicate on the *Elements* as also filling the glaring gap in 'official' (Museum-based) scientific classification and publication, and to find its author leading in them by example.

Textbook Natural History

In consequence, the *Elements* (*ENH*) was no 'copy' or minor 'adaptation' of Cuvier's *Règne animal*. Its defining system of classification, operating as the organisational backbone of Sarah's work, was the essential 'grammar' for all new natural history endeavour. Classificatory tables using Latin nomenclature to summarise the main and subclasses of animals concerned, and culminating with Cuvier's French vernacular terms, head the subsections that are Sarah's longer text descriptions using English-language names. The 'student' thereby learns to identify and name vertebrate animals in three (European) scientific languages. Intercalated illustrations aid larger interest and active memory by varying the otherwise repetitious text format following the order of Cuvier's *Règne animal*. But a second, no less important systemisation underpins and updates the *Elements* (*ENH*) that further corroborates Sarah's particular scientific expertise and authority. While Cuvier began his revolutionary work in the *Règne Animal* to reclassify Fishes as vertebrates (unlike Linnaeus's separate Family including cetaceans), his 22-volume *Histoire Naturelle des Poissons* had yet to be completed (by Achilles Valenciennes). It was Sarah's main classificatory source for the contents of the fourth and final part of her *Elements* (*ENH*, 328–end), which also contains the fewest 'accounts' and supporting authority references. To make its drier listings easier reading for her 'student', key technical terms are bypassed (such as 'en velours', *ENH*, 336), the better to acknowledge the contemporaneity of her scientific information *by its absence*. An unusual footnote states: 'The guide which has been followed for this brief description of Fishes here ceases; for the great work of Cuvier, so worthily continued by M. Valenciennes, does not, as yet, reach further than the sixteenth volume, and a portion of the family of Cyprinoides' (*sic*)

(*ENH*, 422). Sarah's *Elements* therefore provided her 'student' reader in 1844 with the fresh excitements of the very latest (French) science of ichthyology in the making, and as supplied (in her translation) by one of Cuvier's and Valenciennes' main collaborators in it.

The format of the *Elements* (*ENH*) is therefore deceptive in its succinct expert duality, with regard both to its deft integration of Cuvier's two major works of (re-)classification and to their aeration by insertion of the work of other supporting, or alternative, scientific authorities on both sides of the Channel including 'Bowdich' (as also double). The supplementary specialist references collated in Appendix 8 therefore operate as enlightening points of dialogue and exchange, between 'laboratory' and 'field' expertise, and between the technicalities of recent scientific papers, such as Owen's on the platypus – and 'accounts' about natural history as published in periodicals for informed general readers. At no juncture does the *Elements* (*ENH*) compromise or patronise its young student's 'capacity' for understanding scientific contents, formats (including Sarah's use of tables in Latin, and French) and model research *conduct*. Clear 'elements' of the last put directly before the reader's eyes include accurate observation and identification (including drawing), correct naming and clarity of description that enhance application in the field and in the personal and museum collection. Such qualities by pedagogical example thus define and develop the student's understanding of natural history as an open, expansive and inclusive subject. No wonder, then, that Sarah's pioneering single-volume *Elements* (*ENH*) became a reference text for British university-level 'students' in Zoology,[19] and saw a revised and augmented second edition including new drawings by Sarah in 1850. The striking dearth of critical attention to this textbook work in British history of nineteenth-century natural history can only be explained by narrowly 'Victorian Studies' approaches,[20] blind to the 'continental' (Cuvierian and Humboldtian) comparative natural history quintessentially underpinning its 'Anglophone' reapplications and developments.

Sarah's unusual command of comparative and concentric circles of contemporary French and British natural history expertise (illustrated in earlier chapters) further defines her much larger importance as a major contributor to pioneering natural history knowledge. The *Elements* (*ENH*) variously reveals how and where. Her scientific familiarity with Cuvier's major works of classification (in the original French[21]) in theory and in practice meant that she could also challenge them – for example on bats (*ENH*, 51), parakeets (*ENH*, 223) and boas (*ENH*, 307) – through galvanising her extensive knowledge of British and other scientific authorities (and her own). Her work in fact cites more than fifty named experts *apart from Cuvier* on the animal(s) in question, collated in Appendix 8 in alphabetical order for the first time, to identify and recognise in these intertextual authorities and scientific sources Sarah's multiple expertise.[22] The majority, including those names most frequently referenced such as 'Gould', 'Waterton' and 'Yarrell' published their work, including book-length studies, in English. In consequence, they are readily accessible to the target 'student' of the *Elements* (*ENH*) as well as to Anglophone history of science. I deliberately excluded from the roster of Appendix 8 references to 'Mr Bowdich the African traveller', because Sarah was herself behind his front as a scientific authority, as well as named overtly as 'the Author' elsewhere in the

Elements (*ENH*). Space does not permit close analysis of Sarah's different reformulations of her some fifty authority sources. A number offer a shorthand allusion, additional reference point or résumé-paraphrase; others a short or longer direct citation (Appendix 8 notes those she quotes directly at length). To account for their main scientific uses instead in the *Elements* (*ENH*) more quintessentially raises important questions about Sarah's position as the more pioneering scientific authority among them.

Names on the first page of Appendix 8 are representative in its (alphabetical) list: Adanson, Audubon, Baird, Bennett, Burchell, Commerson (and Forster), Cook and Duvaucel were scientific travellers of intercontinental reach, whose published 'travel' accounts detail in the regions concerned their observations and discoveries of natural history new to Western science.[23] Gould, Humboldt, Pentland, Pringle, Sir Stamford Raffles, Schomburgk and Waterton also feature centrally in this category of scientific traveller authority. But Sarah's 'book knowledge' of them as revealed in the Appendix bibliography is complemented in many cases by her personal acquaintance; with Humboldt obviously, with Alfred Duvaucel's sister Sophie (and Cuvier's stepchildren), with George Bennett a friend and protégé of Owen,[24] with Thomas Bell also through Owen,[25] and friends of friends. For example, the Introduction to the second volume of Audubon's *Ornithological Biography* acknowledges the recognition of it by 'Mr Neill', Secretary of the Wernerian Natural History Society of Edinburgh. This is the same 'Mr Neill of Cannon Mills' in Appendix 8, whom Sarah thanks (*ENH*, 273) for sending her extensive information (on gulls) by private correspondence. (John) Gould is similarly acknowledged (*ENH*, 257) for his loan to Sarah as a London-based friend in natural history of the relevant descriptions and plates from his *Birds of Australia* for her inclusion of them. Her mention of 'plates' implicitly includes the work uncited in the Appendix of his wife, Elizabeth Gould (1804–1841), as their illustrator.

'Collections-based' natural history experts represent a second category that Appendix 8 reveals, some of whom Sarah again knew personally – Frédéric Cuvier, the director of the Paris Menagerie (and brother of Georges), Geoffroy (Saint-Hilaire) and Valenciennes at the Paris Muséum – Risso in Italy, and in London 'Mr Gray' (John Edward), keeper of zoology at the British Museum from 1840, and 'Professor Owen' at the Royal College of Surgeons. But London-based private collectors such as George Loddiges, famed for his world-class collections of hummingbirds, are also among her sources, as are Sir Stamford Raffles and Sir Humphry Davy, who established the Regent's Park Zoological Society of London in 1826. It is quoted in the *Elements* in observations on for example the 'Rhinoceros' (*ENH*, 127) and the 'Giraffe' (*ENH*, 138–39). Mention is additionally made of the Surveyor of Royal London Parks, Edward Jesse. These names uncover and extend Sarah's circles in London science, and hence her access to its world specimen holdings and their curators with expert specimen knowledge not always found in scientific publications.

In a third category with exclusively British provenance are key representatives of British scientific 'popularisers' in Bernard Lightman's terms,[26] without formal scientific training such as Agnes Catlow, Col. Hamilton Smith and William Yarrell, who produced major compendium works for educated and leisured audiences focussing on a given class of British natural history. That Catlow and Yarrell were clearly also 'naturalists' in David Allen's terms,[27] meaning they had unrivalled knowledge of British fauna in the field, would then include Couch, Lowe, Pennant, and Selby in this list, together with 'sportsmen naturalists', such as Sir Humphry Davy, Captain Scoresby, Mr Scrope and Mr Waterton.

The intersecting scientific echelons and interests of these 50 named experts in Appendix 8 as representative natural history authorities therefore make Sarah's *Elements* in 1844 an important snapshot of the overlapping circles of British 'gentlemanly' and (majority) 'non-gentlemanly' natural history activity, to reveal her largely unacknowledged place within their ranks.[28] Indeed, Sarah's proven expertise in the *Elements*, and overtly in its final section of Fishes, derived from her undertaking of expert (publishable) natural history endeavour in all domains of 'national-level' production – in the field, including overseas, in the museum collection, in scientific natural history publication and dissemination – that her authorities represent. Additionally, Sarah's uses of her authorities in the *Elements* as both technical and amusing-instructive sources then display how her galvanising of their mixed-mode scientific dissemination formats better reveal her own, for example in *The Fresh Water Fishes*. Although not overtly named or referenced in the *Elements*, it is immediately recognisable in the final part in her expansions of classificatory descriptions concerning (British) freshwater fish species that are especially rare, for example the Vendace (*ENH*, 427) 'only found in the lakes of Dumfriesshire, especially Loch Maben. It is said not to live in any other water in the United Kingdoms.'[29]

To understand the *Elements* (*ENH*) for its larger, interconnective *modus operandi* for expert natural history-making means substantially to rethink Sarah's strategic London status in key British scientific networks in the 1830s and early 1840s at first, rather than 'outsider' second hand. Appendix 8 therefore reveals and reconstructs something of Sarah's 'address book' for London natural history in her interlinking circles of scientific friends, interlocutors and informants, because she was their peer as an authority in her own right.[30] Personal loans of published texts and accounts, private correspondence,[31] access to private collections, in-person encounters at lectures and private gatherings all corroborate Sarah's expert, and constantly developing scientific interests informing her 'British' publications of the 1840s and 1850s. Only through the Appendix bibliography triggered by her named authorities can we also count the significant number of their works appearing only in the early 1840s. They are unassailable evidence of Sarah's integrated account in the *Elements* (*ENH*) of the latest state (without nationality) of expert natural history knowledge.

Appendix 8 therefore encapsulates Sarah's pioneering pedagogical work in 1844 in her several abilities to capture the 'the best authorities' (other experts; their publications) in natural history. In turn, these enrich critical understanding of her own many expert scientific 'capacities' including publications among them. Moreover, the extensive 'bibliography' behind the *Elements* (*ENH*) other than the works by Cuvier indicatively demonstrates the enrichment of Sarah's own scientific inquiry after his death. In consequence, this text offers the clearest evidence that her 'Paris' *modes* of natural history knowledge exchange (including visits at least three times yearly[32]) were after 1832 proactively and concertedly remediated in 'London'. But Sarah's foremost scientific 'capacity' to bring succinctness, clarity and rare liveliness to her descriptions in the *Elements* (*ENH*) – their design was to amuse as well as to instruct – also required the authority to bring others' expert scientific accounts to account. We now turn to the scientific 'Author' orchestrating the generic hybridity of the *Elements* (*ENH*) with an especial eye to its curation of inspirational natural history-making.

The Scientific Pitch of the *Elements*: 'Accounts' of Direct Encounter

To sample randomly any three of the many works underpinning the *Elements* listed in the Appendix Bibliography immediately discovers that even the most allegedly technical, impersonal and 'scientific' in 1844 – for example, Owen's work on the platypus – contains what the modern scientific paper designates the 'discussion'. In the overwhelming majority of cited works in Appendix 8, narrative description explains the 'findings' from the 'data' according to the conventions (and genres) of verifiable observation(s), authenticated by a named expert in their field (as in 'Bennett'). We noted the preponderance of renowned scientific travellers and (local) 'British' naturalists among Sarah's main informant-experts. The travel narrative, like the many genres of 'British field notes', presents the 'object' of natural history not in a museum case, but directly within its natural environments and seasons. The viewer-reporter then accounts for 'new' observations, and their contribution to previous (expert) knowledge. Such significant animal encounters, because by a reliable informant, pepper the majority of Sarah's source works for the *Elements* (*ENH*). For the nineteenth-century reader, scientific authority for a 'discovery' rests primarily in the (re-definitional) expertise of the observer/observation, not in the distinction of experts by formal scientific status and qualification, or none. Nineteenth-century natural history works with 'specialist' illustrations – many works in the Appendix bibliography contain plates – then further inspire reader observation whether as leisure pursuit, instruction or new inquiry in the field, or in the (museum or private) collection. Sarah's use of 'amusing and instructive original accounts' throughout the *Elements* (*ENH*) therefore served not only as her means of selecting pithy, representative, important and comparative scientific examples in her many sources. She also used such 'accounts' as textual figures in the manner of 'illustrations' to captivate new interest in an animal *in situ* in the field, the menagerie, the private collection, as well as in the text. We encountered Sarah's arts of natural history fiction in Chapter 6 to capture creatures such as the Inghena new to the protagonists and the reader. Her parallel deployment in the *Elements* (*ENH*) of others' scientific 'accounts' – in three main categories – also teaches by demonstrating entertaining expert observation.

The main category of 'account' is the excerpted natural history 'anecdote', with third-person (published) authority behind its contents, namely its report of new and/or surprising details for familiar, and less familiar, animals. The provenance of these eyewitness descriptions appearing in named scientific travels and natural histories served immediately to verify, characterise and authenticate the creature being described, with the aim of enlivening, exemplifying or putting more individualising form on the dry scientific classification and anatomical description. Such illustrative natural history anecdotes in the third person also further authenticate the authority of the *Elements* as 'textbook' natural history, by providing additional, sometimes contrary, observational examples and latest findings. The reader also gains familiarity with particular leading authorities, for example 'Kirby' for insects, through encountering their specialist anecdotes. The pithy quality of observational content, not the (social) status of the expert informant, remains uppermost in Sarah's selection and insertion of suitable anecdotes into her textbook work.

A second important category constitutes 'accounts' of her 'friends' in scientific endeavour. Quality content again overrides author status. However, this category is especially revealing in the *Elements* (*ENH*) to include those who otherwise leave no published trace in the fields and communities of natural science endeavour it covers. In the description of 'Hyaenas' (*ENH*, 80–81) for example, Sarah elaborates on their various attributes by means of a graft of different eye-witness reports. The first is by the 'Uncle of the Author in Tantrum Querry on the leeward coast of Africa' (Edward's maternal uncle, Mr Hope Smith) prefacing a longer 'account' – illustrating the hyaena's geographical range – by 'Mr Burchell, the South African traveller'. Sarah's expert 'science anecdote review' (to which the next chapter returns) then directs the reader to recent study of hyaenas in the field or in captivity, as exemplified by the work of Frédéric Cuvier on hyaenas at the Paris Menagerie. Last but not least is the 'account' of the hyaena by 'the Author', who turns out also to be the collector of the first and previous one. Her eye-witness observations at once corroborate the others and resituate her (as an authorial and scientific authority) squarely within wider transcontinental African natural history, and its knowledge in science museum collections and print publications.

The seriousness of the natural history anecdote, like the recorded report by the 'friend' ('amateur' in French) of its endeavour, pivots on its main purpose as illustrative 'anecdote of enlightenment'.[33] However, the second category of account from more personal first-person experience could have a recreational as much as a more overtly instructive point. Sarah's 'anecdotes' of different animals at the Paris Menagerie mention not only its curators but also its informed publics: for the 'Jaguar' (*ENH*, 88), her companion in its encounter is 'Mlle [Clémentine] Cuvier'.[34] As with Sarah's deft inclusion of the female gorilla for science, an otherwise invisible model for the (serious) 'student' therefore goes on record here, to underscore the equal enjoyment and appreciation of natural history by women and girls. Sarah's seemingly superfluous footnote on conchology (*ENH*, 194) similarly flags a larger 'story': the only overtly named woman authority in it, Agnes Catlow and her work were representative. Natural history therefore proves altogether at home in the city and in 'distant lands', within 'drawing-room' and 'museum' settings, albeit that the Cuviers' house within the Paris Muséum precincts blurred these usually distinct spheres.[35] What could appear casual name-dropping of important 'friends' in the *Elements* (*ENH*) also returns the more personal insider scientific 'account' of its author. Her work could reference and recognise as examples those overcoming many 'impossible' means to access 'official' natural history in France as well as in Britain. Cuvier, Bowdich and now Owen counted among the friends of natural history pursuit that was not defined by sex, class, nationality or independent 'gentlemanly' means.

Open Conclusions in 1850 to Others' Expertise

In the light of Sarah's dedication and preface to the *Elements* (*ENH*) as models of respectful acknowledgement, her self-promotion as revealed above in her 'pedagogical' authority and various authorships appears bold. It is then salutary to compare her many overlooked contributions and legacies in British and French histories of natural science with Owen's reputation as the 'English Cuvier' in 1844, and in revisionist review: 'The secrets of Owen's professional success were industry, opportunism, self-promotion and ability to synthesize ideas from the works of others'.[36] If a gendered double standard for 'industry' and 'ability to synthesize ideas from the work of others' favours an Owen but not a Sarah, her *Elements* (*ENH*; *ENH2*) as textbook natural history of the 1840s and 1850s directly challenges models and narratives of 'heroic' discovery in the history of nineteenth-century natural history for their injustices to no less superlative interventions by anecdotally important, yet uncited 'outsiders' including women. In Sarah's most prolific last six years of pioneering natural history work, her second revised edition of the *Elements* in 1850 is then its further lesson, especially in what was a pivotal decade for British science exemplified in 'Darwin'. What she kept without change in 1850 therefore makes the changes the more apparent as also evidence of its author's cumulative, collaborative and transnational contributions to natural history. Notably, the dedication to Owen is gone, and none replaced it: the new edition concentrated attention on its new materials.

Sarah's revised and expanded *Elements* (*ENH2*) is therefore an important barometer for leading-edge, that is 'textbook' natural history in the mid-nineteenth century through its more clearly-defined discipline focus in the qualified main title – 'Elements of Natural History; or First Principles of Zoology' – and its new preface, reformatted in Figure 8.3. Reader 'demand' and making the work 'more complete' are, however, augmented by the adept contribution of 'anecdotes [...] most of which result from the personal experience of the author, or [...] her friends.' If 'anecdote' overtly replaced the 'amusing and instructive original accounts' of 1844 as shorthand synonym, Sarah also uses its media of third-person illustration literally and figuratively, to confirm the provenance of 'observations' and the 'cut of the Ingheena' published in 1844 as her own. In consequence, the first edition also counts among contemporaneous coverage of 'admirable writings which take but one portion [...] thoroughly', in order to promote the work of 'naturalists' as opposed to 'manufacturers of classes'. Her simple identification by addition of 'Brit.' to focus and inspire the student to swell the numbers of 'naturalists' culminates in Sarah's admonition by example to other 'synthesisers' (compiler-plagiarists) in 'Zoology', who do not name their 'authorities'. She had encountered the problem personally, in others' plagiarism of her *Taxidermy*. She thus names 'Dr Carpenter' as the main authority for the entirely new final section of the 1850 *Elements* on 'Molluscous Animals', on 'Articulated Animals' and on 'Radiated Animals' (*ENH2*, 469–511). She was already knowledgeable about, and had figured many of them in the two-volume *Elements of Conchology* by 'T. Edward' Bowdich, (see Appendix 1). To tie her earlier work in 1822 very directly by its title to her 'textbook', natural history in 1850 allowed her to display her sole-authored

The first Edition of this Work contained only the Vertebrated Animals, with very slight notices of the three other Sub-Kingdoms; as, however, the demand for it rapidly increases, it appears desirable to render it still more worthy of public favour, and make it more complete by extending these notices as far as the limits of an elementary book will allow. The comparatively few genera of the first Sub-Kingdom have enabled the author to mention them; but the multitude of Invertebrated Animals makes it impossible in their instance to afford each such details. Those who peruse the following pages, therefore, will find a fuller description of the first Sub-Kingdom than any other. Man is placed first, as the highest in intellectual development, and the series gradually descends to those minute beings, whose countless myriads, in many parts of the globe, form the very soil which receives his footsteps.

While treating of Vertebrata, there has been room for several anecdotes which give proofs of sagacity or affection, most of which result from the personal experience of the author, or have been supplied from that of her friends. Few of them exist in any other publication, unless they have been copied from this work; the cut of the Ingheena is also unique. The genera of the Vertebrata have been brought down to the present period, and will probably remain unchanged. It is not among these that we are to expect much novelty; but the discoveries of travellers are yearly swelling the Invertebrata to an extent which alarms the manufacturers of classes, while it delights the naturalist. On the whole it is hoped that this Zoological Work will place before the public that broad and comprehensive view of some of the operations of their great Creator, which may satisfy the general reader, and lead the student to a closer attention to those numerous and admirable writings which take but one portion, and treat it thoroughly.

Wherever the genera are marked *Brit.* they are found in the British Islands.

A publication of this kind must of necessity contain a great deal this is mere compilation, and this amounts to an open theft if not acknowledged. The Author has, therefore, to own the great assistance which she has received in all the additions to this work, from Dr. Carpenter's Zoology. (iii-iv)

Figure 8.3 'Preface to *Elements of Natural History* (1850)'.

developments regarding natural history knowledge of invertebrates and vertebrates. The new 'assistance' of 'Dr. Carpenter's Zoology' for 'all the additions' to her 1844 *Elements* was therefore significant. This was not William Carpenter's *Scripture Natural History* (1833),[37] but the latest two-volume *Zoology* of 1848 by the Unitarian invertebrate zoologist and promoter of women's education in science, Dr William Benjamin Carpenter (1813–1885).[38] The interesting preponderance of 'non-conformist' rather than 'gentlemanly' natural history behind Sarah's *Elements* (*ENH*; *ENH2*) was therefore further confirmed in 1850.

Sarah's clear revision of her authorities and materials in deletions and new insertions in the 1850 *Elements* stands out alongside the main augmentation of animal classes, because the original pagination and illustrations of the first edition remain. For example, gone is the long acknowledgement of Mr Loddiges (on hummingbirds, *ENH*, 209) and direct reference to Owen on the Ingheena (*ENH*, 27). New material on both species names no additional authorities, although for the latter a new footnote states that 'the skulls of an even larger and more powerful animal of this kind have been recently brought to the Museum of the Royal College of Surgeons from Liberia, but nothing more is known concerning them' (*ENH*, 29). A further new note, on the 'Ostrich', specifies that 'Mr Darwin speaks of a second species in South America, which is chiefly found in Patagonia. The feathers come six inches below the knee' (*ENH2*, 239). If this bird is now known as 'Darwin's Rhea', its image and description in the *Second Voyage of the Beagle* (from the Gauchos' alerting of Darwin to it in 1834) was published by John Gould in 1841.[39] We have already noted Sarah's private access to his work and to Elizabeth's illustrations above, to exemplify the importance of London networks of scientific friends. Sarah would have been equally aware of the classification dilemmas of these 'ostriches' if species in Cuvier's view were fixed. Her new preface therefore circumnavigates 'evolutionary' ideas in her use of 'descent' of species (Figure 8.3 above, paragraph one) as a chain of being with a 'Creator' (as very much Carpenter's view), not a proto-Darwinian 'descent with modification'.

The significant further additions to the 1850 *Elements* are its overtly-flagged new coverage of invertebrates. Because their descriptions are mostly without 'anecdote', they follow the same format as the Fishes, and hence also 'disguise' Sarah's two further major contributions. The first is her completion of the 'Vertebrates' through coverage of the orders of classification of Fishes beyond the Cyprinids that end-stopped the 1844 *Elements*, but without recourse to 'Dr Carpenter'. The second volume of his *Zoology* (1848) attending to 'Fishes' was much less substantive than Sarah's 1850 *Elements*. The now completed *Histoire Naturelle des Poissons* in 1848 (volumes 17–22 since 1844) remains the source for Sarah's seamless integration of its materials in 'translation-adaptation' into her augmented 'Fishes'. In light of her long-standing engagement with the latest French scientific works, it is therefore difficult to ascertain whether she is then wholly, or only partially reliant on 'Dr Carpenter' as (sole) authority for the smallest invertebrates. One example proves the point, since these creatures comprise the concluding class and paragraph of the *Elements*:

> The fifth class, Polygastrica, even yet holds an uncertain place in classification, although from their simple organisation, they are now referred to the end of the Radiata. [...] in every drop of water these animalcules exist. [...] Professor Ehrenberg believes that they have many stomachs, whence their name; but in other respects naturalists differ much concerning them. One thing, however, appears to be certain, that they have cilia, whose vibrations brings food to their mouth. In one genus, that of Nassula, bristle-like but firm teeth have been discovered. Their movements, which offer a variety of motion, are supposed to be effected by cilia. Light influences, and too much injures them; some exist in utter darkness; a great degree of heat or cold, or an electric spark, will destroy them. One drop of salt water kills those which inhabit fresh water. Alcohol is destructive to them, sugar annihilates some, and strychnine causes all to perish. Their powers of reproduction are almost incredible, as well as their universal diffusion, which makes it impossible to say where they do not exist. They remain torpid in dry earth, and are thus disseminated in the dust. In Sweden they form a nutritious earth called Bergmehl; and of their crowded abundance some notion may be given by stating, that one cubic inch of animal iron ochre will contain 1,000,000,000,000. (*ENH2*, 510–11)

Sarah's clear acknowledgement of 'Professor Ehrenberg' (Christian Gottfried Ehrenberg, 1795–1876) and close friend of Alexander von Humboldt with whom she was still in contact, is to his work on 'animaux infusoires'[40] in 1839 (in French translation from the German). Consultation of it confirms it as the clearly recognisable source for Sarah's very succinct 'compilation' of salient physiological details here about 'Polygastrica', and his experiments on them, in lieu of an 'anecdote'. Carpenter's coverage of the Polygastrica in his *Zoology* also acknowledged Ehrenberg as his authority, but his description is a much closer verbatim 'translation' of the same French source. That Sarah went to 'Ehrenberg' directly, rather than through Carpenter, is also supported by her second unflagged, but no less important major addition to the 1850 *Elements.* Her drawings of these representative 'animaux infusoires' are copies from Ehrenberg's (multi-image) plates in the 1839 French edition. These drawings are among some *sixty* new figures of animals in all classes to appear in the *Elements* of 1850, their ready insertion being as complements to a given description, or as representative animals in a main class. All derive from Sarah's pen and 'narrative' use of similarly-sized small woodcut formats. A new image can then be inserted into the main text at the relevant place, requiring only minor typesetting reconfiguration of the description and page, rather than extensive repagination. The entirely new plates I–VII, interleaved to face and open each respective class of animal life differently, collate six or eight new woodcut images representative of the section in one 'snapshot'. They directly invite 'student' perusal of their subsequent textual descriptions. If these new plates especially visualise Sarah's final prefatory remark about 'compilation' as copy, the *Elements* in both editions distinguish between 'mere compilation' and works by accomplished 'compositor-compilers' such as William B. Carpenter as *de facto* experts contributing new knowledge, because also an author/authority in some domain(s) of the work. 'The Author [...] she' could not make Sarah's final point in her new preface more clearly. She was no assistant, nor in need of 'assistance' in the *Elements,* whether in 1844 or in 1850, because her work was no 'mere compilation' (or 'popularisation' in Victorian Studies). Her text and the figures for it are by the same expert science writer and illustrator drawing first-hand from the life, and her life in concentrically French and British natural history. We can only be intrigued by why Sarah very unusually ends her 1850 *Elements* on Ehrenberg's large number for the smallest invertebrates in his million million sum, perhaps to include its final full stop. A different answer lies in the (still) singular footnote reference to women's natural history, Agnes Catlow (*ENH*, 194), in 1850. Had Sarah already had private sight of Catlow's illustrated *Drops of Water: Their Marvellous and Beautiful Inhabitants Displayed by the Microscope*?[41] In 1850 Sarah also interestingly omitted to specify, or delimit, the 'capacities' of her 'student' of Zoology. The mix in the *Elements* (*ENH*; *ENH2*) of first-hand expert science, exemplary illustrative reference and scientific 'anecdote' therefore epitomises 'textbook' natural history as best endeavour without bars, such as institutional qualification or class to the undertaking of new and significant work. As also lessons in 'amusing and instructive original accounts' in very different narrative-compilation formats, the *Elements* (*ENH*; *ENH2*) trial textbook rationale and new pitch, to propel Sarah's no less 'textbook' last works in natural history from 1851 to 1856, to which the final chapter turns.

Notes

1 See Kevin Padian, 'Biology in History: The Rehabilitation of Sir Richard Owen', *American Institute of Biological Sciences* 47, no. 7 (July–August 1997): 445–53.
2 Letter from Cuvier to the Bowdichs acknowledging their return of his books.
3 See Orr, 'Keeping it in the Family'.
4 See Orr, 'Catalysts'.
5 See Jacob W. Gruber's vituperative Oxford DNB entry of 2006: 'Too often [Owen] considered a personal attack and invested too much of his ego in the demonstration of error. Whether in a letter to the public press, an anonymous review, or personal encounter, his response went too often beyond the limits of gentlemanly behaviour [...]'. For a different view on Owen's importance in Lancaster and medical science, see Q. Wessels and A. M. Taylor, 'Anecdotes to the Life and Times of Sir Richard Owen (1804–1892) in Lancaster', *Journal of Medical Biography* 25, no. 4 (2017): 226–33.
6 See Yarrell's long letter to William Scrope – on eels – published verbatim in the latter's *Days and Nights of Salmon Fishing in the Tweed* (1843, cited by Sarah in *Elements*, 427), warmly acknowledging his 'friends Mr Clift and Mr Owen' (62) for their expertise on these Fishes.
7 It lies outside this chapter, but Caroline's Commonplace Book (Ref. G30114MS0283) contains evidence not only of her roles as scientific illustrator but also 'connector' of science and society in a letter to her from Joanna Baillie (writer and sister of the anatomist), and 'a pen and ink sketch showing Oxford Street with placards referring to Owen's researches, by Robert Lee' (Sarah's second husband?), and a list of 'books consulted in the progress of the catalogue of the Gallery' at https://www.aim25.com/cgi-bin/vcdf/detail?coll_id=10218&inst_id=9&nv1=search&nv2=
8 See Orr, 'Keeping it in the Family'.
9 It was published in London by G. B. Whittaker (1827–1835), publisher of the 1825 *Excursions*. The *Magazine of Natural History and Journal of Zoology, Botany, Mineralogy and Geology* 6 (Jan. 1833): 432–33 reviewed this translation, but others had appeared by 1844, including those by William MacGillivray and Edward Blyth, curator of the Calcutta Museum, and correspondent with Owen.
10 Thomas S. Kuhn, *The Structure of Scientific Revolutions*. Chicago: University of Chicago Press, 1962.
11 She may not have known the very different textbook by American educator, Amos Eaton (1776–1842). His *Zoological Text-Book Comprising Cuvier's Four Grand Divisions of Animals, also Shaw's Improved Linnean General System*. Albany: Websters and Skinners, 1826, is in work-book format with 'directions' and exercises (for example, 281). Sarah's work also differs from Henri Milne-Edwards's (1066-page) *Elemens de Zoologie ou Leçons d'anatomie, de physiologie, la classification et les mœurs de animaux*. Paris: chez Crochard, 1834, also including human anatomy and invertebrates, or Milne-Edwards's work with Achille Comte, *Cahiers d'histoire naturelle à l'usage des collèges, des institutions religieuses et des écoles normales primaires* vol. 1 cahier Zoologie. Paris: Fortin, Masson et Cie, 1840 first edition, and 1844 second edition. The target audience was young boys, because primary eduction was not mandatory for girls until 1850.
12 Unlike Mavor's work, Sarah's contains no glossaries of (basic) scientific terms, because specific terminology is introduced in the context of use and examples.
13 See the coverage of fossil vertebrata in T. Ed. Bowdich, *An Analysis of the Natural Classifications of Mammalia. For the Use of Students and Travellers*. Paris: J. Smith, 1821. Sarah had also published on a fossil Nautilus, 'Notice of a Fossil Nautilus found in the Sandstone of the Isle of Sheppey', *Magazine of Natural History* 4 (1831): 137–38.
14 Bernard Lightman's important term and coverage in *Victorian Popularizers of Science: Designing Nature for New Audiences* will be taken up in Chapter 9.
15 In her *Taxidermy* (all six editions) Sarah circumnavigated these knotty problems, both comparative anatomical and racial, by remaining silent on overseas collection, let alone instructions for preservation of *human* body parts. See Orr, 'The Stuff of Translation'.
16 Pratt, *Imperial Eyes*, 208–09.
17 See *An Analysis of the Natural Classifications of Mammalia* (1821), 66. Sarah's drawing of the platypus is Plate XIV, Figure 4.

18 See Richard Owen, 'On the Mammary Glands of the Ornithorhynchus paradoxus' (communicated by J. H. Green FRS 21 June 1832), *Philosophical Transactions of the Royal Society* 122 (1832): 517–83; Richard Owen, 'On the Ova of the Ornithorhynchus paradoxus' (paper communicated by Sir Anthony Carlisle FRS 19 June 1834), *Philosophical Transactions of the Royal Society* 124 (1834): 555–66; Richard Owen, 'On the Young of the Ornithorhynchus paradoxus, Blum' (Paper read 27 May 1834), *The Philosophical Transactions of the Zoological Society of London* 1, no. 3 (March, 1835): 221–28. Owen does not cite the earlier work by Jameson. See Bill Jenkins , 'The Platypus in Edinburgh: Robert Jameson, Robert Knox and the Place of the *Ornithorhynchus* in Nature, 1821–24', *Annals of Science* 73, no. 4 (2016): 425–41.

19 The Natural Sciences Tripos and Moral Sciences Tripos were introduced at Cambridge only in 1848, as discussed by Philippa Levine in her *The Amateur and the Professional: Antiquarians, Historians and Archaeologists in Victorian England, 1838–1886*. Cambridge: Cambridge University Press, 1986, 136.

20 See mention but not further analysis of the *Elements* in Ann Shteir, *Cultivating Women, Cultivating Science: Flora's Daughters and Botany in England, 1780–1860*, 198, and in Gates, *Kindred Nature*, 244. Byrne's 'The Scientific Traveller', Chapter 1 also mentions, but does not analyse the *Elements* to augment a resumé of my own published work on Sarah Bowdich Lee.

21 Only the first volume of Cuvier's *Histoire naturelle des Poissons* (22 vols.) has been translated into English.

22 See Susan Sheets-Pyenson, 'War and Peace in Natural History Publishing: *The Naturalist's Library*, 1833–1843', *Isis* 72, no. 1 (March 1981): 50–72, and in particular 'Table 2: Bibliographic information for *The Naturalist's Library*' (70–71) for the orders and topics covered exclusively by Sarah's compatriot compiler peers, all men. Sarah certainly consulted Hamilton Smith's volumes on dogs (vols. 1, 1839 and 2, 1840).

23 See John Bastin, 'Sir Stamford Raffles and the Study of Natural History in Penang, Singapore and Indonesia', *Journal of the Malaysian Branch of the Royal Asiatic Society* 63, no. 2 (1990): 1–25 for 'publications' that connect 'zoologists' as expert scientific collectors, such as Alfred Duvaucel, Pierre Diard (15-16, 20) and Dr Thomas Horsfield (5–7) behind the work of others such as Sir Stamford Raffles.

24 Lamond Lindstrom sees Bennett as also wronged by Owen as the original collector and publisher of the 'Pearly Nautilus' in 'Sophia Elau, Ungka the Gibbon and the Pearly Nautilus', *Journal of Pacific History* 33, no. 1 (June 1998): 5–27.

25 Thomas Bell became a Fellow of the Royal College of Surgeons in 1844.

26 See Lightman, *Victorian Popularizers of Science*, note 14 above.

27 Allen, *The Naturalist in Britain.*

28 See the Introduction, and Roger Cooter and Stephen Pumfrey, 'Separate Spheres and Public Places: Reflections on the History of Science Popularization and Science in Popular Culture', *History of Science* 32, no. 3 (Sept. 1994): 237–67, and Secord, 'Science in the Pub', 269–315.

29 See Orr, 'Fish', which discloses similar uses by Sarah in her later *Anecdotes of the Habits and Instincts of Birds, Reptiles and Fishes* (London: Grant and Griffith, 1853), but not in the *Elements* (*ENH*; *ENH2*) before it.

30 See Orr, 'Women Peers'.

31 I am very grateful to Peter Lincoln for sharing his discovery with me on 19 Feb. 2021 of a letter (DRO138F/F?) from Sarah to Mary Buckland (dated 27 Oct. 1827, concerning the funeral of Clémentine Cuvier in Paris).

32 See *Anecdotes of the Habits and Instincts of Birds, Reptiles and Fishes*, 47.

33 I borrow this qualifying feature of my term 'scientific anecdote' as framed in the eighteenth century in the title of James R. Wood's recent *Anecdotes of Enlightenment: Human Nature from Locke to Wordsworth*. Charlottesville: University of Virginia Press, 2019.

34 Clémentine's premature death in 1827 (which so devastated Cuvier that he shut his Saturday salon) makes her mention in the *Elements* commemorative as well as pedagogical for 'the Author'.

35 For a fuller discussion of the 'maison-musée' phenomenon in France, which Cuvier's home within the Muséum precincts connected to its galleries earlier modelled, see Bertrand Bourgeois, *Poétique de la maison-musée 1847–1898*. Paris: Harmattan, 2009.

36 See Padian, 'Biology in History', note 1, 449.

37 William Carpenter, *Scripture Natural History: Containing a Descriptive Account of the Quadrupeds, Birds, Fishes, Insects, Reptiles, Serpents, Plants, Trees, Minerals, Gems and Precious Stones Mentioned in the Bible.* Lincoln: Edmands & Company, 1833. It lies outside this study, but Carpenter's work is indicative of others by men who published in the same areas that Lightman, *Victorian Popularizers of Science*, note 14 above, claims were the domain of 'maternal popularizers'.

38 *Zoology: A Systematic Account of the General Structure, Habits, Instincts, and Uses of the Principal Families of the Animal Kingdom.* London: Wm. S Orr, 1848. For the biography of Carpenter, see Roger Smith's entry for the *DNB*. https://doi.org/10.1093/ref:odnb/4742.

39 See the Darwin Correspondence Project online facsimile of Charles Darwin's *Narrative of the Surveying Voyages of His Majesty's Ships* Adventure *and* Beagle *between the years 1826 and 1836, Describing their Examination of the Southern Shores of South America, and the Beagle's Circumnavigation of the Globe.* London: Henry Colburn, 1839 at http://darwin-online.org.uk/content/frameset?itemID=F10.3&viewtype=text&pageseq=545.

40 Christian Gottfried Ehrenberg, *Recherches sur l'organisation des animaux infusoires.* Paris: J. B. Baillière, 1839.

41 Agnes Catlow, *Drops of Water: Their Marvellous and Beautiful Inhabitants Displayed by the Microscope.* London: Reeve and Benham, 1851.

Chapter Nine

'JUST AN ANECDOTE'? PIONEERING PERSPECTIVES FROM THE LIFE IN *ANECDOTES OF THE HABITS AND INSTINCTS OF ANIMALS* (1852), *ANECDOTES OF THE HABITS AND INSTINCTS OF BIRDS, REPTILES AND FISHES* (1853) AND *SIR THOMAS THE CORNISH BARONET* (1856)

In light of the '"Accounts" of Direct Encounter' discussed in Chapter 8, Sarah's generic classification of both her *Anecdotes of the Habits and Instincts of Animals* (1852) and *Anecdotes of the Habits and Instincts of Birds, Reptiles and Fishes* (1853) as 'anecdotes' was doubly unambiguous. These companion volumes overtly adopted the term that she had strategically deployed in the revised *Elements* (*ENH2*), to replace the synonymous 'amusing and instructive original accounts' of its first edition. 'Anecdotes' therefore provided 'proofs of sagacity or affection' in the animals concerned and 'proofs' of the tellers' expertise: '[f]ew of them exist in any other publication, unless they have been copied from this work' (*ENH2*, iii–iv). Sarah's phrasing therefore exemplifies the first and second definition of 'anecdotes' in the OED: '**1.** Secret, or hitherto unpublished narratives or details of history. **2.** The narrative of an interesting or striking incident or event'. Because many of the same underpinning expert authorities for the *Elements* (*ENH; ENH2*) informed her *Anecdotes* (*AnecA*; *AnecBRF*) as Appendices 8 and 9 clearly demonstrate for the first time, their names similarly endorsed the *scientific* objectives and expertise of her double-volume *Anecdotes*, yet also question its novelty. Its reader already familiar with the *Elements* (*ENH; ENH2*) in 1852–1853 or today might reasonably suppose that Sarah simply deleted its drier 'textbook' components and remixed the 'anecdotes' that she had already used in it for the same animal. The *Anecdotes* (*AnecA*; *AnecBRF*) were then adroit abridgements – the clearly reduced coverage of mammals in the first volume – and simple supplementation, for example concerning 'reptiles' in the second. On closer inspection (also clarified by comparing Appendices 8 and 9), Sarah's *Anecdotes* (*AnecA*; *AnecBRF*) offer no reheat of the anecdote materials used in the *Elements* (*ENH; ENH2*). Rather, the major importance of 'anecdotes' lies in their careful 'selection', as Sarah states in her three-page preface to the two-volume *Anecdotes* states in 1852 (reformatted in Figure 9.1). As this chapter sets out, expert (natural) selection seriously raises the stakes for the genre of the scientific anecdote itself in her pioneering hands in 1852–1853, and hence for natural science communication today.

In making a selection of anecdotes, those have been assembled which were supplied by me to other works, and in most instances have received considerable amplification; others have been given which never before were printed – perhaps not even written; while all which have been transferred from other pages to mine have received the stamp of authenticity. Besides those whose names are already mentioned, I have to thank several friends who have drawn from their private stores for my advantage, and thus enabled me to offer much that is perfectly new.

Dry details of science and classification have been laid aside, but a certain order has been kept to avoid confusion; and, although endeavours have been made to throw as much interest as possible over these recorded habits and actions of the brute creation; I love the latter too well to raise a doubt by one word of embellishment, even if I did not abstain from principle.

The intentions with which this work was commenced have not been carried out, inasmuch as materials have crowded upon me beyond all calculation; and, although a large portion has been rejected, the anecdotes related go no farther than the Mammalia, while almost all animals were to have been included.

With regard to the remaining orders—if the present work should meet with a favourable reception, I shall hope next year to present the public with touching and amusing proofs of the sagacity and dispositions of birds, and of "hair-breadth scapes" from reptiles, etc., some of which will, like those in the present volume, be carefully selected from the works of travellers, from the resources of friends, and from my own experience.

To the pleasing task of enlightening those, who, shut up in close cities, have no opportunity of observing for themselves, and to the still higher enjoyment of directing young minds to an elevating pursuit, the naturalist adds a gratification even better than all, by making known the hidden wonders of nature; and leaving to those who delight in argument, the ever unsolved question of where instinct ends and reason begins, he sets forth the love of the great Creator towards all His creatures, and the ways He takes to show His wisdom. (iii–v)

Figure 9.1 'Preface to *Anecdotes of Animals* (1852)'.

As the vehicles of pithy, noteworthy yet untechnical information, the immediate apprehensibility of anecdotes also defines what makes their importance paradoxically (not) self-evident in modern critical reception of nineteenth-century science writing. Because anecdotal evidence is problematic for its subjective, potentially unverifiable viewpoints, the 'anecdote' genre does not count as 'serious' report according to modern models and formats for either 'expert' science or 'non-expert' (citizen) science communication. Twenty-first century peer-reviewed scholarship by Michael Dahlstrom, for example, therefore claims that '[a]lthough narratives have a long history of scholarly study […], their integration within a science context is fairly recent. As such, existing discussions within the field of science communication may benefit from an inclusion of narrative constructs.'[1] Clearly, the 'Two Cultures' made famous by C. P. Snow in his 1959 Rede Lecture, which were 'Literary intellectuals at one pole—at the other scientists, and as the most representative, the physical Scientists'[2] have so entirely parted company that science professionals today know next to nothing about the important histories and literatures of their discipline precisely for its interconnecting 'narratives'.

The (scientific) anecdote already set long eighteenth-century precedent for the clear benefit upon which Dahlstrom then concluded: 'The plural of anecdote is engaging science communication'.[3] Its nineteenth-century developments were as the staple format for fresh, expertly informed '(participant) observation' in major published scientific field discovery, notably by Humboldt as indicative, in travelogues, in 'local' expert natural history accounts and in the reporting of well-known and rarer species in the periodical press. Sarah's *Excursions* was in fact unusual by comparison for its paucity of 'anecdotes'.

But historians of science also ignore at their peril the malleable narrative arts also at work in 'anecdotes' to convey their scientific contents in the above range of expert genres, rather than (only) in Victorian popularising forms especially in female hands. Bernard Lightman's significant consideration of 'maternal popularizers' included his particular note of Sarah's *Anecdotes* as 'composed almost entirely of them' as a 'pastiche [...] glued together by Bowdich Lee's short narrative [...]', because she 'viewed the anecdote as an important source of evidence in natural history.'[4] In his concern therefore to better situate women such as Rosina Zornlin (1795–1859) and her contemporaries including Sarah (Bowdich Lee) within Victorian scientific publication endeavour, Lightman overlooked the fact that most male authors of scientific repute also used anecdotes for 'evidence' and for effective communication. The clear precedent of Sarah's *Elements* (*ENH; ENH2*) then overturns Lightman's contention that in preparation of the two-volume *Anecdotes* Sarah was 'unsure of current conventions':

> Requiring assistance in finding new material, and unsure of the current conventions surrounding the anecdote, she wrote to Richard Owen while working on the first book. Declaring her wish to 'supersede the hackneyed stories now current,' she inquired if he would loan her those books that he owned that contained juicy quotes. She also asked him 'what sources do you advise me to draw from when the anecdotes are not private?' Finally, she solicited his opinion on the issue of whether or not it would 'injure my reputation as an author if I repeat my own anecdotes?' Bowdich Lee decided to include anecdotes about her own interactions with animals, anecdotes told to her by acquaintances, and anecdotes from books and journals. In *Anecdotes of the Habits and Instincts of Birds, Reptiles, and Fishes* she included stories from natural history journals, such as the *Naturalist* and *Naturalists' Magazine*, and from the general periodical press, for example, the *Northampton Mercury*, *Saturday Magazine*, Chambers's *Edinburgh Journal*, and the *Edinburgh Literary Gazette*. Among the naturalists quoted were Philip Gosse, Darwin, and Charles Waterton. One of her own anecdotes recounted the story of how she caught a twelve foot shark [...].[5]

Our different evidence in Chapter 8 clarifies that Sarah was writing to Owen for his expert opinion on her measures to authenticate the *science* of her undertaking in the *Anecdotes*, especially when reusing scientific anecdotes in the public domain, including her own. As we will explore in one key example below, her request for loan of 'books that he owned' was precisely because she was aware that 'the hackneyed stories now current' too-readily collocated with broadsheet, omnibus, or 'pick and mix' publication formats such as the *Penny Cyclopedia*, which frequently recirculated and reprinted material without permissions or acknowledgement. But Sarah was also eliciting new 'anecdotes' nowhere else in print from Owen and his circle, conscious of the particular sensitivities for *their* scientific reputations that would be her acknowledging of them publicly in print. For example, Appendix 9 tellingly reveals for the first time that Owen and his wife supplied private 'accounts' to Sarah on (their) dogs. That the serious 'point' of new observation in the scientific anecdote hides in plain sight of the wider genre's ubiquity, familiarity and accessibility in mass print press publications for informed middle-class readers including female,[6] explains why it eludes modern critical notice as 'science proper'. At best for cultural critics and historians of Victorian science, anecdotes collocate with the popularisation of science rather than the communication of *new* science, due to their 'untechnical' content as mistakenly inferring lesser scientific knowledge value and authorship. Their problematic status is then compounded by what Joel Fineman claims for anecdote as 'the literary form that uniquely *lets history happen* by virtue of the way it opens an opening into the teleological, and therefore timeless, narration of the beginning, middle and end.'[7]

In Sarah's hands, the 'literary' advantage of anecdotes for science 'proper' was rather their ability to disrupt accepted knowledge and narrative of 'things as they are'. In such contexts, let us recall her specific use of the word 'anecdote(s)' already in the introduction to the *Memoirs* (discussed in Chapter 3): 'Unwilling to incur the risk of confusion, by mingling too much *anecdote*, either with my narrative of events or description of scientific and legislative labours, I have divided the present volume into four parts […]; and the fourth will be chiefly confined to those *anecdotes* which will best illustrate his character as a man' (*MBC*, 6, emphasis added). The original and authoritative provenance of her 'anecdotes' in 1833, and punctilious acknowledgement of others' first-hand scientific anecdotes in the *Elements,* explained her especial opprobrium in 1850 of unscrupulous authors passing off others' (her) anecdotes as their own. By contrast as her preface states, her anecdotes were 'carefully selected from the works of travellers, from the resources of friends, and from [her] own experience' in her two-volume *Anecdotes.* By probing Sarah's overt genre choice of 'anecdotes' for *new* science in her two-volume *Anecdotes,* this chapter investigates her unassumingly innovative, even distinctive use – 'selection' – of anecdotes in both contents and form in her two-volume work, including its strategic pedagogical developments of the revised *Elements* of 1850.

The target market and listings for Sarah's last works in their Publisher catalogues were equally unambiguous: for 'Juveniles'.[8] Her pioneering approaches to scientific observation 'from the life' as our chapter title intimates included well-chosen 'anecdotes' that further hallmark her 'field'-defining ventures in natural history. In consequence, her deployment of such original material was to capture both the interest and the striking *event* of her subject(s) in *natural* history thus warranting recording,[9] because the observation was vouchsafed by expert, not second-hand authorities. Sarah therefore overtly sought out for her *Anecdotes* those that constituted 'recognised' science in published books and journals by leading exponents. As their peer, she could also include her own in the best traditions of scientific observation as practised by field naturalists such as Gilbert White. In recent reconsideration of anecdote in his work, Melissa Sodeman noted his legacy in '[s]ubsequent collections of animal anecdotes and general works of natural history […] using the form to delineate animal behavior, to illuminate providential design and to appeal to those new to scientific study', and cited Sarah's *Anecdotes* for their 'evidentiary function [...] to capture behaviors thought characteristic of each species.'[10] What Sodeman overlooks is that Gilbert White and Sarah both use scientific anecdote to report what is also *not* 'thought characteristic of each species', to challenge perceptions about natural (or 'providential') design, and hence previous scientific knowledge concerning well-known, as well as rarer species. Sarah's 'anecdotes' in the *Anecdotes* did not derive from 'mass market' sources, whether writers or print formats. Although some appeared in periodical publications as noted by Lightman above, they originated as anecdotes within book-length studies by species or habitat 'experts', who constitute her main authorities, whether scientific explorer naturalists in all continents of the globe, or renowned national and regional specialists. Sarah's *Anecdotes* are then no 'pastiche [...] glued together' as Lightman claims, whether in his etymologically correct use of the word 'pastiche', or in its more pejorative meaning in literary-critical ears of unoriginal imitation, bricolage or potpourri. Rather, her 'selection' as a collection of *suitable* anecdotes concerned how they could best define a given creature accurately and engagingly, so that *newcomer* ('juvenile') naturalists could identify and understand its singularity in observable species habits and habitats. Sarah's challenge in her *Anecdotes* was therefore to rescope 'textbook' natural history for those without any prior interest or knowledge.

The hallmark of the scientific anecdote in Sarah's preface above and in her 'selection' in the *Anecdotes* was the uncommon capacity of the form to convey known and new scientific interest freshly, and to make natural history a 'living' pursuit, observable to all who know how and where to look, and what to see. Well-chosen scientific anecdotes thus supply multifarious experiential report of the wonder and diversity of the natural world, especially when sourced from 'naturalists' of renown. Indeed, Sarah's significant and rare use of 'naturalist' in the last paragraph of her preface differentiates the long-standing field expert by degree from the enthusiast, the dilettante or the amateur,[11] because the (expert) 'naturalist' is constantly vigilant for new detail in respect of the already known, and for the uncommon sighting. The *Anecdotes* therefore provide expert report that circumvents the zoology 'textbook' form(at) of '[d]ry details of science and classification' (in her preface), to engage instead in 'the pleasing task of enlightening those, who, shut up in close cities, have no opportunity of observing for themselves', yet who can be directed 'to the still higher enjoyment of [...] an elevating pursuit'. By positioning the scientific anecdote optimally as specialist expertise in familiar form, Sarah clearly confronted and addressed in 1852 the problem that John Durant articulates 150 years later: 'What is meant by the "the public understanding of science" in Britain, by "la culture scientifique" in France, and by "scientific literacy" in the United States?' by stating that '[s]cientific literacy should not be taken to mean the knowledge of a lot of science, but rather the understanding of how science really works.'[12] Sarah's carefully chosen anecdotes demonstrate such 'science-in-action' in the two-volume *Anecdotes*, through offering individual and collective exemplar testimonial reports on a wide selection of different animal creation, pitched to urban target audiences with the least direct access to it.

The two-volume *Anecdotes* therefore turn on their 'naturalist [...] he' of the last paragraph. This figure is not only the intermediary 'enlightening' natural history, but also the 'director' to 'elevating pursuit', and to larger questions of 'the love of the great

Creator towards all His creatures, and the ways He takes to show His wisdom'. But precisely because *her* 'own experience' concludes paragraph three as the frame for this figure and 'his' roles, Sarah deftly points up the named and published woman scientific author(ity) on the covers of the *Anecdotes* (*AnecA*; *AnecBRF*) to figure the place of the more unusual and innovative 'naturalist [...] *she*' within them. If she orchestrates previous expert natural history knowledge and its making, to 'offer much that is perfectly new', by adding to such content and insights as a woman peer, her 'selection of anecdotes' went further. In 'those never before printed' and 'drawn from their private stores for my advantage', Sarah also pivotally refitted the scientific anecdote form and genre for new purpose 'towards all His creatures', rather than only to 'higher' animals. Her *Anecdotes* could not be farther from the saccharine animal tales published by women writers for their moral content in the nursery.[13] Sarah's 'selection of anecdotes' for their scientific content and strategic scientific form overtly promoted field natural history as emancipatory for 'those [...] shut up in cities', especially the 'young minds' that include girls'.[14] The expert evidence and observations of the 'naturalist [...] she' immediately challenged the 'natural' orders and knowledge of things, to include the lesser known, and the lesser spotted.

Sarah's first-hand experience and authority in publications on the natural history of both Britain and West Africa directly informed and guided her 'selection of anecdotes' in the two-volume *Anecdotes.* The 33 animals covered in volume 1 (*AnecA*) stand out against those chosen for compilations made by book-learning alone (by 'popularisers' of both sexes), because they are selectively representative, rather than comprehensive in coverage. The volume focusses on some dozen specifically 'African' mammals – 'Monkeys'; 'Hyaenas'; 'Lions'; 'Leopards, Panthers, Etc.'; 'Elephants'; 'Hippopotamus'; 'Rhinoceroses'; 'Camels. – Dromedaries'; 'Giraffes'; 'Antelopes' – alongside indicative 'familiar' northern and southern hemisphere quadrupeds in 'Bears', 'Wolves' and 'Llamas' on public view in London and other zoological collections in British cities.[15] The non-British animals here thus supplement important male-authored national natural histories, for example Thomas Bell's *A History of British Quadrupeds* to which we return,[16] by highlighting 'common' creatures in specifically British habitats also found across the world. In the order of the table of contents, 'Bats', 'Moles', 'Hedgehogs', 'Badgers', 'Weasels', 'Otters', 'Dogs', 'Foxes', 'Cats', 'Squirrels', 'Rats', 'Mice', 'Hogs', 'Horses', 'The Ass', 'Deer', 'Goats', 'Sheep' and 'Oxen' in Sarah's first volume are 'familiar natural history'. All would have been immediately recognisable to the (young) urban British reader in 1852 if only in an ABC or book about a farm.

Small and large – the Elephant follows the Mouse – are therefore selected not as Bible or 'animal fable' narratives, despite the seemingly 'Christianising' culmination of Sarah's preface,[17] but to exemplify her main point. To promote natural history knowledge through representative animals and their 'anecdotes' clearly if indirectly by 'a certain order [...] kept to avoid confusion' (in her preface above), focusses attention on the observable scientific interest and wonder of *all* creatures, not on exoticism or the anthropomorphising or moralising aspects of (only some) animal 'habits and instincts'. The reader of the *Anecdotes* (*AnecA*) – whether 'young', 'proficient' or 'specialized'[18] – thus apprehends animal classification and orders not through dry disquisitions, but directly by indicative and 'familiar' example: Bats illustrate 'Cheiroptera', Moles and Hedgehogs 'Insectivora', Cats 'Carnivora', Squirrels 'Rodentia' and finally Llamas and Oxen as world 'Ruminantia' explain scientific orders differently found in Cuvier (in the *Règne Animal*) and in Bell's British zoology. The important object was to encourage readers' nascent (specialist) interests in natural history by whetting their curiosity to know more. The two-volume *Anecdotes* and their anecdotes therefore strategically connect urban readers to natural worlds within and farther from reach, by conveying through common exemplification the otherwise alien worlds of expert (dry) zoological terminologies, given only as bracketed 'afterthought' throughout. Commencing with Monkeys, Sarah's *Anecdotes* (*AnecA*) like her *Elements* clearly separates human from 'brute creation'[19] in a selection of similarly vertebrated representative animals, while concomitantly inviting the reader in accounts of their multifarious 'sagacity' to contemplate afresh 'the ever unsolved question of where instinct ends and reason begins' (in her preface). Upon this question hung the ultimate distinction, or its lack, between (sexed) human and animal nature, and hence the moot religious and scientific understandings of creation(s), and a Creator of 'all His creatures'.[20] But there is then nothing overtly Christian, anti-religious, philosophical or, indeed, 'maternal' in Sarah's selection, collection and recompilation of the hundreds of anecdotes in both volumes of her *Anecdotes*: a Bishop (Heber) was one among her some sixty named authority sources for the first volume as Appendix 9 confirms. All prove her (unchanging non-conformist) scientific view as also Humboldt's: mankind is not of foremost or determining interest in creation, whether it was a singular event or plural in occurrence. All the selected and attributed scientific anecdotes in the *Anecdotes* therefore served 'to throw as much [specialist] interest as possible' (Sarah's preface above) on the natural world.

To qualify for inclusion, the 'selected' scientific anecdote(s) therefore had to do much more than entertain or illustrate the more important information delivery (as in the *Elements*). As the primary vehicle for understanding, anecdotes had to provide direct – that is, exemplary, representative, illustrative, instructive – scientific insight, because 'direct' also conferred *in situ* apprehension of observable facts with the facility to pinpoint new particulars of discovery in an immediately enlightening short account. The double requisite of the 'direct' qualifying Sarah's 'selection' of anecdotes also dynamises the sheer variety of the genre's differently expert observation-reports as multiply fresh perspectives in natural history-making. It was this variety of anecdote exponents that could then not only captivate but also maintain the interest of newcomer urban readers of the *Anecdotes*. So many demonstrations from the field were suddenly found to be within the reader's grasp in the volume before them as the spur to discovery of the richer *diversity* of natural history close to home and farther afield. A 'Bat' for example, reveals its many different kinds. The importantly dual economy of Sarah's 'carefully selected' scientific anecdotes – at once to encapsulate knowledge and to be telling – permitted each and all to disclose the many and variously known, as well as the yet-to-be-told wonders of the natural world, because its observation proved possible for all-comers. Indeed, every expert naturalist behind the selected anecdote as new 'event' for science reflected the always (and one-time actual) 'beginner'.[21] In bringing expert natural history in accessible terms to the as yet unformed urban reader, Sarah's *Anecdotes* therefore also 'select' cameos of observational best practice and report as individual and representative in a world-(class) library of scientific travel accounts and zoological works. In briefly examining her use of anecdotes from those by the zoologist (and dental surgeon) Thomas Bell (1792–1880) to exemplify her selection criteria of 'authenticity', we also discover Sarah's larger point for 'anecdotes'/her *Anecdotes* as expert natural history from the life.

Bell's *A History of British Quadrupeds* of 1837, with its extensive Preface (vii–xii) and 'illustrated by nearly 200 woodcuts', provides a model 'comparative anatomy' of integrated expert scientific anecdote by which to compare Sarah's. His work was both larger in scope (numbers of quadrupeds covered, including *Cetacea*), yet much more delimited by its 'locality' of *Fauna*, for which justification was apparently unnecessary because '[t]he advantages [...] are too generally understood and acknowledged to require any lengthened proof or illustration' (Bell, Preface incipit, vii). Bell had become Professor of Zoology at King's College London in 1836. This volume was the first of his 'popular works written for John Van Voorst [...], compilations [...] open to criticism for their inclusion of domestic animals, and for important omissions, as well as for their lack of any original research.'[22] In 1837, experts in British natural and earth sciences were also powering organisations such as the British Association for the Advancement of Science (BAAS). Bell therefore saw in his Preface the further advantage of

> the cultivation of the Natural History of our own country [as] the means which are thus offered to multitudes of persons who are restricted by circumstances from engaging in the study of the higher departments of the science, of obtaining a rational and never-ceasing enjoyment; and to the young especially, of opening an exhaustless source of amusement, at once healthful to the body, and favourable to the elimination of the best qualities of the heart and understanding. (vii–viii)

Sarah's preface clearly dialogues with Bell's. But in outlining animal classification, he then situated his work squarely in the history of the best 'former British Naturalists' (viii–ix) in Pennant, 'The History of British Animals by Dr Fleming', but especially 'The Manual of British Vertebrate Animals by the Rev. Leonard Jenyns' of 1835. Bell concluded his own revision and addition to their work by thanking named friends (all-male), including Yarrell, Hogg and Blyth for communications and loans of specimens. A Mr Dickes and a Mr Vasey were acknowledged for the illustrations (xii).[23]

Sarah's careful choreographing of anecdotes from Bell's *Quadrupeds* therefore adheres in her *Anecdotes* (*AnecA*) to 'his' order of animal classification in altogether scientific, albeit highly selective and summative ways. In picking out only 'Long-eared Bats' (Bell, 53–57), she would have perused the 22 descriptions with accompanying 'vignette' images to find in 'one of the most common of our British Bats' (Bell, 54) her 'specimen' anecdote. Her particular choice then reset his identificatory description, including its 'small sharp cry' (Bell, 55), as an inhabitant of 'towns and villages' and 'towers of churches' (Bell, 56) for the budding urban naturalist also to make a first 'discovery' through the ears as well as the eyes in her *Anecdotes*. This bat's cry being the instantly memorable feature for its identification, even if the Bat was itself 'common', avoided all belittling of the new discoverer: s/he learns that Bats operate at night by their ears (echolocation), and immediately remembers the name, 'Long-eared Bat', in its attributes. Sarah's version in 1850 therefore offered comparative reappraisal for her readers also familiar with Bell with its accompanying first 'vignette' (Bell, 62) of this scene. It illustrates two little girls at play beside their mother's skirts, excitedly seeing a bat outside the window of their (urban) drawing room, with their father and older brother (with family dog as comic detail) looking on. This and other vignettes in Bell only underscore the paternalistic viewpoints writ large in his preface and text regarding hierarchies of (gendered) 'ability' in natural history, and of animal nature as inferior to human.

If such reading here of (natural) 'selection' by Sarah concomitantly to point up and overturn 'deselection' in natural history including female by Bell (as indicative of 'naturalists') appears too modern, then the case of the 'Bat' alerts the reader to others. For the 'Badger' (Bell, 122–28) as sole British representative of the 'Ursidae', Sarah did not take up the 'barbarous and dastardly sports' of badger baiting in Bell (124, also illustrated in a vignette, 128). Rather, her focus aligned with her unchanging abolitionist views to underscore from Bell the centrality in British culture of Badgers in 'common expression' – 'A person who is beset by numerous assailants is said to be "badgered"' (Bell, 125) – and in philology (Anglo-Saxon 'Broc', Bell, 127). More significantly concerning Bell's description of the 'Common Weasel' (Bell, 141–47), Sarah picks out from the longer description only his 'anecdote' concerning its 'sagacity' as the natural prey of hawks. Because the weasel is itself a predator, the 'top' predator's relation to similar prey does not always conform to the expected outcome in the 'order' of nature 'red in tooth and claw', as Tennyson famously put it also in 1850:

> But the following fact shows that violence and rapine even when accompanied by superior strength, are not always a match for the ingenuity of an inferior enemy. As a gentleman of the name of Pinder, then residing at Bloxworth in Dorsetshire was riding over his grounds, he saw, at a short distance from him, a kite pounce on some object on the ground, and rise with it in its talons. In a few moments, however, the kite began to show signs of great uneasiness, rising rapidly in the air or as quickly falling, and wheeling irregularly round, whilst it was evidently endeavouring to force some obnoxious thing from it with its feet. After a short but sharp contest, the kite fell suddenly to the earth, not far from where Mr Pinder was intently watching the manoeuvre. He instantly rose up to the spot, when a Weasel ran away from the kite, apparently unhurt, leaving the bird dead, with a hole eaten through the skin under the wing, and the large blood-vessels of the part torn through. A similar anecdote is related in Loudon's Magazine of Natural History, where the *dramatis personae* were a Stoat and an Eagle; but the truth of it appears not to be vouched for by the narrator. Of the accuracy of the present fact there is, however, no doubt, as I knew Mr Pinder well, and have often heard the circumstance related. (Bell, 145)

In quoting Bell's account in full in the *Anecdotes*, Sarah allowed the force of this 'familiar' natural history from the life to work its de-familiarisation effects of observational drama for the reader. The immediacy of attention to unfolding detail pivots on the reader's positioning as if 'Mr Pinder' in the observation of the event, in the act of discovery of an otherwise uncertain outcome between two top predators. But the memorable illustration that this scientific anecdote epitomises – most obviously 'the ingenuity of an inferior enemy' – also underscores its several lessons in Sarah's hands for science 'proper'. First, alertness to the common and uncommon in any natural history encounter keeps the 'naturalist' open to new surprises in the field. Its experience is always different from the controlled, ordered (urban) environment of the drawing room or zoology collection. Second, not all accounts count equally. Bell's anecdote distinguishes its main animal protagonists – but not the sex of either – as well as its 'accuracy' from other versions, for which 'the truth [...] appears not to be vouched for by the narrator'. In Bell's words here are Sarah's own concerns for 'authentic' report in her requests of Owen for expert sources that will assure the science of *her* work. But Sarah also 'out-anecdotes' Bell by being herself the source of original anecdotes (in *AnecA*), where he was not. If Bell's return more largely in Sarah's second volume (*AnecBRF*) as the authority on the reptile and Galapagos turtle specimens from Darwin's *Beagle* expedition, and as a resource also of anecdotes in its sections on 'Reptiles' and particularly 'Batrachians',[24] the final part on 'Fishes' surpasses his models of expert 'zoology' (including from the life). I have earlier drawn attention to how extensive were its materials from her *Fresh-Water Fishes of Great Britain*.[25] But this last part of the two-volume *Anecdotes* without accompanying illustration (by Harrison Weir) was of such challenge to the (scientific) anecdote genre itself that Sarah drew reader attention precisely to it:

> Nearly destitute of affection; almost deprived of the power of uttering a sound; scarcely knowing even a relative duty towards each other; their chief object, that of eating; it would at first sight seem very difficult to find any anecdotes concerning fishes, which may be calculated for a work like the present. The exceptions, however, to the above facts, are so extraordinary to us; our great Creator has provided so many curious contrivances for their preservation; He has made some of them so valuable to man; He has endowed so large a portion with extreme beauty; He has caused others to be so singular; others again to be so terrible; and He has ordered some so to step from their general character, so to surpass our finite comprehension, so to cheat us or our ideas of fixed laws, that I flatter myself this last portion of my work will afford the same interest to the general reader, as those which have had the precedence. (*AnecBRF*, 276)

Sarah could not have been on more familiar ground. As herself the (source) of foremost natural history knowledge of British fresh-water species, and of anecdotes concerning them, this 'naturalist [...] *she*' demonstrates once more how pioneering were her observational and representational abilities in this field, as Chapters 1, 2 and 7 have all differently confirmed. Sarah's practised use of 'anecdotes' derived from her own earlier work in the *Fresh-Water Fishes of Great Britain*:

> Dr Tench, for such is his appellation in several parts of the country, is said to heal his wounded companions, by shedding over them the mucous secretion with which he is so abundantly provided. He himself keeps alive longer than most fishes, and is a fleshy, clever fellow, hiding himself under mud when danger comes, and when he gets a snug place to live in, not at all liking to come out of it. Let not the portrait painters of fishes be tempted

> by his rich green and gold coat to make a resemblance of him; for when they come to the small scales on his fat sides, they will repent of their undertaking. It so happens, that I have copied him fifty-six times over, and I now cannot look at him without a feeling of weariness. (*AnecBRF*, 298)

Encapsulated here are the 'telling' details of the scientific anecdote in its observational acumen to convey the creature in question in a few identificatory, but memorable 'brushstrokes' and in its appeal as knowledge-in-the-making. But this representative self-insertion (of many) by Sarah in the *Anecdotes* as their eminent 'naturalist [...] she' also look beyond well-known 'animals' (mammals and birds), to include all classes of 'clever' natural history, and the otherwise overlooked experts in the field in word and *image* who had brought them to attention. Sarah's meticulous acknowledgement and precious dissemination of the work, expertise and authority of others' scientific 'anecdotes', including by rarer women in the *Anecdotes* (*AnecA*; *AnecBRF*), constituted their almost more important point for posterity. It chimed with her recording in West Africa of indigenous female 'stories' (in Chapters 4 and 5). Her 'selection' of 'anecdotal' voices for natural history also respected persons who otherwise left little or no trace in 'official' (British and French) natural history papers and transactions, despite their authors being important interlocutors in the communities of specialist natural history knowledge in their time.

By raising the bar for the powers of scientific anecdote to recalibrate the 'deselection' story (interests and politics) of natural history endeavour itself, Sarah's *Anecdotes* therefore reconfigure the scientific anecdote as a genre more inclusively, to encourage future corrective contributions to accepted Views of Nature (to pick up Humboldt's title in translation). Her extraordinary wording above, 'He has ordered some so to step from their general character, so to surpass our finite comprehension, so to cheat us of our ideas of fixed laws' (*AnecBRF*, 276), dethrones the human animal and denotes the 'lesson' through 'anecdotes' in her *Anecdotes* for institutional, gentleman and even parson naturalists, and those ignorant of natural history in the field. The genre's more profound respect for the multifarious 'intelligence' of the natural world also promotes its better 'direct' observation, whether of creatures that are common or rare, at home and abroad. (Expert) natural history is within the reach of all.

Sarah's two-volume *Anecdotes* therefore offer a fascinating display-case and object lesson today of natural history knowledge-making in the field in the first half of the nineteenth century. The powers and impetus of scientific anecdotes to develop *telling* new observations – that is, singular to that creature and worthy of remark – also distinguishes those by the 'naturalist' worthy of the name from travellers' and other reports that are tall tale.[26] Although Sarah's promise in her preface above of 'hair-breadth scapes' connotes the sensationalist, as also reflected in situations summed up in the titles for the six commissioned images by Harrison Weir for each volume,[27] her focus was on animal encounters where animal, not human 'sagacity' variously prevailed (as his illustrations endorse). Indeed, in reporting a range of animal subjects and their often unexpected behaviours in the natural history anecdote Sarah, like Gilbert White, reopened the case for telling observations 'from the life' rather than in the natural history collection. By example, her *Anecdotes* (*AnecA*; *AnecBRF*) inverted the assumed status of the genre as minor, and for minors in science-making, with contents that everywhere demonstrated expert provenance and *interest*. They therefore exemplify, and even surpass in her 'Dr Tench' in word and image, the criteria in the major call in 2020 for 'anecdotes in animal behaviour' by the editors of *Behaviour*, now revealing the provenance of our chapter title:

> 'It is just an anecdote', is an often-heard critique to scientists describing animal behaviour that has not been reported before. But what if this anecdote is a qualitative, rich observation, showing a truly unique behaviour or event that can change the way we think about a species? If an ethologist or another animal expert observes unforeseen behaviour that may be of interest, it is a loss for the field if this observation remains unreported and as a result forgotten. This is especially the case for very rare behaviours or unusual circumstances under which certain behaviours occur. [...] Nowadays, such narrative accounts of unique behaviours are on a stark decline (Ramsay & Teichroeb, 2019). In order to safeguard valuable anecdotes in animal behaviour and prevent them from oblivion, at *Behaviour*, we deem it highly valuable to make them available again to the scientific community.[28]

Sarah's astonishing virtuoso lesson in natural history-making is then through 'selection' of the scientific anecdote genre, because it can promote the seemingly common, as well as the 'rare'. The scientific anecdote therefore already challenges modern 'naturalists' and historians of science to rethink what they assume about the provenance, forms and hierarchies of scientific knowledge-making as long settled (today) by the conventions, standards, methods and publications of expert (official) science. By the same token, Sarah also distinguishes the anecdote's telling report of new scientific facts ('faits') from the differently arresting news story ('fait divers') reporting a singular phenomenon or human event,[29] and what James Wood recently terms the 'anecdote of enlightenment' in eighteenth-century literary studies:

> Whether they came from near or afar, anecdotes prompted philosophers, essayists, travel writers, and poets to rethink what they believed they knew about human nature. Writers were drawn to anecdotes of people (and occasionally animals) who seemed to differ markedly from themselves: tales of hunchbacks and housekeepers, polytheists and parrots, savages and slaves. Anecdote opened paths leading out to the perceived peripheries of the human world. But anecdotes also tended to unsettle conventional notions of what was central and what was peripheral in human life, frequently pointing thinkers toward the conclusion that both the norm and exception obey the very same set of law.[30]

Clearly, if *human* nature and enlightenment form the object of the (literary) 'anecdote' for Wood, the genre's alleged unsettling 'of conventional notions' rather problematically turns on his examples: 'tales of hunchbacks and housekeepers [...] savages, slaves' as almost-humans fail to question the 'law' of human civilised ('Enlightenment') society. By contrast, Sarah's *Anecdotes* more unsettlingly promote the scientific anecdote as the genre 'from the life' that neither anthropomorphises animals nor animalises humans. As the culmination and prism of her work, *Sir Thomas, or the Cornish Baronet* can now be examined afresh for its view of scientific 'anecdote' as also a remarkable purveyor of challenging truths for European natural science and civilising missions.

Coda: Unsettling Report from the Fields of Natural History Overseas

To promote the work(s) by unheralded women like Sarah in natural history before 1860 also means in this book neither to lionise nor overpromote them unequivocally, especially where a corpus proves uneven in quality. Sarah's last fictional works of 1855 and 1856 for 'Juvenile' readers are at first sight neither serious 'adventure fiction', nor serious natural history, including 'natural history fiction' discussed in Chapter 6. The reader of *Sir Thomas* without knowledge of Sarah's earlier scientific or fictional works could only compare it very unfavourably with adventure writing (for boys) of the period (by a Captain Marryat, or a G. A. Henty). For the reader familiar with the *Elements* or the *Excursions*, Sarah's last novel has lost all serious plot and interest in natural history. Yet the colonial mission and its critique that I discussed in Sarah's *Playing at Settlers* (1855)[31] is even more overt in *Sir Thomas*. Is critique the new direction of her scientific travel? From the first paragraph of its two-paragraph Preface (see Figure 9.2), the novel sets up the known failure of its fatal 'adventure'. This strategic prefatory 'anecdote' in light

> As the writer of the following story was one day walking on the Scorpion Hill mentioned therein, she saw several blocks of large stones, cemented together, lying about in various directions. She inquired of the Governor of Cape Coast what they were, and he then told her they were the last remains of Sir Thomas's Folly, or so the Tower built by him had been termed. "And who was Sir Thomas?" she asked. Her uncle replied, that a Sir Thomas somebody, for even his name was forgotten, had come some years before his time to Cape Coast, and was, with his servants and four-footed animals, swept away by the unsparing climate. [...] but a tradition still existed in the minds of the current generation, that he had been an outlaw, and had come to that country in utter ignorance of its climate and inhabitants, and with an exaggerated idea of its freedom and resources. In less than twelve months all the party were in the grave.
>
> All the events described, actually occurred although, perhaps, not to Sir Thomas; and they present a faithful picture of the manners and customs of the people of Fanti, as they existed a few years back. But better times have reached this benighted land, instruction and Christianity are marching through it, hand in hand; a better system of medical treatment has been introduced, and it is now possible for Europeans to return from thence. Steam vessels now ply between the Gold Coast and England, bearing with them comforts and refinements. Lawless and rapacious characters have, in several instances, become respectable members of the community; and the word of salvation is rapidly spreading. The whole continent is opening to us, and enterprising travellers are revealing its secrets to us; so that a few years hence, it is to be hoped that Africa will no longer be a land of mystery. (iii–v)

Figure 9.2 'Preface to *Sir Thomas* (1856)'.

of Sarah's two-volume *Anecdotes* may then have more to reveal, precisely in its melding of verifiable (un)common report and authorial expertise. In neatly demanding and commanding reader attention since this is what anecdotes do, this prefatory anecdote also clearly situates the 'author [...] she', and *Sir Thomas* as a thin autobiographical disguise of her first encounters with 'Ashantee' (see Chapter 5). Does this 'potboiler' in the lineage of *The African Wanderers* tell a cautionary tale for imperial science in its 'misadventure' story?[32] Sarah's unusually direct criticism in *Sir Thomas,* especially for a woman author in the 1850s, therefore importantly raises the bar for interdisciplinary scholars balancing current debates on decolonialisation, diversity and inclusion with contextualised understandings of pre-1860s imperial and colonial science narratives. Did Sarah's double 'Paris-London' view of mission overseas bring unusual comparative and contemporary critique of it? By way of concluding this first book-length study of Sarah's extraordinary natural history corpus in the diversity of its genres and pioneering perspectives, this final chapter section takes up the scientific anecdote in Sarah's hands as a tellingly far-reaching, untrivial pursuit.[33]

The main set piece episodes in the novel, as clearly represented in the initial illustration by the renowned John Gilbert, 'The landing of Sir Thomas', consistently display the eponymous protagonist's mistaken imperial values and their re-examination: there can be no mistaking the novel's wider ideological critique, especially because Chapter 4 frames Sir Thomas's many subsequent misadventures:

> Before dinner, at dinner, and after dinner, the strangers [Sir Thomas's party] were discussed by the Governor [of Cape Coast Castle] and his guests. [...] 'It is impossible that they can stay here,' said the President of the Council; 'and if such be really their intention, it must have been formed in utter *ignorance of the climate of this country*, or *utter ignorance of the casualties which they must encounter.*'
>
> 'The captain of the brig told me,' resumed Mr. Anderson, 'that he had bought her, and had on board nine horses, four cows, and a pack of eighteen dogs.'
>
> 'Mad! mad! mad!' ejaculated the Governor; who then added, 'in common humanity we must try and persuade him to go to Accra.'
>
> But the Governor little knew the disposition of the man whom he talked of persuading. Sir Thomas had been told that the land at Cape Coast was free to every one; that the forests in the immediate vicinity teemed with those beasts which would afford noble sport and maintenance; that gold was so easily and plentifully collected there as to be the currency of the country, and that he might procure enough to purchase back his ancestral halls, without the humiliation of suing for them [...] Sir Thomas was never mistaken, especially in matters of importance. (*ST*, 44–45, emphasis added)

The critical narratives on several intercultural levels here also challenge the seeming claims for 'civilising mission' as contained and articulated in suitably triumphalist language in the second paragraph of Sarah's Preface in Figure 9.2.[34] How we read it and the excerpt above as indicative of the novel's critique in the 1850s (and today) pivots on the initial one-paragraph 'anecdote', and how the reader classifies it and its 'writer [...] she'. For those familiar with Sarah's larger corpus, this incipit directly mirrors 'the naturalist [...] s/he' in her preface to the *Anecdotes.* But the unusual first-hand scientific experience and publications of the 'Mrs Lee' on their title pages and inside covers is also unequivocal. Sarah's forms of third-person self-naming and authority deftly develop the prefatory conventions and uses of anecdote that she had strategically deployed in her *Memoirs.*

Most unusually for the adventure novel in 'African' mode – based on African Travel Accounts as exemplified in Appendix 6, and as reconfigured in autofictional mode in Sarah's *African Wanderers* (see Chapter 6) – *Sir Thomas* consistently reframes the allegedly gender-neutral narrative optics of such genres in the active verbs 'she saw', '[s]he inquired (of the Governor)', 'she asked' in its prefatory anecdote. It is precisely on this 'explorer' information and its perspectives that the narrator(-she) conducts her different 'mission', the reconstruction of the larger story of Sir Thomas from the very limited 'intelligence' given in the first report (by the Governor). Sarah not only further self-attributes this 'she' to her several parts in Edward's *Mission* but also underscores in this 'anecdote' the important fact overlooked in knowledge-making in non-Western cultures: oral narrative and its traditions. The eponymous protagonist's half-remembered 'story sensation' in its time importantly lives on in *local* memory through the powers of anecdote in African story-telling traditions. Sarah's collection and reworking of them, examined in Chapter 5, allowed her to reconfigure *Sir Thomas* also for British *imperial* memory. Hers are therefore the wider 'events described' and 'faithful picture of the manners and customs of the people of Fanti' above through her resituating of this 1856 novel around 1820. Its dating lies in the small but significant detail in *Sir Thomas* in its fictional 'Journey Out' (mirroring Sarah's own, see Chapter 5): 'When off Cape Palmas, a small fleet of canoes surrounded them, and the chief came on board, requesting to know if the master of the brig wanted to buy any bulls for the ship's provisions' (*ST,* 38). Sir Thomas's declining of the request demonstrates his lack both of material need of 'bulls' and necessary savvy, as supplied in the telling aside-explanation-translation: '"By bulls he meant slaves," said the sailing master' (*ST,* 39). The explanation would bring the reader up short in 1856 and today; people-smuggling and modern slavery remain a reality.

But the nub of the prefatory 'anecdote' and its larger 'gloss' in the excerpt from chapter four above is Sir Thomas's larger ignorance of the overseas realities of British civilising mission, and his related cultural arrogance. Despite the (black) comic preposterousness of his plans for setting out with his daughter Blanche and son Alan, and for setting up life in his West African 'Folly', they are also a parody and critique of settler manuals and other colonial propaganda, such as Charles Dickens's 1851 'Emigrant Letters' that Sarah had taken to task in her *Playing at Settlers* (1855).[35] But the very ruins we first encounter in *Sir Thomas* that are his 'Folly' (material and metaphorical) strongly indicate the novel's larger counter-narrative of the Enlightenment project. As an 'anti-Robinsonade',[36] Sarah's views from the outset of *Sir Thomas* allow her to entertain the *unsettling* of her juvenile readers' preconceived imperial notions in the novel's white colonial 'mis-adventure' in British overseas territories. How she adeptly re-choreographs imperial scripts results from her one significant change to the ending of the original 'anecdote': 'In less than twelve months all the party were in the grave.' A 'not' qualifies the 'all'.

Powering Changes to Advance Female Positions

The many perils and real dangers of African (adventure) enterprise replicate in *Sir Thomas* the mainstays of 'swash-buckling' adventure fiction – ships boarded by pirates, dangerous and difficult environments and inhabitants (animal and human), obstacles of all kinds to overcome by heroic acts, escape from 'enemy' capture and near-fatal tropical diseases – with two main differences. First, the novel's 'Adventures in Fanti-land'[37] are determined, like *The African Wanderers*, by the cycle of an *African* year, its hot-dry and rainy seasons directly defamiliarising the four of temperate-zone Europe. Real environmental geographies and their consequences thus determine the plot, and offer a double-reality mirror particularising the especial twice-repeated ignorance of Sir Thomas highlighted in the Preface. The ill-conceived Folly is his 'Little England' transported (by brig) and re-erected, without cognisance of the native climate, 'productions' and cultures – economic and social – of West Africa. Readers therefore know from the start and from first reactions of the Governor above that Sir Thomas's trebly 'mad' enterprises will fail very badly, but not how or for whom. The task of the knowing narrator-she is then not to spare, nor domesticate the fuller West African reality blows to Sir Thomas's 'Folly', in both his material estate and his person. His stubborn will, ego and arrogance – key qualities of heroic imperial masculinity – all undergo scrutiny and challenge by means of their direct impacts not only on Blanche and Alan but also on the Governor of Cape Coast Castle and the supplanted 'natives' of 'his' new land. But it is the country and land itself – 'Scorpion Hill' in the preface on which his 'Folly' sits aptly names its features – that supplies the largest blows by its very differences from English soil. The ravages and renewals of dry and wet seasons in their natural causes and effects cannot but guarantee infestations (of ants and scorpions), shortages of food and related disease for 'non-native' animals – Sir Thomas's horses and cows – and humans alike. Humboldtian principles of 'plant geography' (discussed in Chapter 6) therefore govern the 'scientific' plot and detail in *Sir Thomas* in Asante's unsuitable 'fodder' for cows but not goats, which Sir Thomas did not bring, his failings to prepare for natural famine (the 'hungry' season) and his gross miscalculation of African heat of two extreme kinds, dry and wet. The seasoning fever he and all in his party suffer because they cannot do otherwise, and not all overcome, is compounded by his failure to include chimneys and fires in his Folly to expel the damp (*ST*, Chapter XII). His early forays to explore his new country are also only to hunt its big game for sport (*ST*, Chapter VI), without understanding the larger significance of the leopards he kills as sacred to the 'Fetish Men' (these animals represent tribal families in Fante understandings of natural-human orders). Of his dogs not stolen as delicacies for local chiefs' tables, only those that are relocated to Accra survive. The scientific and cultural messages of *Sir Thomas* are one. To survive let alone 'settle', the white incomer's only future is not to rule, but rather accept and adapt to local natural and cultural conditions (exemplified in the 'Governor'). To impose intransigent British imperial values and to refuse to compromise them is to perish. In short, the starkly unsettling message of *Sir Thomas* is that 'little England' imperial and settler mentality is bound to fail, and fatally. The clear lessons of natural disaster for Sir Thomas's family and (live)stock through not respecting the climate and land also earth *as fictions* the expansionist Victorian 'worldviews' of the second paragraph of the preface.

But the second main difference is the more startling, because it is also clearly flagged from the outset. By means of 'classic' adventure story tropes, the clearly identified writer-she recounts the genre's blind spots as integral to the cautionary tale of the eponymous Sir Thomas. Only his daughter Blanche successfully survives his/its (mis-)adventure and its 'happy ending' after his death. *Sir Thomas* therefore closely follows the conventions of travel writing and adventure-travel fiction, including standard use of chapter synopses, but with an integrally parallel view, encapsulated by the first:

> DESCRIPTION OF FITZOSBORNE CASTLE. CHARACTER OF SIR THOMAS FITZOSBORNE. – LADY AGATHA RANDOLPH. – MARRIAGE OF SIR THOMAS FITZOSBORNE AND LADY AGATHA. PERSECUTION BY EARL RANDOLPH. – BIRTH OF BLANCHE. DEATH OF THE DOWAGER LADY FITZOSBORNE. – RENEWAL OF EARL RANDOLPH'S PERSECUTION. – BIRTH OF AN HEIR. – DEATH AND FUNERAL OF LADY AGATHA.

The 'strange land' of the novel is as much 'home' (in the opening and closing chapters) as 'abroad', to compare and contrast being an 'outlaw' or 'exile' in both. Sir Thomas's 'Folly' is therefore both his Cornish and his 'Cape Coast' 'Castle' home, dependent for their future as in their past upon a (rightful) male heir. More importantly, exile positioning is experienced in unequal measures by the sexes, as indicated by the almost equal number of synopsis references to 'Lady Agatha' here. Her pivotal bit parts in the family plot (as wife and mother) conclude in her early death and removal from the tale upon producing an heir, but are quintessential to enabling Blanche's motherless development as the firstborn daughter. In 'classic' adventure fictions by contrast, especially those set in Africa, chapter synopsis references to women in any role are almost unknown: white women simply do not set foot there; 'native' women supply only coincidental interest alongside other indigenous animals. Very unusually in *Sir Thomas* – and not only because Blanche emerges as its main protagonist – chapter synopsis references to white women are therefore many, but also interconnect with those of often *named* indigenous women: 'Alan Relates how he and his Sister saw a Headless Body thrown into the Sea' (*ST*, Chapter VI, the 'Body' is female), 'Prah-Prah Woman' (*ST*, Chapter VII), 'Custom for Quamina Bwa's Mother' (*ST*, Chapter VIII), 'A Woman going to be sacrificed as a Witch, who caused the Famine. – Blanche summons aid from the Castle. – The Woman Saved.' (*ST*, Chapter IX), 'They [Blanche and her white servant Mary] are nursed by Sarah, the great Mulatto Woman' and 'Description of the Coloured Woman' (*ST*, Chapter XII). Before commencing the story proper, a 'juvenile reader' (boy or girl) perusing these chapter synopses could only be struck by the novel's differences from 'rugged hero' norms. In that these are (relentlessly) and repeatedly challenged – as exemplified in the (treble) 'Mad! mad! mad!' story and response to Sir Thomas above culminating in his death – the greater consequences are for the 'weaker sex'. Although the novel commences with its no less normative clichés – white female health is inevitably 'delicate'; Lady Agatha dies young as consequence of childbed and provision of the 'heir' – the narrator-she ensures that the stronger 'heiress' firstborn is already cliché-challenging in an opening cameo: 'her father's disappointment at having a girl was made up to him by her providing to be a brave little creature, full of life and courage, even at that early age, and, seated on his shoulder, she was early accustomed to scenes which were thought to be more fit for a boy than a girl' (*ST*, 6). For a girl reader, Blanche exemplifies her social place and roles, yet already indicates her distinguishing character ('courage') to come. Moreover, despite the name 'Blanche' typifying white upper-class stock, liability to 'blanching' (realised in *ST*, Chapter XVI) and requisite heroine purity, she embodies the 'blank canvas' for women braving exploits overseas (and at home), like Sarah herself. By the time of Blanche's departure for Africa as decided by her father, she has already encountered the death of her grandmother and mother, and the strains of extreme grief and financial straits. Despite then being delimited by the many constraints of female dependency and caring roles – because a daughter and hence surrogate in female roles of wife, mother, nurse – these only the more underscore Blanche's 'courage' in the plot to come in different West African settings. They prove a site for greater independent female agency. If the eponymous Sir Thomas peels off white male masks to reveal the darker imperial '(un)settler tale', the impacts on his daughter find her thrust proactively into its 'adventure', and how she then overcomes.

But Blanche is not the only important female figure in *Sir Thomas* of stature and acuity in the hands of the narrator-she. As the synopsis summaries quoted above underscore, 'Sarah, the Great Mulatto Woman', plays more than a tokenistic role as 'servant' nurse in the novel's pivotal Chapter 12 (describing Alan's death from fever), or as black foil for 'Mary' (the white maid Sir Thomas also brought to Africa):

> Both Mary and Sarah were judicious enough to suffer a continuance of Blanche's weeping, and the latter sat down by the side of the white woman's bed, with an expression of tenderness and compassion, which shewed her kind disposition, and from that moment Blanche felt her value and recognised her as a friend. She was a tall and handsome woman; her hair which was black and silken, was combed from her face into a kind of cone at the top, and ornamented with gold. She was rather stout, and was not young; but her eyes were full and dark blue, her features straight and delicate, her teeth beautiful, and her form exquisitely proportioned. She was one of the great women of the country, and would have been called a princess among civilised people [...] Blanche had often looked with deep interest on these coloured women, some of whom, and Sarah was one, were as fair as many English women, and extremely pretty. Their manners are engaging and quiet, and they are greatly attached to the European husbands, and especially their children. Sir Thomas, however, had forbidden his daughter to have any acquaintance with them [but] being in general very shrewd women, they easily understood the reason why Blanche and Mary kept such a distance [...]. (*ST*, 188–90)

After Alan's funeral as Blanche's comforter 'friend', the fictional Sarah's good auspices also bring Blanche through fever, her father's death and subsequent removal to the Governor's Castle, and hence the happy ever after ('reader, I married him') of the last two chapters. In lending her own name to this key 'great Mulatto Woman', Sarah (Bowdich Lee) reflects back for readers of *Sir Thomas* of both sexes that 'shrewd women' inhabit skins of every colour. This further prominent depiction of 'native' women as among the hallmarks of Sarah's scientific works and colour plates, now deploys the novel form to immortalise the real historical personage (discussed in Chapter 5) behind the fictional character.[38] In the quotation above, Sir Thomas's prejudices, deep ignorance and arrogance are once again upbraided and overturned by clear counter-example in the more civilised 'Sarah the Great Mulatto Woman'.

But what, then, should the astute reader in 1856 and today make of the several scenes in *Sir Thomas* that investigate 'uncivilised' local customs? That all directly involve women is significant since the narrator-she, despite her altogether more informed insider viewpoints, appears to recount them from British imperial points of view. The beheaded body is a human sacrifice (*ST,* Chapter VI), a practice condemned by the 'Governor'. The stoning of the native woman cast as a witch and scapegoat for the famine (*ST,* Chapter IX) is rescued by Blanche. The disturbing episode (*ST,* Chapter X) of her near abduction by a 'Black Man' is saved by her presence of mind to sound the alert, resulting in the Governor's demand that he be flogged and drummed round the ramparts. That Sarah herself lived these uncomfortable episodes, as her *Fragments* report (see Chapter 5), is part of their 'anecdote' veracity. But the second paragraph of Sarah's preface to *Sir Thomas* is then salient concerning the relative values of barbarism and civilisation constitutive of the (inter)cultural history of the novel. The stoning of witches can only remind the reader of their burning in Europe; the beheading conjures the French Revolution. Condemnation of all such practices is therefore held up by the 'enterprising traveller(-she)' authoring *Sir Thomas* as part of its larger cautionary tale. To refuse to paint them as 'savage', or as African local colour in standard scientific travels, Sarah's works of natural history and adventure fiction mirror back similarly barbaric acts at home.

Of equal significance is then Blanche's marriage to the aptly-named, because more enlightened Governor Hope (modelled on Edward Bowdich's maternal uncle, Hope Smith). Theirs is a rare love story along with 'Samba' (Chapter 5) in Sarah's fictional travel writing and natural history fiction. In Blanche is therefore her identification of the blank space in 'classic' explorer and adventure genres upholding the rugged hero plot: it writes individualist conquest of new territory, never his settling down in domesticity in his own. But Blanche is no fictional alter-ego even if she memorialises Sarah's own story of happy marriage to Edward. Blanche's return to Sir Thomas's restored estates in Cornwall managed by 'Hope' only sees her once more subsumed as the new 'Lady Agatha'. Sarah's return to England a widow upon her second voyage to West Africa instead impelled this 'enterprising traveller(-she)' to rise to the life-challenge of what she did *next.* The novel then differently tells Sarah's much larger 'love story' with West Africa and its natural history in her multiple, corpus-long, commitments to make them better known.

Sarah's open condemnation in *Sir Thomas* is therefore of the larger 'Folly' (barbarism) of the civilising mission as a colonial misadventure, including scientific. It disproportionately and critically affects 'native' women, abroad and at home. In the 2020s, the novel also offers an unusually chilling and prescient cautionary tale of ignorance and arrogant disrespect for natural environments through exploitation, non-adaptation and implantation of non-native species. To take up Sarah's differently pioneering perspectives on West Africa's legacies is now also to see her more informed natural history-making in the field; it harnessed alternative non-conformist views, practices, expertise and forms of relaying knowledge of the natural world. Scientific 'anecdotes' in oral histories now stand out as a significant and telling multi-cultural form that Sarah had already recorded in the 'Notes' to her *Stories of Strange Lands*. Such 'anecdotes' remain outside the scope, knowledge and imagination of the recent call in *Behaviour* above. But the second paragraph of Sarah's preface to *Sir Thomas* also reveals the further legacies with which this study of her corpus began. Thanks to the unprecedented and enterprising decisions of Edward and Sarah Bowdich to include their young family in their pursuit of natural history, wherever it would take them – Paris, Lisbon, Madeira, the Gambia, London – their children had a different preparation and understanding of its integral undertaking 'from the life'. When Eugenia Keir had nursed and then buried her mother in 1856, she knew from experience something of what lay ahead in emigration and settlement in Australia with her husband, Dr Swayne. It was among the many highlights of the journey of this book that I could meet Sarah's great-great-great granddaughter from that union, and see through her, and therefore Sarah's eyes the Gambia and the Burton Street, Camden, that I knew so well from my research.

Notes

1 Michael F. Dahlstrom, 'Using Narrative and Storytelling to Communicate Science with Nonexpert Audiences', *PNAS* 111, suppl. 4 (16 Sept. 2014): 13614–20. See also the pejorative sense of 'anecdotal evidence' in Alfred Moore and Jack Stilgoe, 'Experts and Anecdotes: The Role of "Anecdotal Evidence" in Public Scientific Controversies', *Science, Technology & Human Values* 34, no. 5 (Sept. 2009): 654–77.
2 Stefan Collini, ed., *C. P. Snow: The Two Cultures* (Canto Classics). Cambridge: Cambridge University Press, 2012, 4.
3 Dahlstrom, 'Using Narrative', 13618.
4 Lightman, *Victorian Popularizers of Science*, 130–31.
5 Lightman, *Victorian Popularizers of Science*, 131.
6 See Susan Bruxvoort Lipscombe, 'Introducing Gilbert White: An Exemplary Natural Historian and his Editors', *Victorian Literature and Culture* 35 (2007): 551–67 (553) as an example of critical emphasis in British history of science on the revolution of the steam press for its causality in the dissemination of knowledge.
7 Joel Fineman, 'The History of the Anecdote: Fiction and Fiction', in H. Aram Veeser, ed., *The New Historicism*, 49–76 (61 emphasis in the original). New York and London: Routledge, 1989.
8 In the publisher's catalogue printed at the back of *Anecdotes* (1853), Grant and Griffith advertise and list her late works on Animals for pre-teenage children: '*In super-royal* 16*mo, beautifully printed, each with Seven Illustrations by* Harrison Weir, *and Descriptions by* Mrs Lee, 6d *plain*,1s *coloured* 1. BRITISH ANIMALS. First Series. 2. BRITISH ANIMALS. Second Series. 3. BRITISH BIRDS. 4. FOREIGN ANIMALS. First Series. 5. FOREIGN ANIMALS. Second Series. 6. FOREIGN BIRDS.* Or bound in One Volume under the title 'Familiar Natural History,' see page 4' (8).
9 See Paul Fleming, 'The Perfect Story: Anecdote and Exemplarity in Linneaus and Blumenberg', *Thesis Eleven* 104, no. 11 (2011): 72–86 for the importance of the 'event' in the anecdote (74). Sarah's capturing of *animal* habits and instincts as scientific anecdotes very differently challenges Fleming's discussion of contingency as modus operandi of the genre.
10 Melissa Sodeman, 'Gilbert White, Anecdote, and Natural History', *SEL Studies in English Literature 1500–1900* 60, no. 3 (2020): 507–28. Her list includes Thomas Brown's *Anecdotes of*

the Animal Kingdom (1834) and Jane Bourne's *A Winter at De Courcy Lodge; or, Anecdotes Illustrative of Natural History* (1837).

11 Sarah's usage of 'naturalist' on a sliding scale is in contradistinction to the person/status of the informed 'amateur' outside professional bodies. See Philippa Levine, *The Amateur and the Professional: Antiquarians, Historians and Archaeologists in Victorian Britain, 1836-1886*. Cambridge: Cambridge University Press, 1986. See also S. Shapin and A. Thackray, 'Prosopography as a Research Tool in History of Science: The British Scientific Community 1700–1900', *History of Science* 12, no. 1 (1974): 1–28, defining (i) those with publications in science, (ii) those active in scientific societies and institutions or who taught and disseminated scientific knowledge and (iii) 'a large mixed bag of cultivators of science who patronized, applied or disseminated scientific knowledge and principles, but who themselves neither published science, taught science, nor actively associated themselves with scientifically-oriented institutions' (12–13).

12 John Durant, 'What is Scientific Literacy?' *European Review* 2, no. 1 (1994): 83–89 (83).

13 Talking animals featured largely. See, indicatively, Tess Cosslett, *Talking Animals in British Children's Fiction, 1786–1914*. Farnham: Ashgate, 2006.

14 Natural history work is without sex for Sarah (as Cuvier). See Orr, '*Adventures in Australia* (1851) by Mrs R. Lee'.

15 In heading her work with 'Monkeys' as representative of 'Quadrumania' and thus distinct from the other 'quadrupeds', Sarah also omits from her selection both Marsupials and Cetaceans. As will be shown, specific representative naming matters.

16 Thomas Bell, *A History of British Quadrupeds*. London: John Van Voorst, 1837.

17 See the works of Ann Pratt or William Carpenter's 1833 *Scripture Bible Natural History* and their differences from serious 'textbook' works by William B. Carpenter, *Zoology: A Systematic Account* ..., which Sarah referenced overtly in the preface to the revised *Elements* (1850).

18 These usefully distinguishing, yet multi-audience categories are Matthew D. Eddy's in his 'Natural History, Natural Philosophy and Readers', in Bill Bell, Stephen W. Brown and Warren McDougall, eds., *The Edinburgh History of the Book in Scotland. Vol. 2 'Enlightenment and Expansion'*, 297–308. Edinburgh: Edinburgh University Press, 2012.

19 There is no pejorative sense here. This formulation meaning 'animal creations' is common to late eighteenth-century works on natural history, for example by William Smellie. See Stephen W. Brown, 'William Smellie and Natural History: Dissent and Dissemination', in Charles W. J. Withers and Paul Wood, eds., *Science and Medicine in the Scottish Enlightenment*, 191–214. East Linton: Tuckwell Press, 2002: 'he asserted that a chief attention of natural history should be to the study of the "instincts of brute creation" because such an acquaintance was essential to 'minutely investigating the principles of human nature' (199–200).

20 Sarah's practice here follows her treatment of humans and great apes in the *Elements*.

21 Because outside 'official', 'gentleman' and 'professionalising' natural history as a woman, Sarah's works make the major socio-political point that the bars of class, sex and even race do not debar the (major) pursuits of science.

22 R. J. Cleevely. 2004. 'Bell, Thomas (1792–1880)'. Oxford Dictionary of National Biography. Oxford: Oxford University Press.

23 See R. B. Williams, 'The Artists and Wood-Engravers for Thomas Bell's *History of British Quadrupeds*', *Archives of Natural History* 38, no. 1 (2011): 170–72.

24 Thomas Bell, *A History of British Reptiles*. London: John Van Voorst, 1839.

25 See Orr, 'Fish', 233–34.

26 See, as an indicative study of such report, Jan Bondeson, *The Feejee Mermaid and Other Essays in Natural and Unnatural History*. Ithaca: Cornell University Press, 1999.

27 Respectively in *Anecdotes* (1852) these are 'The Monkey Painter', 'The Shepherd's Dog and Cur', 'The Bear and her Cubs', 'The Fox and the Hares', 'Leading the Blind Rat' and 'Wild Horses and Wolves' and in *Anecdotes* (1853) 'The Eagles and Sheep', 'The Swallows and the Cat', 'The Raven and the Dog', 'The Guinea Hen and Duckling Brood', 'The Duck rousing the Dog' and 'The Snake and the Ichneumon'. It lies outside this study, but the engagement by Grant and Griffith of Harrison Weir as the illustrator of Sarah's *Anecdotes* is a further indicator of the importance of her work in the scientific anecdote genre.

28 Mariska E. Kret and Tom S. Roth, eds., 'Anecdotes in animal behaviour', *Behaviour* 157 no. 5 (2020): 385–86. Instructions on how to collect and submit 'serious' anecdotes for submission conclude the call.

29 It lies outside this chapter, but the 'fait divers' was a recognised section of 'sensationalist' report in the nineteenth-century French press.
30 Wood, *Anecdotes of Enlightenment*, 3.
31 See Orr, 'Rethinking the Pioneering Text'.
32 This question develops the arguments of my 'Rethinking the Pioneering Text' and '*Adventures in Australia* (1851) by Mrs R. Lee'.
33 My formulation recalls the 1980s board game, 'Trivial Pursuits', based on general knowledge in six areas: geography, entertainment, history, arts and literature, science and nature, sport and leisure. *Sir Thomas* examines all six, as the excerpt from its Chapter 4 exemplifies. Space does not permit their thematic investigation.
34 In the Governor's repetitions of Sir Thomas's 'ignorance' as 'mad', see Chapter 5 for the similar views of the people of Liverpool concerning Sarah herself.
35 See my 'Rethinking the Pioneering Text', 144–45, which frames the seemingly overlong preamble of the first two chapters of *Sir Thomas* to 'explain' his wild decision to take his motherless children to Africa to make his/their fortune (due to the 'persecution of Earl Randolph').
36 Chapter 6 records the response of reviewers to Sarah's *African Wanderers* as an exemplary Robinsonade.
37 This is the posthumous title for *Sir Thomas* in various reprint editions on both sides of the Atlantic from 1881 (London: Grant and Griffith; New York: E. P. Dutton).
38 In 1856 this description is no 'whitewashing' of the historical 'Sarah'. Rather, the 'European' blood of these native women – black 'silken' hair and blue eyes of previous Portuguese, Danish, Dutch and other Western settlement of the Gold Coast – is the outcome of male colonial 'misadventure' in the taking of local 'wives' or, worse, rape behind the gloss (for juvenile audiences) of European (mixed race) progeny.

SARAH BOWDICH LEE AND PIONEERING PERSPECTIVES IN NATURAL HISTORY: LESSONS FOR TODAY

This first study of the major contributions by Sarah née Wallis, Mrs T(homas) Edward Bowdich, then Mrs R(obert) Lee (1791–1856) to new knowledge of natural history has focused on her book-length publications concerning West Africa, including its preparations in the field for her undertaking of *The Fresh-Water Fishes of Great Britain*. The nine chapters and their lessons therefore bring to serious critical attention Sarah's multiple contributions to cross-Channel natural history-making in her published works from 1825 until her death. They also inspire the recuperation of other women in the first half of the nineteenth century at work and publishing in science in at least two language cultures. Chapter findings individually and collectively, as gathered together below, then only magnify the main questions of this book. Why and how, indeed, has Sarah remained so firmly in the blind spots of expert Anglophone and Francophone critical inquiry in the disciplines covered in the introduction, and despite gender, transnational and interdisciplinary lenses? Her works plainly added to major discipline fields, and in the case of (modern) ichthyology and anthropology were in their vanguard. The imprimaturs of Cuvier and Humboldt affirmed their first-ranking, original qualities. Sarah's works also eminently proved the rule that a woman could exist, thrive and regularly publish (expert) natural history in the period. But her corpus also demonstrated why the best modern (inter-)disciplinary inquiry in the history, geography and cultures of nineteenth-century British, French and European science will fail to accommodate its expertly multi-genre, intermedial hybridity. Set assumptions govern and determine who and what constitutes 'serious' contribution to science of the period, also informing science and its histories today.[1] The introduction highlighted the benchmarks that consolidate scientific endeavour in the first half of the nineteenth century by which Sarah cannot be seen. They are modelled by 'genteel' and national(istic) standards compounding the (professionalised) thrall of modern discipline distinctions imposed retrospectively upon 'serious' natural history and earlier 'naturalists' as its makers in consequence. The nine chapters individually and together challenge modern critics to pay much closer attention to pre-1850 context(s) for broad-church natural history on the one hand, and on the other hand to the latter's foundational basis for 'new' nineteenth-century scientific specialisms, such as ichthyology, anthropology and ethnography in which Sarah was a remarkable forerunner irrespective of her sex.

As the nine windows of this study then additionally disclosed, Sarah's sole-authored corpus from 1825 roundly identifies the further benchmark for 'serious' scientific endeavour in the *form* that it should take, precisely because none of her book-length works deliver in it. The scientific paper ('mémoire') read at a learned scientific society or national institution, and published in its 'transactions' was central to the professionalisation of European sciences gathering pace from the early 1830s. If scientific travel writing continued to challenge its mould, especially in light of works by a Humboldt and a Darwin, a woman could be their translator.[2] She would not rank among scientific explorers herself, however, until the later nineteenth-century European 'high' colonial period saw women of independent means, such as Isabella Bird (1831–1904), Mary Kingsley (1862–1900) and Alexandra David-Néel (1868–1969), undertake intercontinental journeys of discovery. Sarah was more than their (occluded) precursor, both as a pioneering scientific traveller and as a scientific travel writer, because of her ground-breaking dual citizenships at the forefronts of French-international museum and field natural history. Her story immediately rewrites the 'facts' of history of science: Mary Kingsley was not the first woman explorer of West Africa or collector of its Fishes.

Almost more trailblazing than were Sarah's remarkable contributions to natural history in their contents before 1860 are the multi-genre formats in which she published them, as concertedly explored for the first time in chapters above. Although a summary of Sarah's many pioneering 'firsts' for natural history is therefore important – in the manner of Table C in chapter 2 – any tabulation of her achievements also needs to address their multiform modes of dissemination. They constituted the scientific paper and template (male) scientific travelogue in alternative, no less expert guises. Ahead of Cuvier's *Histoire naturelle des poissons* and other national studies of freshwater ichthyology, Sarah's contributions to knowledge of West African ichthyology, and European fresh-water species discussed in Chapters 1 and 2 have proved second to none in text and image. Her formats and outlets for this work are retrospectively the more significant in consequence. The imprint of Levrault for the French *Excursions* in 1826 with its Appendix of her drawings annotated by Cuvier made Sarah's annexed contribution a French scientific paper ('mémoire') in all but name, including in the endorsements of its contents by Muséum reviewers for its *Bulletin*. Her *Fresh-Water Fishes* similarly reset Paris Muséum standards for *illustrated* specialist ichthyology in the higher-quality Plates from living rather than pickled specimens. But Sarah also capitalised on her coverage of a first (inter)national Natural History of Freshwater Fishes (read *Histoire naturelle des poissons d'eau douce*) documented from the life. Her 'research expedition' recovering Britain's Fishes for the new field of modern ichthyology also commanded their highest quality collectability for the expert fisher and (world) armchair *aficionado* of both sexes. The Gift Book format therefore differently extended the reach of her work's specialist interest beyond that of Cuvier's 22 volume *Histoire naturelle des poissons*, to ensure specialist, connoisseur and amateur readerships. With the hindsight of the final section of this book, her project already prepared the fundamental principles of her *Elements* and *Anecdotes*: expert natural history clearly described and illustrated was potentially within the reach of all, with access to it available through any and every entry-interest point. The material format and experience of the *Fresh-Water Fishes* already exemplified such durability of future interest and use by specialist and newcomer alike. Its numbers and their Plates were designed in the quality of the paper, bindings and illustrations for integrally 'bound' aesthetic-scientific appreciation that invited scientific and artistic reordering and re-evaluation at once, and upon the collector-subscriber acquiring the complete set. No other modern ichthyology could touch Sarah's highest bars for fish science until the work of Jonathan Couch.

Reinsertion of the French term, 'mémoire', for a scientific paper above then highlights the pioneering nature of Sarah's *Memoirs/Mémoires* determined in Chapter 3 as 'a first science biography "from a female pen"'. But her deft refiguration of the very format – the French *Éloge* – that Cuvier had made his own to memorialise him for and outside French (Muséum) science, further distinguished her choice of the *Memoir/Mémoire* form. It took science biography into unprecedented new territories by accounting for Cuvier's 'scientific life' as private and public, and these for his reputational reach in the international 'fields' of her treble publication the same year in cross-Channel, and Transatlantic markets. The further hallmark of her 'accomplished' work for a US reviewer in 1834 was that it remained the 'official' biography of Cuvier in these multinational contexts until the 1960s.

Although more hidden in plain sight, a similarly long pedigree pertained to Sarah's first (historical, 'plant') Geography of the Gambia, discussed in Chapter 4. In displaying the 'Bonplandian' reaches of its aquatic fields to offset Humboldt's 'mountain-eye' overviews, Sarah's first intercontinental geography and geology of the Gambia offered a (new) blueprint for such endeavours in content and form. I use the word 'blueprint' advisedly for its reassessment within current Blue Humanities research, and how the latter may learn important lessons from Sarah's precedent. Hers connects transcontinental 'plant geology' integrally with people*s*. Although unfinished, fragmentary and compressed her 'Narrative' of the Gambia also set out a 'prospectus' of what the Bowdichs' natural history of Sierra Leone might newly have accomplished. But Sarah could also single-handedly transform her 'Narrative' and 'Zoological Appendix' in light of their model scientific field 'Notes'. Their more radical remake in her (already pioneering) Gift Book stories restitched their science into the illustrated *Stories* and *Fragments* of 1835. Chapter 5 also discovered in this first 'collection' of *West African* (Gambian, Ghanaian, Gabonese) natural history and 'ethnography' her larger parts in Edward's *Mission* of 1819, during *her* first Voyage to 'Ashantee'. *The African Wanderers* not only broadened the review of the natural history of West Africa to include *her* brief sojourn in Sierra Leone (where Edward never set foot). As Chapter 6 contended, her first 'natural science fiction' also focused her increasingly critical view of the British civilising mission overseas and at home, especially against the backdrop of new legislation to abolish world slavery. The Gift Book story (and its collection), and a West African natural history novel, allowed her a no less public platform for her early 'environmental' views that had yet to find space in the political and pro-abolition pamphlets of the times.

Sarah's multipy expert figures of natural history then also showed much more than she might tell through her mastery and developments of the conventions for (French and British) science illustration. Chapter 7 newly recorded her 'doubly expert pen and ink' and 'foremost uses of (water) colour' to include her ground-breaking representations of

West African human subjects, particularly women. We could then identify her unmissable hand in one of the most iconic Figures according to British Imperial, Commonwealth and Colonial Studies. But Chapter 7 also revealed for the first time the greater wealth of her expert art history productions, to include 'text book' natural history in Chapter 8. To have identified a further hundred drawings constituting Sarah's 'parallel' corpus in the *Elements* is therefore to establish in this book her hitherto unacknowledged place in nineteenth-century art history and history of scientific illustration. '*S. Bowdich del*' and 'SL' was an illustrator second to none before 1860 for the quality and range of her scientific drawing formats 'from the life.'

Chapter 9 then undertook a similar generic re-examination and task of restitution. Sarah is less an exemplary 'maternal popularizer' in Lightman's terms, and more the unheralded major exponent of the scientific 'anecdote' form. In drawing serious attention – the function of the anecdote – to the rich hybridity of observational, critical and corrective 'account' (including her own), Sarah could illuminate and enliven all 'fields' of natural history and its reporting. In remaking the anecdote specifically to (re-)train the eye of 'juvenile' (and urban) natural historians, Sarah also challenged in her two-volume *Anecdotes* the very expert authority forms and formats for natural science report – the specialist paper, the scientific expedition, the subject-defining Muséum taxonomy, the expert natural history – that by definition foreclosed next-generation access to scientific discovery for all (boys and girls) outside its privileged ranks. Sarah's expert extraction of key anecdotes precisely from these sources but also her own for the *Elements* and the *Anecdotes*, now comes to light for the first time through our Appendix matter for these chapters. Her doubly preeminent achievement in 'textbook' natural history behind both the *Elements* and the *Anecdotes* resides in the polymathy of its informed understanding, including recent 'expert' natural history for the widest range of newcomer and specialist readerships, and as eminently inspired by the author-family educator herself on the cover. It was her deft ingenuity, however, that could bring first-hand expert 'observation from the life' to life in form as well as content in these masterful exercise books for expert and newcomer 'naturalists' alike. Sarah was therefore no minor scientific or fiction writer for minors. Ultimate respect for natural habitats in all their plethora is the moral of her *Sir Thomas* and her integrally developed scientific and literary corpus.

To identify what all nine chapters together endorse as 'subject-defining' in the contents and forms of Sarah's natural history work(s) would then be the tallest order of all were it not for the models, parameters and standards by which she produced them. They are located in this book (as in my earlier publications) in her all-important Paris mentors, Cuvier and Humboldt, as illustrative of international engagement in specialist museum and field natural history respectively. The first two parts of this study then clarified her 'canvassing' of Cuvier and 'harnessing' of Humboldt in her work(s), to highlight in the third her further 'access[ing]' of their mutual commitments to the communication of expert science to wider audiences. In applying their best models to the science of her natural history work in text and image, Sarah's corpus arguably far surpassed theirs precisely in its creative multi-genre range and the multi-media expertise of its no less pioneering *formats*. Natural history in her hands then expressly opened doors to all-comers, because overtly designed to attract those that 'official' (British and French) science excluded by sex, class or creed, and hence education, but also natural interest.

To adjudicate on this 'arguably' to rank Sarah's corpus and its perspectives as pioneering is, therefore, to take with renewed seriousness the historical and comparative (Anglophone and Francophone) contexts in which she worked and published. To quote more extensively from my earlier work (on the *Fresh-Water Fishes*): 'Scientific advances, achievements and vital contributions cannot be measured, or evaluated, without attention to their many intersecting histories of knowledge. Claims to first 'discovery' of objects, or ideas in natural history, and to originality of contribution must therefore be carefully contextualized and authenticated with reference to first publication, and/or public dissemination in print'.[3] In 1828, Cuvier had already identified Sarah's as 'talens distingués' (distinguished talents) when he published the blueprint for his subject-defining *Histoire naturelle des poissons* with the words

> Natural History is a science of facts, and the number of facts it encompasses is so large that no man (*sic*) could gather or verity by himself the totality of those forming any one of its branches; it can therefore only be studied fruitfully by consulting all authors devoted to it, and by comparing the evidence between them, and with nature; but to consult such writers fruitfully, to be able to appreciate the degree of confidence to be placed in any one of them, to distinguish what they report according to their own observations, and what they have garnered in the works of their predecessors, *it is essential to know the circumstances under which they worked; the historical period when they lived; the state in which they found science; the enablement provided to them by their personal position, the help of their friends and protectors, or the cooperation of their students.*[4]

Sarah's works amply hold up her consultation of all the best 'authors', and her 'distinguished' observations by their lights (as Appendices 5, 7, 8 and 9 further endorse). Her 'lights' were foremost representative exponents in their different ambits for science in the 1820s. As my earlier work confirmed, 'Sarah's authority [is] as an *internationally* renowned (British) ichthyologist first, and as the first (British) *woman* in its larger history second'.[5] This book makes the larger case for her place in nineteenth-century natural history as an internationally renowned 'naturalist' first, and woman naturalist second to none in the period before 1860 by the measures of Cuvier's blueprint above.

Cuvier and Humboldt had differently privileged access to (German) scientific education and knowledge. Both pursued their distinctive intercontinental work with all the means and necessary resources at their disposal. In their awareness of such privilege, however, including how they extended it to mentor an 'unlearned person' (*MBC*, 3) who was British and female, resided their shared and unusual view for the time that science had no sex, a view that Sarah upheld in doing it in the primary roles that women could play in science (by men). To return Cuvier and Humboldt more prominently in this book to Sarah's 'science from the life' is therefore also to challenge scholarship on gender and science to discover other major men of science seconding – in the meanings of to endorse and to promote – women's work as 'peers' in their fields.[6] In that he was not her mentor, however, Chapter 8 calls new attention to Richard Owen as a further 'seconder', but also beneficiary of Sarah's French-international natural history expertise. To identify pivotal, expert national and international deuteragonists in women's scientific work is therefore also to differentiate the key *seconding* roles that they play, because more largely *outside* the woman's 'intimate' life.[7] Alongside unsung deuteragonist 'heroes', this larger story for women's scientific work contains multiple 'villains' masquerading as their 'friends of science'. Cuvier's blueprint also already identified them: those who do *not* 'consult [women] writers fruitfully, [or] distinguish what they report according to their own observations, and what they have garnered in the works of their [female] predecessors.' As I have shown in the 'un-heralding' of Sarah's ichthyology by contemporaries,[8] such figures play primary roles in ensuring the larger erasure of rival peer-pioneers, irrespective of their gender.

Sarah's primary place in Cuvier's and Owen's circles in Chapter 8 illuminates the further major lacuna in national-monolingual histories of science and geography. In the nineteenth century, the international vernaculars and cultures of science mattered concerning who circulated new natural history knowledge and in what formats. Cuvier's blueprint therefore returns us to what it did not identify, to illuminate Sarah's corpus as the more creatively pioneering: he practised and automatically accessed all the forms of 'official' science, the '*mémoire*', the '*Éloge*', the multi-volume '*Histoire naturelle*', the Muséum '*Bulletin*'. Humboldt's treatise, *Kosmos*, more comprehensively brought together his expert comparative scientific travel writing and its underpinning *Naturphilsophie*.[9] Creative ('literary') outlets for science were automatically excluded by both, however. By contrast, this book has investigated Sarah's book-length publications for her re-crafting of their generic outside and inside for disseminating new natural history. In overturning Donald deB. Beaver's claims that she wrote 'natural history for survival', Chapters 6 and 8 of this book have also mapped how her writing of Gift Book stories to support her family and her independent natural history work(s) developed into her 'natural history fictions'. In their wake, there is now a much larger case to be made for expert 'nature poetics' as undertaken by other authors outside scientific circles, who nonetheless represented the latest findings in natural history by also eschewing travel adventure, 'animal' tales, and other popular science formats. Sarah's adept, corpus-wide, developments of the 'scientific' anecdote form defied and refused models of 'two cultures' pitting sciences against arts and insiders against outsiders by national, linguistic or other qualifications for 'expert' science publication. In amplifying what was pioneering and *subject-defining* about the contents and forms of Sarah's work(s) in this book's chapter titles, we can now elucidate her pioneering perspectives on natural history and natural history-making.

The imagined, but historically verifiable, profile of Sarah in a contemporary Gambian woman's eyes opening this book provides their framing device. The most important of all Sarah's sole-authored works from 1825 were her unusually comparative perspectives *on* West Africa's abundant natural history (new to Western science), and *from* the several countries in its region that she lived and worked in *before* as well as *after* her pivotal training in international natural history in Paris. Her first voyage to join Edward in 'Ashantee' laid the groundwork for her expertise in 'natural productions' (botany and zoology) and 'manners and customs' (of the Fante and other peoples) in Ghana and the Gabon. The step-change of the Bowdichs' mentoring under Cuvier and Humboldt added to these fields new knowledge of geology, of invertebrates and of the lesser-known vertebrate classes of natural history. Where Edward set forth to measure and map the reaches of the River Gambia, Sarah made its larger West African (geo-)ichthyology her own, and as integral to the region's better comparative understanding. Her variously West African perspectives for natural history-making then accommodated indigenous knowledge and naming of species new to Western natural history.

To see natural history-making through the perspectives of (West African) Fishes in Chapter 1 was then also to see history 'from below' newly turned for ichthyological sciences. Sarah's additional pioneering insights on (the hitherto hidden) interconnected and distinguishing geographical habitats of Fishes also identified their different collecting worlds. There were no 'centres' (or peripheries) of expertise, because her findings in Madeira and the Gambia in the field inflected knowledge of similar species 'at home' (in Britain), including in major national collections (pickled and dried specimens) in London and Paris. As Chapters 1 and 2 demonstrated, Sarah was denied her pioneeringly connective perspectives and findings at the forefront of (world) ichthyology of 'West Africa' only because the patronymic naming conventions of Western science – confirmed by the gold standard of specimen holotypes and permitted neotypes in last resort, and stacked against *local* (first-)naming histories and geographies – overruled her indisputable evidence. The case to re-establish the rightful '*S.* Bowdich, 1825' for 'Ethamalosa fimbriata' is not only made by this nomenclature clearly in recent specialist *Francophone* ichthyology. Her naming also derived from identification in the field, and *ahead* of Cuvier's 'definitive' new ichthyological taxonomies for the *Histoire naturelle des poissons* with vast collection resources behind its preparation. To prove the 'science of facts' in Cuvier's blueprint for Natural History wrong (in his *Histoire* and its successors, such as FishBase, and CLOFFA) is to right fundamental error according both to international rules for nomenclature, and to the ethics of correct attribution and citation for all *scientia* worthy of the (rare woman's) name.

But Sarah's pioneering perspectives on Fishes in the field already overcame her larger unrightful exclusions from modern discipline historical record. She set no such limits in her books as multiple outlets for suitable 'specialist' dissemination of natural history knowledge. Her 'non-specialist' showcase for ichthyology therefore included her lengthy expert glosses of Cuvier's *Histoire naturelle des poissons* in the *Memoirs* (Chapter 3), her report of its advancing volumes in the *Elements* of 1844 and 1850 (Chapter 8) and synopsis of her own work in the field in the second volume of her *Anecdotes* (Chapter 9). Fishes also illuminated her astonishing quality contributions to specialist scientific illustration.

The interconnected 'collecting worlds' of natural history, including Fishes (in Ghana, the Gambia) among many others in Sarah's works, also revealed her unprecedented access for a (white) woman in 1817 and 1823 to local, and especially women's knowledge of *flora* and *fauna*. Her collection of specimens then almost pales before her collection of oral histories of West Africa's natural history that illuminate the 'Notes' to her *Excursions* and *Stories* (elucidated in Chapters 4 and 5). Her many reports in word and image of encounters with 'native' women translate into her pioneering perspectives on and portrayals of them. We chose 'Costume of the Gambia' as the literal, but also metaphorical cover for this book to capture Sarah's pioneering perspectives, with herself unseen yet fully observing. There was no overtly 'white', 'European' or 'black' knowledge in her view, although the singularities of her position as the first white woman (also accompanied by her children) that her interlocutors had encountered demonstrably added in Chapters 4 and 5 to what she as a woman could collect in terms of specimens and 'stories'. Chapter 6 (and 8) could then also identify Sarah's especial acuity in collecting the female of the species as the vitally missing component of Western natural history. Hers were extraordinary first representations – in text and image – of the female gorilla in the unprecedented 'gorilla episode' of *The African Wanderers* of 1847, and in her 'cut' of the 'Inghena' for the *Elements.*

Sarah's larger independent 'cargo in return', not the disaster of her crates in 1824, has now revealed her West African (re)collections as catalysts for the 'continuance' of her work in natural history across her multiform corpus after 1825. From her West African travels also came her incomparable comparative Paris–London perspectives on expert French and British natural history endeavour, as a participant observer outside and inside their official organs. The various good auspices of Cuvier and Humboldt after 1824 epitomised in their 'Notes' to her *Excursions* then go far to explaining why and how she could continue her expert work nonetheless. Cuvier's death would then have dealt the deciding blow to her Muséum access after 1833, and ended Sarah's pioneering natural history publications at that point, except that Chapters 5 to 9 demonstrate the contrary. Only her tried-and-tested non-conformist convictions, stances and perspectives can explain *how* she successfully negotiated the many exigencies of her insider-outsider natural history-making, encapsulated in her self-repositioning in 1824 and 1834, despite the closed doors of London and Paris science. Her prefaces across the board are then masterclasses in pioneering negotiation of deeply ingrained social bars and protocols not only for female authorship, but also for *leading* international female scientific authority. The indicative reviewers of her *Memoirs* can only admire the foremost rank of her (non-conformist) 'woman's pen'. In these anonymous critic-readers is then the sternest lesson for the best experts in the domains of (gender and) historical geography, history of science and travel, and also comparative literary-cultural studies in the introduction to this book. The utmost seriousness should now be paid to the astute *self*-perspectives encapsulated in the conventions of female-authored prefaces, and hence to the work(s) that such a woman-authored preface fronts. We have seen in various chapters how Sarah's prefaces and the works they frame operated between the bars for women's science.

Sarah's unchanging and quintessentially non-conformist views therefore enabled her to forge an independently entrepreneurial and creative life in natural history by her own pen after 1825 until her death. Generic inventiveness and intermedial expertise went hand in hand with her book-length scientific contributions in its fields. Her pioneering perspectives for natural history-making could then more firmly uphold the view propelling it, that such endeavours could be pursued by all with informed observational acumen like herself, and take a myriad of suitable 'anecdote' forms. Their gamut in her corpus clearly ranged from the unusual insider biographical view of Cuvier (in Chapter 3), the serious 'Narrative' in the scientific travelogue (Chapter 4), 'Description' in specialist ichthyology (Chapter 2) and 'textbook' entries (Chapter 8) to the oral cautionary 'legend' that is 'Adumissa' (Chapter 5) or 'Sir Thomas' (Chapter 9), because the scientific 'anecdote' is both a genre and a manual for new, and corrective, observation skills when understood as their selected (re)collection. The charge of anecdote to question things and knowledge 'as they are' also shone through Sarah's command of colour in scientific illustration forms concerning vertebrate animal and human subjects.

Behind all Sarah's pioneering perspectives above remained the sharpening focus of her material situation in natural history (as a pivotal consideration in Cuvier's blueprint for new science above), and of her pen. That she was an unusual widow with three children to support was no excuse or cause for pity, but rather the stimulus for her different pursuit of expert natural history, whatever it would take to resource and develop in content and in form. This book has therefore identified in Sarah the very early model for combined scientific endeavour and childcare that is entirely familiar to women in STEM(M) today. But Sarah's successful support and upbringing of her children as mutually accommodating the pioneering pursuits of her science, and as determining her resilience when hard-won doors closed for her, are no further crosses and models for 'things as they are' that women in STEM(M) continually bear. Necessity and discrimination had already spurred and mothered invention for the Bowdichs in 1816 as a more unusual, non-conformist, scientific couple automatically including their children in their field endeavour, whether the Paris of the Muséum or in tropical West Africa. Edward's secondments in both more largely shine a spotlight on *his* pivotal deuteragonist roles to forge natural history-making *en famille*, fully inclusive of his partner-in-science-wife and their children together. This pioneering perspective of 'Bowdich, 1825', almost more than any other for expert natural history-making, breaks the two immutable moulds that are the 'lone' Chair-Gentleman in his laboratory, and the 'rugged' scientific explorer, when their epitomes in Cuvier and Humboldt also unusually mentored 'outsiders', including women. Sarah's best lesson in non-conformist natural history-making for women in STEM(M) today is the need to know, and to seek out, the best seconders at all stages of a 'career', and hence avoid obstructors masquerading as 'collaborator friends' or 'co-authors'. Science 'en famille' can then also offer unforeseen openings to knowledge foreclosed to establishment experts, who have never had to more boldly go.

The obscurity and alleged impossibility of women like Sarah to exist in natural history in the period was explained in the introduction by the educational and other historical odds against them, underpinning the collective focus of the various contemporary critical frameworks by which she was 'invisible'. They differently upheld 'things as they are' within national, monolingual, modern discipline and even interdisciplinary studies. By contrast this book could identify how and why Sarah was a very large presence in natural history-making in France and Britain (1825–1856). Her non-conformism, her comparative and bi-cultural perspectives, her multi-subject corpus also published in two languages, each and all challenge current (inter)disciplinary, intermedial and intercultural inquiry, including with gender lenses. Our chapters thus overtly focused on and promoted the comparative 'Western' wealth and breadth of her pioneering contributions to, and perspectives on, natural history-making in the first half of the nineteenth century.

Her work(s) thus newly interconnected and inter-reflected (her) views of West Africa and Western Europe in forms that expanded the bounds of 'scientific', 'literary' and 'artistic' productions. If our nine chapters discovered these major new lines of inquiry for their fruitfulness, they also leave much to unpack. There was not enough space to do fuller justice to her *Anecdotes of Birds, Reptiles and Fishes, Adventures in Australia* (as a larger development of her West African natural history fiction),[10] or her *Trees, Plants and Flowers.* Each in different ways, and most clearly the first in light of Chapter 9, exemplifies the motor of Sarah's multiform natural history and the 'how?' of her innovating productions. They demonstrate that 'Natural History is a science of telling observations', to recast Cuvier's template for it above. To enlist her perspectives for women in STEM(M), and for research on the transnational histories and cultures of nineteenth-century natural history that ground modern science is to take a leaf from her published books. Sarah's angles of pioneering vision were consistently informed, comparative, multilingual, intermedial, polymathic, curiosity-driven, creative and committedly non-conformist.

To frame Sarah's pioneering perspectives, this book has promoted a methodology that combines intercultural nineteenth-century French Studies and comparative history of science, because its application first identified and cross-connected the importance and potential significance of the 'Madame Bowdich' on the inside cover of a French scientific travel work in 1826 endorsed by none other than Cuvier and Humboldt and printed by Levrault. The lenses of Cuvier and Humboldt that inspired my article publications underpinning this book, and that identified why Sarah has remained largely invisible in French and British history of science, have withstood the test of the nine new chapters here, particularly those in part 3. But it was particularly important, in consequence, that I also stress-tested the larger story that this book now tells about Sarah's pioneering natural history work. The 'narrative of its continuance' lay in the double-edged production of her corpus – in French and English; in text and image; in scientific and literary genres – and in the continuing (lack of) resources at her disposal in museum and field natural history, whether human, specimen or material. As my means to frame and introduce all of these and Sarah herself, when no known portrait exists of her, I wrote a version of the opening imaginary 'profile' in 2013 while in Banjul in the Gambia with Sarah's great-great-great granddaughter. We found ourselves largely corralled in a tourist hotel that made larger independent exploration of Sarah's 'Bathurst' difficult for white women not in the Gambia for package holiday sun and 'fun'. But we did take an organised river 'excursion' to the former slave market at Albreda via 'Kunte Kinte Island' (the much eroded 'Fort St James Island') that pressed the story of transatlantic, but not transcontinental African slavery. We also broke out for sufficient time to visit the environs of 'Government House', Hannah Kilham's Mission and Banjul's markets. Its trading women vied (as in Sarah's day) to sell us their wares of fabrics and local fish, a quizzical, colluding smile on their faces. The call in *Behaviour* in 2020 for 'anecdote' (in Chapter 9) then prompted me to think again about what 'proper' academic and scientific publication can encompass for telling effect. My original narrative perspective had been little Eugenia's in the mirrors of her great-great-great granddaughter's. These remain, but in the larger knowing gaze of the inquiring woman trader 'with parasol' observing and presenting 'Mrs Sarah' directly for your view.

This book therefore open-ends where it started, having now seen what Sarah could do *next* through how she set out indeed in 1824, to leave to posterity her unusual legacies, extensive corpus and lessons in natural history-making. The inspirational case of 'Mrs Sarah' in profile for her times and today presents a preeminent international naturalist first, and (fore)mother second, although her presence with her children on the Bathurst quayside (and behind her stowed barrels and crates) had yet to disclose her already adroit, double pen. In consequence, this book calls for renewed study of others at work between the bars of Western European and West African natural history-making from 1800. The pioneering perspectives of a 'Mrs Sarah' therefore matter for twenty-first-century cross-cultural critical endeavour and for women in STEM(M) as a model of already prescient and interconnected local-global knowledge-making with committedly comparative environmental views.

Notes

1 See Eileen M. Byrne, *Women and Science: The Snark Syndrome*. Bristol: Falmer Press, Taylor & Francis, 1993. If the 'snark syndrome' sums up all such received wisdom, assumption, beliefs and prejudices as tantamount to superstitions, because sound empirical research demonstrates their fallacy, among the problems with Byrne's term and (circular) argument is that the 'snark syndrome' keeps feeding what she calls the 'snark effect'. These are the same internalised assertions (with no ground) that form the basis of education and policy decisions. The lack of historical contextualisation for portmanteau terms is a further problem for their usage beyond identifying a pattern.

2 See Martin, *Nature Translated*, and Joy Harvey, *'Almost a Man of Genius': Clemence Royer, Feminism, and Nineteenth-Century Science*. New Brunswick and London: Rutgers University Press, 1997.

3 Orr, 'Fish', 208–09.

4 Orr, 'Fish', 220–21, emphasis added. The translation is mine (from the incipit to the first volume of the *Histoire naturelle des poissons*, 1–2). I discussed Sarah and her 'talents' (205) as the model Cuvier may have had in mind for the final sentences of this quotation.

5 Orr, 'Fish', 235.

6 See Orr, 'Fish', 212 and 'Women Peers'.

7 I re-inflect the keywords of the important work by Abir-Am and Outram discussed in the introduction.

8 Orr, 'Fish', 235, and more recently Orr, 'Catalysts'.

9 The first volume of this multi-volume work was published in 1845 and rapidly translated. It did not include fiction among its 'literary' elements, a distinction that evades historians of science. See, as recently as indicative (and within Anglophone contexts), Alice Jenkins, 'Alexander von Humboldt's "Kosmos" and the Beginnings of Ecocriticism', *Interdisciplinary Studies in Literature and Environment* 14, no. 2 (Summer 2007): 89–105 at https://www.jstor.org/stable/44086615.

10 See Orr, '*Adventures in Australia* (1851) by Mrs R. Lee' for the French natural sciences behind it.

APPENDIX 1: PUBLICATIONS AUTHORED BY SARAH BOWDICH THEN LEE, 1825–1856

A. Anonymous and Unacknowledged (Joint) Publications and Translations, 1819–1824

(Where 'Sarah Bowdich del.' is the published illustrator, the number of her drawings in the volume is recorded in bold)

Anon. *Taxidermy: On the Art of Collecting, Preparing and Mounting Objects of Natural History. For the Use of Museums and Travellers*, **3**. London: Longman, Hurst, Rees, Orme and Brown, 1820.

Anon. *Taxidermy: On the Art of Collecting, Preparing and Mounting Objects of Natural History. For the Use of Museums and Travellers*. Second Revised Edition. London: Longman, Hurst, Rees, Orme and Brown, 1821.

Anon. *Taxidermy: On the Art of Collecting, Preparing and Mounting Objects of Natural History. For the Use of Museums and Travellers*, **3**. Third Revised Edition. London: Longman, Hurst, Rees, Orme and Brown, 1823.

Anon. *Taxidermy: Or the Art of Collecting, Preparing and Mounting Objects of Natural History. For the Use of Museums and Travellers*. Fourth Revised Edition. London: Longman, Rees, Orme, Brown and Green, 1829.

Anon. *Taxidermy: Or the Art of Collecting, Preparing and Mounting Objects of Natural History. For the Use of Museums and Travellers*. Fifth Revised Edition. London: Longman, Brown, Green and Longmans, 1835.

Bowdich, Mme. *Excursions dans les Isles de Madère et de Porto Santo*. 2 vols., **40**. Paris: F. G. Levrault, 1826.

Bowdich, T. Edward. *Mission from Cape Coast Castle to Ashantee*. London: John Murray, 1819.

———. *Travels in the Interior of Africa to the Sources of the Senegal and Gambia; performed by Command of the French Government in the Year 1818 by G. Mollien*. London: Henry Colburn and Co., 1820.

———. *The British and French Expeditions to Teembo, with Remarks on Civilization in Africa*. Paris: J. Smith, 1821.

———. *An Essay on the Superstitions, Customs and Arts Common to the Ancient Egyptians, Abyssinians and Ashantees*, **19**. Paris: J. Smith, 1821.

———. *An Analysis of the Natural Classifications of Mammalia. For the Use of Students and Travellers*, **120**. Paris: J. Smith, 1821.

———. *An Introduction to the Ornithology of Cuvier. For the Use of Students and Travellers*, **261**. Paris: J. Smith, 1821.

———. *Elements of Conchology, Including the Fossil Genera and the Animals. Part 1: Univalves With Upwards of 500 Figures*. Paris: J. Smith; London: Treuttel and Würtz, 1822.

———. *Elements of Conchology, including the Fossil Genera and the Animals. Part II: Bivalves. Multivalves. Tubicolae*, **500**. Paris: J. Smith; London: G. B. Sowerby and H. S. Tutchbury, 1822.

———. *A Geognostical Essay on the Superposition of Rocks in both Hemispheres by Alexander von Humboldt*. Translated from the original French. London: Longman, Hurst, Rees, Orme, Brown & Green, 1823.

———. *History and Description of the Royal Museum of Natural History*, translated from the French of M. Deleuze. Paris: L. T. Celliot, 1823.

———. *An Essay on the Geography of North-Western Africa*. Paris: L-T Celliot, 1824.

———. *An Account of the Discoveries of the Portuguese in the Interior of Angola and Mozambique from Original Manuscripts. To Which Is Added a Note by the Author on a Geographical Error of Mungo Park in His Last Journey into the Interior of Africa*. London: John Booth, 1824.

B. Sole-Authored Publications by Sarah Bowdich (Mrs T. Edward Bowdich), then Mrs R. Lee, 1825–1856.

Bowdich, T. Edward. *Excursions in Madeira and Porto Santo, during The Autumn of 1823…*, **40**. London: George B. Whittaker, 1825.

———. 'Adumissa'. In *Forget Me Not; A Christmas, New Year's and Birthday Present*, edited by Frederic Shoberl. pp. 233–53. London: Ackermann and Co., 1826.

———. 'Amba, the Witch's Daughter'. In *Forget Me Not; A Christmas, New Year's and Birthday Present*, edited by Frederic Shoberl. pp. 9–39. London: Ackermann and Co., 1827.

———. *The Fresh-Water Fishes of Great Britain. Drawn and Described by Mrs T. Edward Bowdich*. London: R. Ackermann, 1828–38.

———. 'The Booroom Slave'. In *Forget Me Not; A Christmas, New Year's and Birthday Present*, edited by Frederic Shoberl. pp. 37–77. London: Ackermann and Co., 1828.

———. 'Anecdotes of a Tamed Panther'. *Magazine of Natural History* 1 (1828): 108–12.

———. 'Eliza Carthago'. *The Gentleman's Magazine and Historical Chronicle* 98 (July–Dec. 1828): 347–49.

———. 'On the Natural Order of Plants, Dicotyledòneœ Anonàceæ'. *Magazine of Natural History* 1 (Jan. 1829): 438–41.

———. 'Eliza Carthago'. In *Forget Me Not; A Christmas, New Year's and Birthday Present*, edited by Frederic Shoberl. pp. 57–64. London: Ackermann and Co., 1829.

———. 'Anecdotes of a Diana Monkey'. *Magazine of Natural History* 2 (1829): 9–13.

———. 'Attachments formed by Animals…'. *Magazine of Natural History* 2 (1829): 62.

———. 'Some Account of the Progress of Natural History during the Year 1828, as reported to the Academy of Sciences in Paris by the Baron Cuvier'. *Magazine of Natural History* 2 (1829): 409–28.

———. 'Going to Sea, and the Ship's Crew'. *Friendship's Offering: A Literary Album* (1829): 361–72.

———. 'The Life of a Hero'. In *Forget Me Not; A Christmas, New Year's and Birthday Present*, edited by Frederic Shoberl. pp. 325–37. London: Ackermann and Co., 1830.

———. 'The Voyage Out'. *Friendship's Offering: A Literary Album* (1830): 85–96.

———. 'Notice sur une panthère apprivoisée' par madame Sophia (*sic*) Bowdich'. *Gazette Littéraire* (8 avril 1830): 301–02.

Lee, Mrs R. (late Mrs Bowdich). 'Some Details regarding the Garden of Plants and the National Museum of Paris'. *Magazine of Natural History* 3 (1830): 22–26.

———. 'Two Poodles from Milan'. (signed Sarah Lee). *Magazine of Natural History* 3 (1830): 290–91.

———. (formerly Mrs Bowdich). 'Agay, the Salt Carrier. An African Tale'. *Friendship's Offering: A Literary Album* (1831): 137–53.

———. 'Notice of a Fossil Nautilus found in the Sandstone of the Isle of Sheppey'. *Magazine of Natural History* 4 (1831): 137–38.

———. (formerly Mrs Bowditch (*sic*)). 'Jacqueline'. *The English Annual* 1 (1832): 245–56.

———. (lately Mrs Bowdich). 'La Mère des Soldats'. In *Forget Me Not; A Christmas, New Year's and Birthday Present*, edited by Frederic Shoberl. pp. 212–23. London: Ackermann and Co., 1832.

———. 'A Night Alarm'. In *Forget Me Not; A Christmas, New Year's and Birthday Present*, edited by Frederic Shoberl. pp. 117–27. London: Ackermann and Co., 1832.

———. 'Jacqueline'. *The Court Magazine and Belle Assemblée* (Sept. 1832): 117–21.
———. 'A Visit to Empoöngwa; or a Peep into Negro-Land'. *Friendship's Offering: A Literary Album* (1833): 294–304.
———. *Memoirs of Baron Cuvier.* London: Longmans Rees Orme Brown Green & Longman, 1833.
———. *Memoirs of Baron Cuvier.* New York: J. & J. Harper, 1833.
———. *Mémoires du Baron Georges Cuvier*, trans. Théodore Lacordaire. Paris: H. Fournier, 1833d.
———. (Formerly Mrs Bowdich). 'A Scene in Negroland'. *The Court Magazine and Belle Assemblée* (June–Dec. 1834): 95.
———. (formerly Mrs T Edward Bowdich). *Stories of Strange Lands and Fragments from the Notes of a Traveller.* London: Edward Moxon, 1835.
———. 'A Night Alarm'. In *Forget Me Not; A Christmas, New Year's and Birthday Present*, edited by Frederic Shoberl. pp. 117–27. London: Ackermann and Co., 1835.
———. 'The Busy Friends: A Spanish Story, Founded on Facts'. In *Forget Me Not; A Christmas, New Year's and Birthday Present*, edited by Frederic Shoberl. pp. 321–38. London: Ackermann and Co., 1836.
———. 'The Caterer. A Fragment from the Notes of a Traveller'. In *Forget Me Not; A Christmas, New Year's and Birthday Present*, edited by Frederic Shoberl. pp. 211–22. London: Ackermann and Co., 1837.
———. 'The Coat of Arms'. In *Forget Me Not; A Christmas, New Year's and Birthday Present*, edited by Frederic Shoberl. pp. 259–67. London: Ackermann and Co., 1838.
———. 'Hammer and Nails'. In *Forget Me Not; A Christmas, New Year's and Birthday Present*, edited by Frederic Shoberl. pp. 175–93. London: Ackermann and Co., 1839.
———. 'The Tapestried Chair'. In *Forget Me Not; A Christmas, New Year's and Birthday Present*, edited by Frederic Shoberl. pp. 343–59. London: Ackermann and Co., 1840.
———. 'Florence; or the Sutherland Family'. In *Forget Me Not; A Christmas, New Year's and Birthday Present*, edited by Frederic Shoberl. pp. 269–88. London: Ackermann and Co., 1841.
———. *The Juvenile Album or Tales from Far and Near.* London: Ackermann and Co, 1841b.
———. 'Scenes in Wales'. In *Forget Me Not; A Christmas, New Year's and Birthday Present*, edited by Frederic Shoberl. pp. 59–81. London: Ackermann and Co., 1842.
———. *Taxidermy: Or the Art of Collecting, Preparing and Mounting Objects of Natural History. For the Use of Museums and Travellers.* Sixth Revised Edition. London: Longman, Brown, Green and Longmans, 1843.
———. 'The Pirate'. In *Forget Me Not; A Christmas, New Year's and Birthday Present*, edited by Frederic Shoberl. pp. 63–80. London: Ackermann and Co., 1843.
———. 'Emily. A Portrait from the Artist's Studio'. In *Forget Me Not; A Christmas, New Year's and Birthday Present*, edited by Frederic Shoberl. pp. 149–166. London: Ackermann and Co., 1844.
———. (formerly Mrs T. Edward Bowdich). *Elements of Natural History. For the Use of Schools and Young Persons. Comprising the Principles of Classification, Amusing and Instructive Original Accounts of the most Remarkable Animals.* London: Longman, Brown, Green and Longmans, 1844.
———. *The African Wanderers or the Adventures of Carlos and Antonio. Embracing Interesting Descriptions of the Manners and Customs of the Western Tribes and the Natural Productions of the Country.* London: Grant and Griffith, 1847.
———. *The African Crusoes.* New York: Dick and Fitzgerald, 1847.
———. *Willie Fraser or the Little Scotch Boy and Other Tales.* London: Harvey and Darton, 1850.
———. (formerly Mrs T. Edward Bowdich). *Elements of Natural History or First Principles of Zoology. For the Use of Schools and Young Persons, Comprising the Principles of Classification, Interspersed with Amusing and Instructive or Final Accounts of the most Remarkable Animals.* New Edition, Revised and Enlarged. London: Longman, Brown, Green and Longmans, 1850.
———. *The African Wanderers or the Adventures of Carlos and Antonio. Embracing Interesting Descriptions of the Manners and Customs of the Western Tribes and the Natural Productions of the Country.* Second Edition. London, 1850c.

———, *Adventures in Australia; or the Wanderings of Captain Spencer in the Bush and the Wilds. Containing Accurate Descriptions of the Habits and Natural Productions and Features of the Country.* London: Grant and Griffith, 1851.

———. *Anecdotes of the Habits and Instincts of Animals.* London: Grant and Griffith, 1852 (Second Edition, 1853).

———. *Anecdotes of the Habits and Instincts of Birds, Reptiles and Fishes.* London: Grant and Griffith, 1853.

———. *Adventures in Australia; or the Wanderings of Captain Spencer in the Bush and the Wilds. Containing Accurate Descriptions of the Habits and Natural Productions and Features of the Country.* Second Revised Edition. London: Grant and Griffith, 1853.

———. *Anecdotes of the Habits and Instincts of Animals.* Philadelphia: Lindsay & Blakiston, 1853.

———. *Familiar Natural History: with Descriptions.* London: Grant and Griffith, 1853.

———. *Twelve Stories of the Sayings and Doings of Animals.* London: Grant and Griffith, 1853.

———. *The African Crusoes.* Philadelphia: Lindsay and Blakiston, 1854; †1860.

———. (formerly Mrs T. Edward Bowdich). *Trees, Plants and Flowers: Their Beauties, Uses and Influences.* London: Grant and Griffith, 1854.

———. *Playing at Settlers or The Faggot-House.* London: Grant and Griffith, 1855.

———. *Sir Thomas, or the Adventures of a Cornish Baronet in North-Western Africa.* London: Grant and Griffith, 1856.

Lee, Mrs. R. *The African Crusoes.* Boston: Lee and Shepard, 1873; 1876.

———. *Adventures in Fanti-Land.* London: Grant and Griffith, 1881.

———. *Adventures in Fanti-Land.* New York: E. P. Dutton, 1881.

APPENDIX 2: (FRENCH) DATABASE NAME/NAMING SEARCH

- *Ethmalosa fimbriata* : Fowler, 1928 a : 38 | Hornell, 1929 : 1 | Fowler, 1936 a : 175 | Cadenat, 1947 : 28 | Steven, 1947 : 1 | Blanc, 1948 : 45 | Brown, 1948 : 563 | Cadenat, 1950 a : 127 | Lozano-Rey, 1950 : 7 | Postel, 1950 : 49 | Salzen, 1958 : 1388 | Doutre, 1959 : 513 | Postel, 1959 : 230 | Longhurst, 1960 a : 276 | Longhurst, 1960 b : 1337 | Gras, 1961 : 575 | Monod, 1961 : 506 | Boeseman, 1963 : 5 | Longhurst, 1963 b : 1 | Watts, 1963 : 235 | Longhurst, 1964 a : 686 | Daget & Iltis, 1965 : 47 | Whitehead, 1967 a : 86 | Whitehead, 1967 b : 585 | Daget & Stauch, 1968 : 25 | Reizer, 1968 : 229 | Zei, 1969 : 105 | Boely & Elwertowski, 1970 : 182 | Reizer, 1971 : 85 | Scheffers, 1971 : 1 | Scheffers et al., 1972 : 1 | Scheffers, 1973 : 1 | Fagade & Olaniyan, 1974 : 248 | Albaret & Gerlotto, 1976 : 113 | Gerlotto, 1976 : 1 | Leveque & Paugy, 1977 : 13 | Daget, 1979 : 59.
- *Meletta senegalensis* : Valenciennes, 1847 : 370 (junior synonym) | Whitehead, 1967 a : 86 (junior synonym) | Whitehead, 1967 b : 588 (junior synonym).
- *Harengula forsteri* : Valenciennes, 1847 : 299 (junior synonym) | Whitehead, 1967 b : 587 (junior synonym).
- *Alausa dorsalis* : Valenciennes, 1847 : 418 (junior synonym) | Dumeril, 1858 d : 264 (junior synonym) | Whitehead, 1967 b : 589 (junior synonym).
- *Alausa aurea (non Steindachner)* : Dumeril, 1858 d : 264 (junior synonym).

- *Alausa platycephalus* : Bleeker, 1863 a : 123, pl. 26, fig. 2 (junior synonym) | Günther, 1868 : 438 (junior synonym) | Boeseman, 1963 : 5, pl. I, fig. 1 (junior synonym) | Whitehead, 1967 b : 590 (junior synonym).
- *Clupea dorsalis* : Günther, 1868 : 438 (junior synonym) | Steindachner, 1870 b | Steindachner, 1870 e : 566 (junior synonym) | Dambeck, 1879 : 414 (junior synonym) | Boulenger, 1905 g : 187 (junior synonym) | Ehrenbaum, 1915 : 20 (junior synonym) | Kahsbauer, 1962 : 159 (junior synonym).
- *Clupea senegalensis* : Steindachner, 1870 b | Steindachner, 1870 e : 569 (junior synonym) | Dambeck, 1879 : 414 (junior synonym) | Rochebrune (de), 1883 : 143 (junior synonym) | Steindachner, 1894 b : 81 (junior synonym) | Osorio, 1895 a : 133 (junior synonym) | Osorio, 1898 : 199 (junior synonym) | Pellegrin, 1905 g : 137 (junior synonym) | Pellegrin, 1907 i : 85 (junior synonym) | Pellegrin, 1914 l : 13 (junior synonym).
- *Clupea setosa* : Steindachner, 1870 d : 311 (junior synonym) | Steindachner, 1882 a : 14 (junior synonym) | Whitehead, 1967 b : 590 (junior synonym).
- *Clupea aurea (non Steindachner)* : Rochebrune (de), 1883 : 142 (junior synonym).
- *Clupea eba* : Pellegrin, 1907 i | Pellegrin, 1914 k : 13 (junior synonym) | Pellegrin, 1914 l | Pellegrin, 1922 i.
- *Ethmalosa dorsalis* : Regan, 1917 c : 303 (junior synonym) | Fowler, 1919 : 265 (junior synonym) | Pellegrin, 1923 b : 84 (junior synonym) | Chabanaud & Monod, 1926 : 238, fig. (junior synonym) | Monod, 1927 : 658, fig. (junior synonym) | Fowler, 1928 a : 38 (junior synonym) | Pellegrin, 1928 e : 6 (junior synonym) | Pfaff, 1933 : 298 (junior synonym) | Frade et al., 1946 : 368 (junior synonym) | Cadenat, 1947 : | (junior synonym) | Trewavas & Irvine, 1947 : 110, fig. (junior synonym) | Blanc, 1948 : 45 (junior synonym) | Monod, 1949 e : 53-65, fig. 18-30 (junior synonym) | Roux & Collignon, 1950 : 26 (junior synonym) | Maclaren, 1952 : 79 (junior synonym) | Poll, 1953 a : 29 (junior synonym) | Bainbridge, 1957 : 874 (junior synonym) | Watts, 1957 : 539 (junior synonym) | Bainbridge, 1960 : 1 (junior synonym) | Cadenat, 1960 : 1372 (junior synonym) | Bainbridge, 1961 : 347 (junior synonym) | Bainbridge, 1962 : 262 (junior synonym) | Bainbridge, 1963 : 270 (junior synonym) | Daget & Iltis, 1965 : 47 (junior synonym) | Longhurst, 1965 a : 302 (junior synonym) | Whitehead, 1967 a : 86 (junior synonym) | Blache et al., 1970 : 139, fig. 366 (junior synonym) | Longhurst, 1971 : 353 (junior synonym).
- *Sardinella eba (non Regan)* : Metzelaar, 1919 : 203 (junior synonym) | Chabanaud & Monod, 1926 : 239 (junior synonym).
- *Clupea fimbriata* : Fowler, 1936 a : 175 (junior synonym) | Zolezzi, 1938 : 213 (junior synonym) | Whitehead, 1967 b : 586 (junior synonym) | Talwar & Whitehead, 1971 : 80 (junior synonym).

Attributs

- DISTRIBUTION: Lien Faunafri
- Anatomy : Monod, 1961 .
- Fisheries : Salzen, 1958 | Scheffers, 1973 | Scheffers & Taylor-Thomas, 1974 .

- Food habits : Bainbridge, 1957 | Bainbridge, 1963 | Monod, 1927 .
- Growth : Gerlotto, 1976 .
- Migrations : Blanc, 1948 | Gerlotto, 1976 | Reizer, 1971 .
- Reproduction : Albaret & Gerlotto, 1976 | Bainbridge, 1961 | Longhurst, 1963 | Reizer, 1971 | Scheffers, Conand & Reizer, 1972 | Watts, 1963 .
- Types :
    - "Alausa dorsalis Valenciennes in Cuvier & Valenciennes, 1847: 418. Type locality: "Côte d'Afrique". Lectotype MNHN n° 3175 (larger one); paralectotype MNHN n° 3175 (smaller one) - designated by Whitehead, 1967 b."
    - Alausa platycephalus Bleeker, 1863 a: 123, pl. 26, fig. 2. Type locality: "Ashantee, Ghana". Holotype RMNH n° 3310.
    - Clupea fimbriata Bowdich, 1825: 234, fig. 44. Type locality: "Porto Praya, Cape Verde Islands". Most likely Gambia, fide Whitehead, 1967 b: 590. Putative neotype BMNH n° 1900.6.28: 25 (Whitehead, 1967 b).
    - "Clupea setosa Steindachner, 1870: 22. Type locality: "Mazatlan, Mexico"; true provenance Liberia and Gaboon fide Steindachner, 1882. Holotype NMW n° 76999."
    - Harengula forsteri Valenciennes in Cuvier & Valenciennes, 1847: 299. Type locality: "San Jago, Cap Vert". No type material, based on drawing.
    - Meletta senegalensis Valenciennes in Cuvier & Valenciennes, 1847: 370 (non Alosa senegalensis Bennett, 1831). Type locality: "Sénégal". No type material.

http://vmcloffa-dev.mpl.ird.fr/table/taxon2/view?idtaxon:int=1107, gray-shading added, to highlight the first provenance and naming of this Fish (captured 11/03/2016, re-verified 21/06/2022, single view pagination, 10/08/2023).

APPENDIX 3: ORIGINAL PROSPECTUS FOR THE *BRITISH ICHTHYOLOGY*, ED. W. JERDAN ESQ.

Mr. S Pether* was our leading counsel and guide in regard to the fine art departments of drawing and engraving the fishes, which Sir John, at first, estimated at thirty, but when the list came to be made, we found only the following:—British fresh water fishes: 1. pike; 2. perch; 3. carp; 4. tench; 5. trout; 6. barbel; 7. grayling; 8., gwynnard, in Bala Pool, North Wales; 9. char, in Windermere; 10. chub; 11. bream; 12. roach; 13. dace; 14. pope; 15. bleak;

The cherished plan of a publication between Sir J. Leicester and I was thus announced:—

Will be published in One Volume Quarto.

BRITISH ICHTHYOLOGY.

EDITED BY

W. JERDAN, Esq., F.A.S., M.R.L.S., &c.

WITH OCCASIONAL REMARKS BY

AND ILLUSTRATED WITH ENGRAVINGS EXECUTED
UNDER THE IMMEDIATE INSPECTION

OF

SIR J. F. LEICESTER, Bart., H.M.R.D.S., H.M.R.C.I., &c.

FROM CORRECT DRAWINGS IN HIS POSSESSION.

The idea was more specifically developed in the annexed suggestions in the handwriting of my respected coadjutor.

"In One Vol. Quarto,

BRITISH ICHTHYOLOGY;

Illustrated with engravings of the principal Fish of Great Britain, and others frequenting its shores, from drawings taken from Nature by Sir J. F. Leicester and some of the first artists.

16. eel; 17. gudgeon; 18. loach; 19., minnow; 20. miller's thumb; 21. stickleback: river fish: 1. salmon; 2. smelt; 3. flounder; 4. lamprey eel, in the Severn.—Total 25.

* Painted for Sir John, who suggested the subject, "A Caravan overtaken by a Sandstorm in the Desert."

The cost of the drawings and engravings was estimated at 200 guineas, of which the moiety was to be paid for the former by Sir John, and the second hundred for the engravings, to be repaid by the publication if it succeeded, any surplus to be mine. There was to be a popular octavo edition, after the quarto, with wood-cuts, as in Salter and Walton; when I got assent to the corporeal introduction of my small-fry friends, the minnows, sticklebacks, loaches and miller's thumbs. The plates were to be about the size of those in old Albin's "History of *Esculent* Fishes," 1794; and we had a bream executed as a specimen by Mr. Clarke, recommended by Pether. The other engravers spoken with were Griffiths, the engraver of Cuvier, in which the dory was beautifully executed; Curtis, the engraver of Franklin; Milan, Marsh, Swaine and John Scott, all able artists, and competent to do ample justice to the undertaking; which became quite a pleasant hobby, and led to many an amusing letter, as we discussed tails, scales, gills, fins and localities where the finest specimens were to be caught, including journeys to the English lakes, Scotland, Wales and Ireland!

My part was to write an introduction, and a page or leaf of description, with the various generic names given by different authors, and modelled on Block's "Ichthyologia" (3 vols, folio, Berlin, 1795), with remarks on particular species, and specimens extraordinary for size, or other peculiarity. In one of his notes, my co-author was pleased to observe that he had no doubt my share of the work would be very complete, if he could persuade me not to go quite so far as the cookery. Paintings of an enormous jack and a carp, taken out of the Tabley waters, by Thomson, R.A., were admirably copied by Sir John, and with several other drawings of his own of perch and roach, dispatched to me in London, wherewith to commence proceedings, with the instruction, "I am fully persuaded that nothing can be had *excellent* and *cheap*; if the first is not obtained in a very *high degree*, I shall not like to have a hand in it." (ibid. pp. 145-6, emphasis in the original)

APPENDIX 4: LIST OF SUBSCRIBERS TO *THE FRESH-WATER FISHES OF GREAT BRITAIN*

Thomas B. Aveling Esq.
Robert Barclay, Esq
R. Bland, Esq.
Mrs Branthwayte,
Messrs Calkin and Budd
J. G. Children,
E. Cooper Esq.
Alex. Copeland Esq.
R. Davies Esq.
Edward Downes, Esq.
Mrs Erskine
Dr. Franck
H. Gamble Esq.
Hudson Gurney Esq
Revd. Barnard Hanbury
F.W. Herschel,
C. Hodgson Esq
Mrs Holland
J. Hudson, Esq
J. Marryat, Esq M.P
George Marryat
Mrs Marryat
Miss Marryat
R. Murchison, Esq
J. Murray, Esq.
Gerard Ralston Esq.
Rev Oliver Raymond
J. Rump, Esq
W. Sturt, Esq
Charles Tennant, Esq.
Thomas Tooke Esq.
Miss Tooke
Miss Waddington
George Waddington
Charles Webber Esq,
Dr Williams

Duke of Sussex
Duke of Devonshire
Marquess of Lansdowne
Earl Gower
Baroness Grey d Ruthyn
Sir Humphry Davy

APPENDIX 5: 'HUMBOLDT' IN REFERENCES AND FOOTNOTES FOR THE *EXCURSIONS* (1825)

Reference or Note + p. number	*Trigger term*	*Humboldtian Source (as given in the* Excursions*)*
Reference and Note, pp. 31–32	*sida carpinifolia* and *symplocos alstonia*	*Plantes équinoxiales*, t. I, p. 185. See https://gallica.bnf.fr/ark:/12148/bpt6k61310k/f4.itemp. 159.
Reference and Note, p. 47	Peak of Teneriffe (*sic*)	*Voyage de Humboldt*, I, c. 3. See *Voyage aux régions équinoxiales du nouveau continent fait en 1799, 1800, 1801, 1802 1803 et 1804.* Paris: F. Schoell, 1814, ch. 2. See https://gallica.bnf.fr/ark:/12148/bpt6k97281w
Reference and Note, p. 64	salt in (volcanic) tufas	In clay in the Cordillieres nearly 13,000 feet above the sea' *Relation historique*, t. 2, ch. 5.
Note, p. 68	[Madeira was not created by a marine volcano]	*Humboldt's Relation historique*, Supplement, p. 640.
Note, p. 82	[shells in sandstones]	*Voyage*, t. 2, c. 5.
Note, p. 84	[system of winds in this part of the Atlantic]	*De l'influence de la déclinaison du soleil sur le commencement des pluies équatoriales.* Annales de Chimie, 1821, p. 179. See *Annales de Chimie et de Physique*, par MM. Guy-Lussac et Arago. T. 8. Paris: chez Crochard, 1818: 179–190.
Reference and Note, p. 86	Impressions like *arcae.*	Humboldt, *Relation historique,* t. 2 c. 5.
Reference and Note, p. 89	Calcareous tufa [...] which is a tertiary formation	Humboldt, *Relation historique,* t. 2 c. v.
Reference and Note, p. 90	Shelly limestone of [Portugal] and Baxo	Humboldt, *Rel. Hist.* I, c.ii and *Supplement*, p. 641. 4to.
Reference and Note, p. 92	Dracoena Draco	*Tableaux de la Nature. Physionomie des Végétaux,* t. 2, p. 110.
Note, p. 104	[mosses]	Baron de Humboldt observes, that, in countries near the tropics, succulent plants appear before mosses (no ref.).
Reference only, p. 105	the extreme of Baron de Humboldt's vine region	
Reference only, p. 116	*d. sativa* of Linneaeus	Remarked by Baron de Humboldt (*Essai Politique,* &c. p. 407). See *Essai politique sur le royaume de la Nouvelle Espagne.* Paris: F. Schoell, 1811.
Reference only, p. 117	The *til* has been confounded with the *l. foetens*	Mentioned by Baron de Humboldt under the name *laurus til.*
Note, p. 126	[Temperatures at the equator]	Humboldt, *Mémoires d'Arceuil*, 5. p. 512.

(*Continued*)

(*Continued*)

Reference or Note + p. number	*Trigger term*	*Humboldtian Source (as given in the* Excursions*)*
Reference and Note, p. 127	'A Funchal la temperature des caves…'	Humboldt, *Annales de Chimie*, p. 602. *Rel. Hist.* p. 424, qto. I quote from my MS extracts from these expensive works, and omitted to note the volume.
Reference and Note, p. 137	[artificial horizon] and 'the tact of De Humboldt.'	De Humboldt, *Voyage*, t. 1 p. 63, 64.
Note, p. 147	[highest point of Madeira]	Humboldt's *Voyage*, &c. t. 1. c.1
Reference only, p. 148	the beautiful *inga obtusifolia*	'(first found in Cumana, by De Humboldt and Bonpland)'

APPENDIX 6: NOTES TO '*ADUMISSA*'; THE LONGEST NOTE IN '*AGAY*'

Notes to 'Adumissa'

(a) Note [31]

I have seen many of these anniversaries, both on Adumissa's account and that of others; and nothing can be more stunning or distressing than their noise, except, indeed, that made at the funcral itsclf. My cars were generally saluted at sunrise by a discharge of muskets [...] About ten o'clock the procession usually started from the house of the deceased, the body in its own basket or palanquin [...] the bearers being painted in all the grotesque fashion that could be devised. [...] all screamed to the sound of the most barbarous instruments [...] and, by night, some were so tired, and others so intoxicated with the rum and fermented palm wine [...] that unable to stand, they laid down, kicked and roared. I used to dread giving permission to any of my servants to assist at these ceremonies, for they were generally unable to speak, or even move, for a week [...] The victims, generally bulls or cows [...] accompanied by the priests and slaughterers with large knives, formed a part of the parade, and being killed in the evening, were then dealt out to the hungry mob. Where the English forts stood, thank God! these victims were confined to quadrupeds, with very few exceptions; but, in the interior, hundreds of men and women are butchered on these occasions. Many, however, were the attempts even of the natives on the coast to preserve this ancient custom, but nothing but the exceeding energy and firmness of Mr Hope Smith (Mr Bowdich's uncle), succeeded in wholly putting a stop to this horrible practice under the walls of our fortresses. I was in the country only for three weeks, and was then residing in the fort of Annamaboo; I heard a most unusual noise, and running to the ramparts wit overlooked the town, [...] immediately sunder me I saw some people dragging the headless trunk of a female, which they were going to throw into the sea; her head had been cut off at the grave of her master, and buried with him. Horror-struck, I flew to Mr Hope Smith [...] and he instantly summoned the authorities of the town, threatening to fire the heavy cannon [...] if they immolated any more victims. He thus frightened them into compliance, and a few such decided actions as these, for they well knew that Pynin Smith, as they called him, never threatened in vain, effectually put a stop to these monstrous rites during his government. (*SSL*, 31–32)

(b) *SSL*, 'Note [16]'

All the decorations here mentioned are in strict keeping with the fashion of the country; [...] The blackening of the eyelids is borrowed from the Moorish women, who always keep a little bodkin-like case of powdered lead, or antimony, near them. The soolah tooth-picks come chiefly from Accra, further in the Bight of Benin. [...] The shea tolu, or vegetable butter, comes from a very large tree, first made known to Europeans by the enterprising Mungo Park [...] It bears different names in the various parts of Africa [...] but, in Fantee as the butter alone is met with, it is called Ashantee grease. It is wholly distinct from the tallow tree, and is a new species of Bassia, which, in honour of him who found it, ought to be called Bassia Parkii. It extends over a large portion of the continent, from Jaloff and Houssa to the latitude of Gaboon [...] It is an excellent article of food when quite fresh, and enters into almost all the dishes of the natives. [...] It is one of the finest cosmetics possible, and, without some such aid, the skins of the negroes, constantly exposed to the sun, would crack and peel off in white scales. White people are always obliged to purchase it for their servants [...] When up the river Gaboon, I went to see the tree, which was rather a difficult task for an (*sic*) European woman to accomplish, as the nearest grew in the thickest part of the forest. [...] Unfortunately the tree was not in flower, and the chief advantage I derived from my excursion was the having actually seen the 'fat tree.' A servant from Booroom, to the north of Ashantee, told me, that her people bruise the nut, boil it in water till the oil rises to the top, and then skim this off, and put it into calabashes to cool and harden. The celebrated chemist, M. Chevreuil, analysed it for me, and found it admirably adapted for the manufacture of soap. (*SSL*, 24–26)

(c) Double Account of 'vegetable butter' in T. Edward Bowdich's *Mission* of 1819:

> There is a white grease, which has long been called Ashantee grease by the natives on the coast, who supposed it to be produced in that country. They use it daily to anoint their skins, which otherwise become coarse and unhealthy. The Ashantees purchase it from the interior, and make a great profit by it: it is a vegetable butter, decocted from a tree, called Timkeeii: it is doubtless the Shea butter of Mr. Park.* (*Mission*, 1819, 334)

Chapter XIII, 'Sketch of the Gaboon':

> The vegetable butter (which certainly belongs to the natural order Sapotece) brought to the Ashantee market, is here well known by the name of Onoongoo: it is a large tree, and the nuts are enclosed in a round red pod, containing from four to six: the flower is also red, from description. My servant, a native of Booroom, called the tree Kirrimkoon, and the butter Incoom; the Ashantees call the latter Sarradee; in Mallowa the tree is called Timkeea. The nut is first boiled, and the oil or butter afterwards expressed; in Booroom and Mallowa it is skimmed from the surface. It tasted quite as good as fresh butter before any salt is added, and we relished the meat fried in it exceedingly. Being the rainy season I could neither get a sight of the flower or the pod of this or the odica, but I procured the nuts and produce of both. The curious may compare this butter with the specimen of the Ashantee grease.

Before I understood them to be distinct trees, I concluded the odica and the butter both to be the produce of the cacao-nut, but the butter answers closely to Mr. Park's description of the shea-tolu, though the tree did not resemble the American oak. (*Mission*, 1819, 447)

The Longest Note in 'Agay': Sarah's Account of Herself in 'Note [18]'

([18]) The Ocrahs are supposed to know all the king's secrets; and are bound to perform his commands, whatever they may be; and to die when he does. In return for this, they are exempted form all care for their existence, generally in the Palace. They wear a large gold ornament, generally circular, but sometimes in the shape of a star or wings, round their necks, and few dare to resist the orders of one so decorated.

An Ocrah was sent to Cape Coast by the king, while I lived there; and prompted by an anxiety which taught me to conciliate the Ashantees as much as laid in my power, I yielded to his proposal to dine with me. He insisted on it that Mr Bowdich had desired him to do so; but I knew this to be untrue; and, uninfluenced by the falsehood, I appointed a day for the visit. The hour was to be three o'clock, and I was rather puzzled as to the choice of viands. At ten o'clock in the morning the gentleman arrived, with a retinue of at least fifty persons, some ragged and dirty, and among them the usual chamberlain, a piece of African state which is very absurd, for he bears a large bunch of rusty keys, for which his master has not a single lock. I was obliged to tell the Ocrah that I could not have him all day, and he left me in no very good humour. At the proper hour he re-appeared with his train; but on my further informing him that I was not prepared for so many, he dismissed all but his intimate friend, to the hall below. The two then sat down to the table, and I helped them to fish, which they began to eat with their fingers; but, on observing the use I made of a knife and fork, they begged to be instructed how to handle their's (*sic*). I could scarcely keep my countenance at their attempts, nor at their putting a piece slily into their mouths with their fingers, when they thought themselves unseen; but when Ocrahnameah had eaten half of his fish, he begged permission to send the remainder to his wife. [...] we proceeded to a chicken-pie, but I had been unfortunate in my selection. The fetish had forbidden him to eat fowls [...]; he however devoured mutton and pastry wholesale, and then returned to the fish; he drank wine and porter till he was nearly intoxicated, and I was very glad when he found himself so sleepy that he was obliged to retire. After this it was difficult to keep him at a distance, and he thought himself entitled to come at all hours of the day. The good-for-nothing person returned to Ashantee, saying he had received neither kindness nor presents, but my letters having informed Mr Bowdich of the truth, and all his property being seized by the king, numerous proofs of hospitality and attention were found. And Ocrahnameah was disgraced and stripped of every thing that he possessed.

He was a very unworthy specimen of Ashantee manners for a better behaved person than Adoo Bradie, nephew to the king, never appeared. [A story in a story ensues.] I must instance another instance of native manners, in the person of Yokokroko, the king's ironer. He paid his first visit when I happened to be alone, but although he had never before seen a white woman, he evinced no symptoms of surprise, but lowering his cloth from his shoulder, and waving his left hand in the air (the Ashantee salutation), he drew himself up to his full height, and stood till I requested him to sit down. He seated himself as if he had been used to an arm-chair all his life, but presently rising again, he took from the hands of one of his servants a small sanko (which is now in the British Museum), and unwrapping the cloth which enveloped it, he laid it on the table before he, and commanded the remainder of his presents to be submitted to my examination. He received my thanks with extreme dignity, took the refreshment I offered him without hesitation, and after a visit of a quarter of an hour, he retired; and the sole remark which he had permitted himself to make was a question whether I was of royal birth, the golden studs in my shoes having given him that idea; for none but the kingly descent can wear gold on their sandals in Ashantee. I may safely say, that in no European country have I ever seen a more elegant and self-possessed a person (*SSL*, 175–79).

APPENDIX 7: WEST AFRICAN TRAVEL ACCOUNTS INFORMING *THE AFRICAN WANDERERS* (1847)

1. Key Resources (in English and French, in Alphabetical Order) Informing *The African Wanderers*

Allen, Capt. William and T. R. H. Thomson. *A Narrative of the Expedition Sent by Her Majesty's Government to the River Niger in 1841.* 2 vols. London: R. Bentley, 1848.

Beecham, John. *Ashantee and the Gold Coast: Being a Sketch of the History, Social State, and Superstitions of the Inhabitants of Those Countries: With a Notice of the State and Prospects of Christianity Among Them.* London: John Mason, 1841.

Bowdich, T. Edward. *Mission from Cape Coast Castle to Ashantee.* London: John Murray, 1819.

———, Trans. *Travels in the Interior of Africa to the Sources of the Senegal and Gambia; Performed by Command of the French Government in the Year 1818 by G. Mollien.* London: Henry Colburn and Co., 1820.

———. *The British and French Expeditions to Teembo, With Remarks on Civilization in Africa.* Paris: J. Smith, 1821.

———. *An Essay on the Geography of North-Western Africa.* Paris: L-T Cellot, 1824a.

———. *An Account of the Discoveries of the Portuguese in the Interior of Angola and Mozambique from Original Manuscripts. To Which Is Added a Note by the Author on a Geographical Error of Mungo Park in His Last Journal into the Interior of Africa.* London: John Booth, 1824b.

Depuis, Joseph. *Journal of a Residence in Ashantee: Comprising Notes and Researches Relative to the Gold Coast and the Interior of Western Africa.* London: Henry Colburn, 1824.

Freeman, Thomas Birch. *Journal of Various Visits to the Kingdoms of Ashanti, Aku and Dahomi in Western Africa. With a Historical Introduction.* London: J. Mason, 1844.

Hutton, William. *Voyage to Africa: Including a Narrative of an Embassy to One of the Interior Kingdoms, in the Year 1820; With Remarks on the Course and Termination of the Niger.* London: Longman, Hurst, Rees, Orme and Brown, 1821. [cites 'Bowdich' 1819].

Lander, Richard. *Journal of a Second Expedition Into the Interior of Africa From the Bight of Benin to Soccattoo.* London: J. Murray, 1829.

———. *Records of Captain Clapperton's Last Expedition to Africa.* London: H. Colburn and R. Bentley, 1830.

———. *Journal of an Expedition to Explore the Course and Termination of the Niger.* 3 vols. London: J. Murray, 1832. [immediately translated into French]. The maps and descriptions in these texts inform key 'stretches' of Sarah's fictional 'wanderings'.

Mollien, Gaspar. *Voyage dans l'intérieur de l'Afrique, aux sources du Sénégal et de la Gambie, fait en 1818, par ordre du gouvernement français.* Paris: Arthus Bertrand, 1822.

Park, Mungo. *Travels in the Interior Districts of Africa. Performed under the Direction and Patronage of the African Association in the Years 1795, 1796 and 1797.* London: W. Bulmer and Company, 1799.

2. Episode of the 'White Negro'

Wondo led his friends into the house, saying, 'Now you shall see one of our white men; there he stands', and he pointed to a strange figure close to the door. [...] To the astonishment of Carlos and Antonio, a white negro was before them. The projecting muzzle, large mouth, flat nose and retreating forehead, the characteristics of his race, were much exaggerated in him; but his crisp, woolly hair was almost yellow in colour, his eyes were of a dark blue and from seeing imperfectly in the day-time, they were constantly blinking, and had a vacant expression. His skin was of a reddish-white, and when his cloth fell accidentally from his shoulders, a number of blotches were seen in various parts of his body. He slowly retired as the party entered the house, where they had not been long seated before the inner door opened, and Antonio involuntarily started. The white negro issued from it, carrying a small harp; the frame-work was of a yellowish wood, with a carved head on the upright; the strings were made of the runners of a tree; and when the man struck the first few chords, it gave a deep rich tone. He seated himself, and began a low recitative, as if to preface what followed. At length he burst into a song of defiance, which Wondo said was the war-cry of his native land. He rapidly passed his fingers over the strings, and all seemed bustle and activity; the notes then changed to a mournful air, accompanied by a low wailing. 'Those are the prisoners lamenting,' observed Wondo. The negro then shouted sounds of victory, and at that moment he appeared to be perfectly frantic; he put the harp upon his foot, tossed it up and down, stretched out the arm that was free and performed a number of gestures; the sounds then gradually died away as if they were retreating to a distance, the whole having a highly dramatic character. 'How very extraordinary!' exclaimed Carlos ; 'it is perfect inspiration. Where did he get that harp ?' 'He made it here,' replied Wondo; 'but he tells me there are plenty in his own country. I took him prisoner in a war which we had with' 'Hush!' cried Antonio, 'he is beginning again.' The musician recommenced by imitating the voices of birds; and to this succeeded a soft measure which again changed into a lively strain, as if for dancing. In the midst were heard the names of some of those present. 'He is praising us,' whispered Wondo, 'and he will end by describing a hunting party.' (*AW,* 143–145)

3. Episode of the African Slave Market

[T]hey were marched into a large marketplace, where they took their stand among a great many others, who, like them, were going to be sold as slaves. On casting their eyes around, they saw a display of wealth and industry for which they were not prepared. Natives with cloths round their waists, or short petticoats, were bustling about in all directions; then came their masters with caps on their heads, their feet thrust into highly-wrought sandals, and their figures enveloped in handsome cloths, put on toga fashion, or in robes; and besides these were Mohammedans, with ample tunics and turbans. The appearance of the Europeans created considerable surprise, and many were the questions put to the man who had them in his possession. [...] Several persons, apparently of consequence, offered to buy them; but their price was rejected. [...] At length the crowd which had gathered round them opened, and made way for a Moor, whose rich, brocaded silk dress and jewelled turban bespoke the man of consequence. Many of those around prostrated themselves before him, and others made a salaam to the ground. He stopped and looked at the white men with keen curiosity, and continued to gaze at them the whole time that the slave-trader was telling their history. Then the Moor, scornfully uttering the word 'Nassareen', evidently asked what he was to give for them; their master answered, a little bargaining ensued, and at length, calling an attendant, the Moor desired him to pay a certain number of cowries out of a bag which he carried, and then bidding him take the man with him, probably to deliver to him the remainder of the sum, he ordered that the newly-purchased slaves should be led to his house. 'Thank God!' exclaimed the two friends at the same moment, 'we shall be together'; and, with a light step, they almost joyfully followed their guide even into slavery. (*AW*, 267–8)

APPENDIX 8: NAMED EXPERT SCIENTIFIC SOURCES IN THE *ELEMENTS* (1844)

Name of Scientific Authority	*Animal in* Elements *+ p. ref.*	*Reference to published work as found in the* Elements*)*	*√ = direct quotation (rather than allusion or paraphrase) with page range*
Adanson the French Traveller	Ostrich, p. 237		
Mr Audubon	Vultures, p. 162		
	Shore Larks of Labrador, p. 192		
	American Bittern, p. 247		
Mr Baird	Lemur tardigradus	*Magazine of Natural History* in early number	√ (pp. 38–39)
Mr Bell	Viper, p. 313	*History of British Reptiles*	
	Frog, p. 325		
Mr Bennett	Gibbons, p. 29	*Magazine of Natural History*	√ (pp. 31–32)
Sir Thomas Brown	Moles, p. 56		
Bruguiere (*sic*)	Tenrecs, p. 55		
Mr Burchell the S. African Traveller	Hyaenas, p. 80		
Agnes Catlow	Note, p. 194	*Popular Conchology*, p. 197	
Commerson and Forster	Salarias alticus, p. 398		
Capt. Cook	Kangaroo, p. 97		
Mr Couch	The Pholis, p. 398		
	Sunfish, p. 455		
	Torpedo, p. 466		
Baron G. Cuvier	Dogs, pp. 71–72		√ pp. 71–72
	Humming Birds, p. 208		√ pp. 208–209
M. Frédéric Cuvier	Makes of Madagascar, p. 38		√ anecdote
	Hyaenas, p. 80		
	Horses, p. 129		
Sir Humphry Davy	Salmon, p. 427	*Salmonia*	
M. Duvaucel the Indian Traveller	Parakeet, p. 222		
Mr Ellis	Eels, p. 448		
Colonel Findlay	Crocodiles, p. 289		
The late Captain Fisher of the Navy	Barbary Cow, p. 140		
Mr Freyreise	Bats in Brazil, p. 48		
M. Geoffroy	Bats in Egypt, p. 47		
Gessner	Goldfinches, p. 196		
Gmelin	Silurus, p. 415		

(*Continued*)

(*Continued*)

Name of Scientific Authority	*Animal in* Elements + *p. ref.*	*Reference to published work as found in the* Elements)	√ = *direct quotation (rather than allusion or paraphrase) with page range*
Mr Gould	Satin Bower Bird, p. 178 Nightjar, p. 187 Ortolan Bunting, p. 195 Megoapodius or Jungle Fowl, p. 256 cit. Mr Gilbert Megalpodius and Telegalla of New Holland, p. 233 New Holland Vulture, p. 257	*Birds of Australia*	
Mr Gray	Species he makes a genus, p. 47 Ophidians, p. 304		
Colonel Hamilton Smith	Gluttons, pp. 65–66 Otters, p. 71 Weasels, p. 71	*Mammalia* in *The Naturalist's Library* acknowledged as important source, p. 48, on bats (and other animals)	√
Mr Hancock	Doras Hancockii or costatus, p. 416		
Dr Horsfield	Tupaias, p. 55		
Baron Humboldt	Pogonias?, p. 364 Atherina (Smelt) in Peru, p. 391 Astroblepus, p. 416 Gymnotus, p. 450		
Washington Irvine	Prairie Dogs, pp. 104–107	*Tour on the Prairies*	√ Anecdote
Mr Jesse	Moles, p. 56 Corncrakes, p. 258		
M. Le Court	p. 57		
Mr George Loddiges	Hummingbirds, p. 209		
Mr Low	Conger Eels (and Otters), p. 449		
Captain Lyon	Snow Bunting, p. 194	*Journal of a Naturalist*	√ pp. 194–95
Dr McCullough	? local acc. p. 431		
Montagu	Chough, p. 200 Grebes, p. 265		√ pp. 200–202
Mr Neill of Cannon Mills	Gull, p. 273		
Mr Owen	Specimens at Royal College of Surgeons, p. 27 Monotremes, p. 118		
Pallas the Russian Naturalist	Hedgehogs, p. 54 Comephorios, pp. 399–400		
Mr Pennant	The Pholis, p. 398 Lumpsucker, p. 445		
Mr Pentland the able Naturalist and Consul General of Bolivia	Arges, p. 416 Loricaria, p. 417		

(*Continued*)

(*Continued*)

Name of Scientific Authority	*Animal in* Elements + *p. ref.*	*Reference to published work as found in the* Elements)	√ = *direct quotation (rather than allusion or paraphrase) with page range*
Mr Pringle	African Elephant, p. 124	*A Residence in South Africa*	√ p. 125
	Springboks, p. 139		√ p. 140
	Gnu, p. 141		√ pp. 237–238
	Secretary Bird, p. 165		
	Ostrich, p. 237		
Sir Stamford Raffles	Tupaiais, p. 55		
M. Risso	Tetragonurus, p. 393		
Mr Scarth	Stormy Petrel, p. 272		√ Anecdote
Mr Schomburgk	Jabirus, p. 248	*Travels in Guiana*	√ pp. 248–49
Captain Scoresby	Fulmar Petrels St Kilda, p. 271		√ pp. 271–272
Mr Scrope	Salmon, p. 427	'Days and Nights of Salmon Fishing'	
Mr Selby	Partridges, p. 232		√ Anecdote, p. 232
	Waterhen, p. 258		√ Anecdote, p. 256–59
Spallanzani	Salamanders, p. 327		
M. Valenciennes	Silurus, p. 415		
	Carp, p. 421		
Mr Waterton	Vampires, species called Spectre, p. 49		√ pp. 167–69
			√ pp. 179–80
	Lion, p. 82		√ pp. 266–68
	Vultures, p. 142		
	Owls, p. 167		
	Missel Thrush, p. 179		
	Goat Sucker, p. 187		
	Starlings, p. 199		
	Guillemot, p. 266		
	Cormorants, p. 278 false myth		
Mr Wilson	Green Woodpecker, p. 218		√ pp. 218–19
Mr Yarrell	Reed Warbler, p. 182	*British Birds*	√ pp. 182–83
	Cross-beaks, p. 198	*British Fishes*	√ pp. 198–99
	Rooks, p. 202		√ pp. 202–203
	Creepers, p. 208		√ pp. 246–47 anecdote
	Bee-eaters, p. 212		
	Bittern, p. 246 (cit. Mr Maxwell)		
	Ducks, p. 284		
	Mackerel, p. 383		
	Herring, p. 428		
	Anchovy, p. 433		
	Common Eel, p. 449		
	Selacians, p. 460		
	Amnocetes, p. 469		

Appendix 'Bibliography'

Adanson (Michel), *Voyage to Senegal, the Isle of Goree, and the River Gambia*. London: J. Nourse and W. Johnston, 1759.

Anecdote of African men and boys riding tame ostriches at a French factory at Podor, Senegal (85–87).

Audubon, Mr (John James), Birds of America + Audubon, J. J. and William MacGillivray, *Ornithological Biography, or an Account of the Habits of the Birds of the United States of America* (accompanied by descriptions of the objects represented in the work entitled *The Birds of America*, and interspersed with delineations of American scenery and manners) 5 vols. Edinburgh: Adam Black, 1831–1839: Vultures (vol. 2: 1834, 33–52 & Plate 106); Shore Larks of Labrador (vol:? Plate 200?); American Bittern (vol. 4: 1838, 296–301 & Plate 337).

Baird (W. Esq.), 'Descriptive Notice of a Specimen of *Lemur tardigradus,* Lin., Makes, Cuv. kept alive for some Time at Edinburgh', *The Magazine of Natural History,* vol. 1 (1829): London: Longman, Rees, Orme, Brown and Green, 208–16 (including an image which is the source of Fig. 6 in the *Elements*).

Bell (Thomas), *A History of British Reptiles*. London: John Van Voorst, 1839, 'Viper', 313; 'Frog' is the 'Common Frog', 84–101 (including an image which is the source of Fig. 46 in the *Elements*).

Bennett (George), 'An Account of Ungka Ape of Sumatra', *Magazine of Natural History* (1832): 131–39 (including an image which is the source for Fig. X of the *Elements*).

Brown, Sir Thomas,

Bruguière

Burchell (William, J), *Travels in the Interior of South Africa,* vol. 1. London: Longman, Hurst, Rees, Orme & Brown, 1822, vol. 2. London: Longman, Hurst, Rees, Orme, Brown & Green, 1824, 'hyaenas' vol. 2, ch. viii, plate, 222, and description, 222–9.

Cadet-de-Vaux, Antoine-Alexis, *De la Taupe: de ses mœurs, de ses habitudes et les moyens de la détruire.* Paris: chez L. Colas, 1803, on 'Henry Lecourt Taupier', 7–16; on 'les variétés de Taupes reconnues par Hanry Lecourt', 50–2 and 'Des Piéges (*sic*) de Lecourt', 205–6 and 'De la manière dont Lecourt procède à la préhension des Taupes', 213–8.

Catlow, Agnes, *Popular Conchology or The Shell Cabinet Arranged According to the Modern System*. London: Longman, Brown, Green and Longmans, 1842 (second ed. 1854).

Commerson and Forster,

Cook, Capt. James

Couch (Jonathan),

Cuvier, Frédéric

Davy, Sir Humphry, *Salmonia, or Days of Fly Fishing*. London: John Murray, 1828.

Duvaucel, Alfred

Ellis,

Findlay, Colonel

Freyreise, Mr., quoted in Hamilton Smith, Lieut. Col. Charles, *Introduction to the Mammalia, The Naturalist's Library* vol. xiii. Edinburgh: W. H. Lizars and London: S. Highly, 1842, 123 and Plate 2 Cheiroptera.

Geoffroy (Saint-Hilaire, Isid.), *P. Egyptiacas* in Hamilton Smith, Lieut. Col. Charles, *Introduction to the Mammalia, The Naturalist's Library,* vol. xiii. Edinburgh: W. H. Lizars and London: S. Highley, 1842, 117.

Gessner

Gmelin

Gould (William),

Gray (John Edward) but check George Robert Keeper of the British Museum.

Hamilton Smith, Colonel (Lieut. Col. Charles), *Introduction to the Mammalia, The Naturalist's Library* vol. xiii. Edinburgh: W. H. Lizars and London: S. Highley, 1843, Gluttons, pp. 206–208; Otters, pp. 249–52. Weasels, pp. 185–87.

Horsfield (Thomas, Dr), *Zoological Researches in Java and the Neighbouring Islands.* London: Kingsbury, Parbury & Allen, 1824, 'Tupaias', Plate, 118, *Tupaia Javanica* (description in Latin, 119–20), and in English for 3 varieties, 120–29.

Humboldt

Irvine, Washington, *On the Prairies.* Paris: A and W Galignani and Co., 1835, 'Prairie Dogs', ch. 32, 176–7.

Jesse (Edward), *Gleanings in Natural History. Second Series. To which are Added some Extracts from the Unpublished MSS. Of the Late Mr White of Selborne.* London: John Murray, 1834, 'Moles', 124–29; 'the Corncrake', 139–40.

Le Court (Henry Lecourt, see Cadet-de-Vaux)

Loddiges, George

Low (Richard Thomas)

Lyon, Captain, quoted in Capt. William Edward Parry, *Journal of a Second Voyage for the Discovery of a North-West Passage from the Atlantic to the Pacific.* New York: E. Duyckink, G. Long, Collins and Co., 1824.

McCullough, Dr

Montagu's Ornithological Dictionary

Neill, Mr of Cannon Mills (Patrick)

Pallas the Russian Naturalist

Pennant

Pringle

Raffles, Sir Stamford

Risso (Antoine), *Ichthyologie de Nice, ou histoire naturelle des poissons du département des Alpes Maritimes.* Paris: F. Schoell, 1810. 'Tetragonurus', 347–50, Plate X, fig. 37.

Scarth, Mr (Robert), habits of 'stormy petrel' reported in letter from him to Dr Leach in *Magazine of Natural History* (1833): 160–62.

Mr Schomburgk (Robert Hermann)

Scoresby, Captain (William), 'Fulmar Petrels St Kildas' anecdote cited in Yarrell, *A History of British Birds* 3 (1843): 621–22.

An Account of the Arctic Regions, with a History and Description of the Northern Whale-Fishery 2. vols. Edinburgh: Archibald Constable and Co., 1820, vol. 1, 528–31 details the 'Petrel, Fulmar or Mallemuk'.

Scrope (William), *Days and Nights of Salmon Fishing in the Tweed. With a Short Account of the Natural History and Habits of the Salmon, Instructions to Sportsmen, Anecdotes &tc.* London: John Murray, 1843, 'Salmon', ch. 2, 8–62 (and Parr is young Salmon).

Selby (Prideaux John), *Illustrations of British Ornithology,* 2 vols. Edinburgh: W. H Lizars, 1833; 'Partridges', vol. 1, 1833 second ed. 432–37, Plate LXI;

———, 'Waterhen', 'On the Instinct of the Water-Hen', in *History of the Berwickshire Naturalists' Club, Instituted Sept. 22. 1831.* Edinburgh: Neill & Company, 1834, 84–5.

Spallanzani, L., 'Reproductions of the Legs in the Aquatic Salamander', in *An Essay on Animal Reproductions.* London: Becket and de Hondt, 1769, 68–82.

Valenciennes, Achilles

Waterton (Charles), on 'Vampires', in Hamilton Smith, Lieut. Col. Charles, *Introduction to the Mammalia, The Naturalist's Library,* vol. xiii. Edinburgh: W. H. Lizars and London: S. Highly, 1842, 126 '*Vampyrus spectrum,* Leach' and Plate 2; Cheiroptera, the source of Fig. 7 in the *Elements.*

———, *Wanderings in South America, the North-West of the United States, and the Antilles, in the Years 1812, 1816, 1820 and 1824: With* Additional *Instructions for the Perfect Preservation of Birds &c. for Cabinets of Natural History*. London: J. Mawman, 1825.
———, *Essays on Natural History, chiefly Ornithology. With an Autobiography of the Author, and a View of Walton Hall*. London: Longman, Orme, Brown, Green, & Longmans, 1838.
Wilson
Yarrell (William), *A History of British Birds,* 3 vols. London: John Van Voorst, 1843.
http://darwin-online.org.uk/content/frameset?itemID=F10.3&viewtype=text&pageseq=545
rheas, vampires, cougar/puma

APPENDIX 9: NAMED EXPERT PROVENANCE FOR *ANECDOTES OF ANIMALS* (1852)

Name of Scientific Authority of Anecdote	*Animal in* Anecdotes of Animals *p. ref.*	*Reference to published work as found in the* Anecdotes of Animals
Mr Baird	Lemur tardigradus, p. 30	
Prof Bell	Long-eared Bats, p. 36	*British Quadrupeds*
	Badger, p. 69	
	Weasels, p. 75 Prey to hawks	
	Newfoundland Dog, p. 98	
	Bloodhound, p. 114	
	Setter, p. 123 (Juno)	
	Horses, p. 312 (origins in Egypt)	
	Ass, p. 335 (belonging to his grandfather)	
Mr Bennett	Monkey, p. 18	
Mr Bewick	Dogs, p. 112	
M. Blaise & M. Guilleman	Dogs for sport, p. 129	
Mr Bowdich	Monkeys, pp. 2, 28	
M. Bonpland	Monkeys eaten in S. Amer. p. 28	
Mr Broderip	Elephants, p. 292	'in his delightful Zoological Reflections'
Capt. Brown	Monkey, p. 30	in his 'Popular Natural History'
	Badgers, p. 68	
	Weasels, p. 74	in his 'Popular Natural History'
	Otters, p. 83	
	Dogs, p. 117	corrects his 'Popular Natural History'
	Wolf, p. 167	
	Foxes, p. 17?	in 'Popular Natural History'
	Tigers, p. 222	in 'Popular Natural History'
	Cats, p. 239	'Natural History'
	Goats, p. 366	'Natural History'
	Sheep, pp. 369–70	Wild Buffalos of Hindostan, Lieut White of the 15th N.I. in 'Popular Natural History'
	Oxen, p. 381	
Mr Bruce	Hyaenas, p. 186	
Dr Buckland	Hedgehogs, p. 50 (killing snakes)	
Mr Burchell	Hyaenas, p. 183	
	Elephants, p. 290	
	Rhinoceros, p. 311 (measurement)	

(*Continued*)

(Continued)

Name of Scientific Authority of Anecdote	*Animal in* Anecdotes of Animals *p. ref.*	*Reference to published work as found in the* Anecdotes of Animals
Mr Byam	Hogs, p. 302	'Central America'
	Oxen, p. 384	
Mr Campbell	Lions, p. 202	His second journey to Africa
Mr Carpenter	Vampires, p. 36	
	Rats, p. 246	
Mr Catlin	Buffalo, p. 382	
Baron Cuvier	Dogs, p. 83	
M. Frédéric Cuvier	Makis of Madagascar, p. 30	Essay on Domestication of Animals
	Wolf, p, 165 at Paris Menagerie	
	Elephants, p. 291	
	Llamas, p. 346	
Mr Darwin	Vampires, p. 36	Enemy is the Puma close to the river Gallego
	Mice, p. 266	
	Llamas, p. 346	
Mr Denham	Hyaenas, p. 182	In his Travels
The Earl of Derby	Moles, p. 41	
Capt. Duff in India	Lions, p. 209	
M. Duvaueel (*sic*)	Monkeys, p. 21	
Lieut. Edwards	Monkeys, p. 30	In his voyaging up the Amazon
Capt Fisher of the Navy	Antelopes, p. 359	Pair of fairy-like creatures to England'
Mr Forbes	female monkey, p. 27	
Mr Gilpin	brace of pointers, p. 118	
Mr Gordon Cumming	Lions, p. 189, and p. 194	
	Lioness, p. 207 (her temper)	
	Elephant (p. 275) foot as trophy	
	Hippopotamus, p. 296	
	Rhinoceros, p. 308	
	Springboks, p. 360	
Mr Gosse	Wolves, p. 171	*Canadian Naturalist*
Capt Basil Hall	Pigs, p. 305	
Colonel Hamilton Smith	Otters, p. 71	
	Weasels, p. 71	
	Dogs, pp. 87, 109 incl. greyhound	
Major Harris	Rhinoceros, p. 311 (brain lies under horns)	
Marchioness of Hastings	Lions, p. 190	'Miscellany of Natural History'
Sir Francis Head	Squirrel (in Canada), p. 252	
Bishop Heber	Otters, p. 79	In Matta Colly river
	Pariah dogs of India, p. 150	
	Arctic Fox, p. 186 in India	
	Tiger, p. 222	
	Ass, p. 337 (of Bombay)	
	Goats, p. 368	
Mr Hogg, the Ettrick Shepherd	Moles, p. 41	
	Dogs, p. 103	
Baron Humboldt	Jaguar, p. 234	

(Continued)

(*Continued*)

Name of Scientific Authority of Anecdote	*Animal in* Anecdotes of Animals *p. ref.*	*Reference to published work as found in the* Anecdotes of Animals
Prof Jamieson of Edinburgh	Cougar, p. 208	
Mr Jesse	Rats, p. 261	
Mr Jukes	Dogs, p. 101	'Excursions in and about Newfoundland'
Mr Kohl	Horses, p. 318ff.	Abridged account of Asiatic horses
M. Lamartine	Horses, 324	Account of Arab chief and his horse'
Landseer		"The Monkey who had seen the World"
Sir Thomas Dick Lauder	Dogs, p. 106 St Bernard	
M. Lecourt	Moles, p. 41	French Naturalist Henri (*sic*) Lecourt studied moles
Mr Lloyd the sporting traveller of Norway	Dogs, p. 162	'Field Sports of North of Europe'
Capt. Lyon	Bears, p. 53 Esquimaux dogs, p. 92	Note to Capt. Phipp's Voyage to North Pole
Mr Martin	Dogs, p. 129	His 'clever little treatise on dogs'
Dr McCullough	local acc. p. 431	
Mr McIntyre of Edinburgh	Dog, p. 99	
Capt. Owen	Hippopotamus, p. 296	'while examining branch of the Temby River in Delagoa Bay'
Prof Owen	Ingheena, p. 16	
And Mrs Owen	Dog in Cornwall Terrier, pp. 145–6 (known to him)	
Mr Pringle	Vampires, p. 39 Hyaenas, p. 185 Lions, p. 194 Elephants, p. 283 Antelopes, p. 360 Cape Buffalo, p. 381	Banks of Little Fish river?
Sir Stamford Raffles	Malayan Sun Bear, pp. 64–65	
Dr Richardson	Bears, p. 55 Dogs, p. 93 Esquimaux dogs, p. 173	Canada L'Acadie Hare Indians on Mackenzie River
Mr Rüppell	Cats, p. 239	
Mr Ruxton	Bears, p. 58 Coyotes, p. 172	A trapper names Glass
Mr Ryan	Coyotes of Mexico, p. 172	
Sir Walter Scott	Greyhound, p. 109 called Maida Bullterrier, p. 147 Camp	
Dr Abel Smith	The Straand Wolf, p. 185	

(*Continued*)

(*Continued*)

Name of Scientific Authority of Anecdote	*Animal in* Anecdotes of Animals *p. ref.*	*Reference to published work as found in the* Anecdotes of Animals
Mr Spaarman	Lions, p. 203 Wild Boars of Africa, p. 299	In his African Travels
Capt Stedman	Vampires, p. 36	Who travelled to Guiana from 1772-1777
Mr Steedman the S. African Traveller	Hyaenas, p. 184	
Mr St John	Badgers, pp. 67, 69 Otters, p. 81 Dogs, p. 103 Water Spaniel, p. 130 (Rover) Foxes, p. 176 Cats, p. 238	'Wild Sports of the Highlands' Sluie on the Findhorn Highland Sports
Col. Sykes	Hyaenas, p. 184 procured one in India	
Prof Temminck	Cats, p. 239	
M. Vaillant	Giraffes, p. 354	'ill-treated' thought to have invented them
Mr Waterton	Vampires, p. 48 Lion, p. 82, p. 209	
Mr Wentzel	Cats, p. 246 Deer, p. 351	'Observations on the Language of Brutes' Dog-Rob Indians in pairs to kill rein-deer
Mr Williamson	Tiger, p. 221	'Oriental Field Sports'
Mr Wilson	Ass, p. 338	
Capt. Woodhouse	Lions, p. 209	

Appendix 'Bibliography'

Mr Baird (W. Esq.), 'Descriptive Notice of a Specimen of *Lemur tardigradus,* Lin., Makes, Cuv. kept alive for some Time at Edinburgh', *The Magazine of Natural History* 1 (1829): 208–16 (including an image which is the source of Fig. 6 in the *Elements*).

Professor Bell (Thomas), *A History of British Quadrupeds.* London: John Van Voorst, 1837.

Mr Bennett (George), 'An Account of Ungka Ape of Sumatra', *Magazine of Natural History* 4 (1832): 131–39 (including an image which is the source for Fig. X of the *Elements*).

Mr Bewick (Thomas), *A General History of Quadrupeds.* London: G. G. J and J. Robinson and C. Dilly, 1790. (CUP, 2014)

M. Blaise (*sic*) (Blaze, Eléazar), *Le Chasseur au Chien d'arrêt.* Mons: Leroux Père, 1844.

M. Bonpland (Aimé),

Mr Bowdich

M. Broderip (W. J), *Zoological Recreations* (also dedicated to Owen) First Edition. London: Henry Colburn, 1847.

Capt. Brown (Thomas), *Biographical Sketches and Authentic Anecdotes of Dogs, Exhibiting Remarkable Instances of the Instinct, Sagacity, and Sociable Disposition of this Faithful Animal.* Edinburgh: Oliver and Boyd, 1829.

——, *Popular Natural History, or, the Characteristics of Animals, Portrayed in a Series of Illustrative Anecdotes.* London and Edinburgh: A. Fullarton & Co. 1848, 3 vols.

Mr Bruce

Dr Buckland (William), *Magazine of Natural History* 6 (1833): 457–58.

Mr Burchell (William, J), *Travels in the Interior of South Africa*, vol. 1. London: Longman, Hurst, Rees, Orme & Brown, 1822; vol. 2. London: Longman, Hurst, Rees, Orme, Brown & Green, 1824, 'hyaenas', vol. 2, ch. viii, plate, 222 and description, 222–29.

Mr Byam (George), *Wild Life in the Interior of Central America*. London: John W. Parker, 1849.

Cadet-de-Vaux, Antoine-Alexis, *De la Taupe: de ses mœurs, de ses habitudes et les moyens de la détruire*. Paris: chez L. Colas, 1803, on 'Henry Lecourt Taupier', 7–16; on 'les variétés de Taupes reconnues par Henry Lecourt', 50–52 and 'Des Piéges (*sic*) de Lecourt', 205–06 and 'De la manière dont Lecourt procède à la préhension des Taupes', 213–18.

Mr Campbell (Rev. John), *Travels to South Africa*. London: Francis Westley, 1822. Ox and Lions, 133ff.

Dr Carpenter (William B.),

Mr Catlin (George), *Manners, Customs and Conditions of North American Indians* (1841). 2 vols.

Baron Cuvier

M. Frédéric Cuvier, 'Essai sur la domesticité des Mammifères, précédé de considérations sur les divers états des animaux, dans lesquels il nous est possible d'étudier leurs actions', *Mémoires du Muséum d'Histoire naturelle*, t. XIII. Paris: A. Belin, 1825, 406–55.

Mr Darwin (Charles),

Mr Denham (Dixon), *Narrative of Travels and Discoveries in North and Central Africa in the Years 1822, 1823 and 1824 by Major Denham, Captain Clapperton and the Late Dr Oudney*. 2 Vols. London: John Murray, 1828.

The Earl of Derby

Captain Duff of India (James Grant), *A History of the Mahrattas*. 3 Vols. London: Longman, Rees, Orme, Brown and Green, 1826?

M. Duvaueel (Alfred Duvaucel),

Lieut. Edwards (William H), *A Voyage up the River Amazon*. New York: D. Appleton, 1847, 'white monkey', 84–85.

Capt. Fisher, *Narrative of Travels in North Africa in the Years 1818, '19 and '20*. London: John Murray, 1821.

———, and see Parry.

Mr Forbes (Edward?).

Franklin, John (Captain), *Narrative of a Journey to the Shores of the Polar Sea in the years 1819, 20, 21 and 22*. London: John Murray, 1823.

Mr Gilpin, 'brace of pointers' quoted in Colonel Charles Hamilton Smith, *The Naturalist's Library*, vol. X, 'Dogs'. Edinburgh: Lizars, 1840, 196.

Mr Gordon Cumming (Roualeyn), *Five Years of a Hunter's Life in the Far Interior of South Africa*. 2 Vols. New York: Harper Brothers, 1850.

Mr Gosse (Philip H.), *The Canadian Naturalist: A Series of Conversations on the Natural History of Lower Canada*. London: John van Voorst, 1840, image and description of 'black wolf', 37–39.

Captain Basil Hall, *Fragments of Voyages and Travels*. Edinburgh: Robert Cadell, 1832. 3 Vols. Vol. 2, Jean 'the grunter' pig, 151–60.

Hamilton Smith, Colonel (Lieut. Col. Charles), *Introduction to the Mammalia, The Naturalist's Library*, vol. xiii. Edinburgh: W. H. Lizars and London: S. Highly, 1842, Otters, pp. 249–52. Weasels, pp. 185–87.

———, *The Naturalist's Library*, vol. X, 'Dogs'. Edinburgh: Lizars, 1840, including 'Greyhound', pp. 160–77.

Major Harris (Captain William Cornwallis), *Narrative of an Expedition into Southern Africa during the Years 1836 and 1837 from the Cape of Good Hope*. Bombay: American Mission Press, 1838, 'Rhinoceroses' in Appendix 376–77 (not quite right ref.).

Marchioness of Hastings, see Rhind

Sir Francis Head (Francis Bond Head)

Bishop Heber

Mr Hogg the Ettrick Shepherd (James), 'Class IV Dogs', *Blackwoods Magazine* 15 (1824): 177–83.

Baron Humboldt

Professor Jamieson

Jesse (Edward), *Gleanings in Natural History. Second Series. To Which Are Added Some Extracts from the Unpublished MSS. Of the Late Mr White of Selborne.* London: John Murray, 1834, 'Rats'.

Mr Jukes (Joseph Beete), *Excursions in and about Newfoundland during 1839 and 1840.* London: John Murray, 1842. 2 vols.

Mr Kohl (Johann Georg), *Travels in the Interior of Russia and Poland* (1841) and 'Asiatic horses' reported in 'The Steppes of Russia III' in *Asiatic Journal and Monthly Miscellany* (1842): 18–27.

M. Lamartine (Alphonse), 'The Arab Horse by Lamartine', *The Lady's Pearl* 3–4 (April 1843): 196.

Landseer (Sir Edwin), 'The Monkey who had seen the World' (1827).

Sir Thomas Dick Lauder, 'The St Bernard Dog, Bass' in *The New Excitement or a Book to Induce Young People to Read.* Edinburgh: W. Innes, 1841: 234–38.

Le Court (Henry Lecourt), see Cadet-de-Vaux

Mr Lloyd (Llewelyn), *Field Sports of the North of Europe comprised in a Personal Narrative of a Residence in Sweden and Norway in the Years 1827-28.* London: Henry Colburn and Richard Bentley, 1830. 2 vols. (with vignettes)

Captain Lyon, quoted in Capt. William Edward Parry, *Journal of a Second Voyage for the Discovery of a North-West Passage from the Atlantic to the Pacific.* New York: E. Duyckink, G. Long, Collins and Co., 1824.

Mr Martin (Martin), *A Description of the Western Islands of Scotland.* London: Andrew Bell, 1703.

Dr McCullough (R. J?),

Mr McIntyre of Edinburgh

Captain Owen (W. F. W)

Professor Owen

Captain Parry (William, Edward), *Journal of a Second Voyage for the Discovery of a North-West Passage from the Atlantic to the Pacific; Performed in the Years 1821-22-23 in his Majesty's Ships Fury and Helca under the orders of Captain William Edward Parry.* London: John Murray, 1824.

Mr Pringle (Thomas), 'Antelopes' in 'South African Sketches', *The British Magazine* 1 (Jan.–June 1830): 161–8 and *Narrative of a Residence in South Africa,* 2nd edn. London: Edward Moxon, 1835.

Raffles, Sir Stamford, 'Malayan Sun Bear' in 'XVII'. *Descriptive Catalogue of a Zoological Collection, made on account of the Honorable East India Company, in the Island of Sumatra and its Vicinity, under the Direction of Sir Thomas Stamford Raffles, Lieutenant-Governor of Fort Marlborough with Additional Notices Illustrative of the Natural History of those Countries* in *Transactions of the Linnean Society of London* (June 1821): 239–74. https://doi.org/10.1111/j.1095-8339.1821.tb00064.x.

Rhind, William, 'Feline Species', in *Miscellany of Natural History.* Edinburgh: Fraser and Co, 1834, on the 'Asiatic Lion', 68–72.

Dr Richardson (John), *Fauna Borealis-Americana of the Zoology of the Northern Parts of British America.* London: John Murray, 1829.

'Bears', 'Dogs' and 'Esquimaux Dogs' in order of original contents, but 'Esquimaux Dogs' and 'Hare-Indian Dog' are paraphrased and illustrated in *The Family Magazine* 2 (1834–5), 360 and 389 respectively.

Mr Rüppell (Eduard), 'cats' quoted in Broderip, 197–98. See *Reisen in Nubien, Kordofan und dem peträischen Arabien.* Frankfurt-am-Main: Friederich Wilmans, 1829.

Mr Ruxton (George F.) *Adventures in Mexico and the Rocky Mountains.* London: John Murray, 1847. + trapper named Glass. https://en.wikipedia.org/wiki/Hugh_Glass.

Mr Ryan

Sir Walter Scott, see Brown, 1829: his dog Maida, 85–88 and Bull Terrier Camp, 408.

Dr Abel Smith (Andrew), 'Descriptions of Two Quadrupeds inhabiting the South of Africa, about the Cape of Good Hope', in *Transactions of the Linnean Society of London*. London: Longman, Rees, Orme, Brown and Green, 1828. Part 2, 460–70 + plate Tab. XIX drawing of Hyaena.

Mr Spaarman (Anders), *A Voyage to the Cape of Good Hope Towards the Antarctic Polar Circle and Round the World, but chiefly into the Country of the Hottentots and Caffres from Year 1772 to 1776*, vol. 1, 2nd ed. London: G. G. J. and J. Robinson, 1786.

Captain Stedman (John Gabriel), *Narrative of a five years' expedition against the revolted negroes of Surinam in Guiana, on the wild coast of South America from the year 1772 to 1777, elucidating the history of that country and describing its productions... with an account of the Indians of Guiana and negroes of Guinea...* 2 vols. London: J. Johnson, 1796, vampires + plate V2, ch. xxii, 142–43 and ch. xxiii, 170.

Mr Steedman (Andrew), 'hyaenas', in *Wanderings and Adventures in the Interior of Southern Africa.* 2 vols. London: Longman and Co., 1835. Vol. 1, Part 2: ch 3, 143.

Mr St John (Charles), *Short Sketches of Wild Sports and Natural History of the Highlands.* London: John Murray, 1846, 'Badgers', ch. xxxi; Otters, ch. xii, Dogs (and snake, ch. viii) and ch. xiv, Foxes, ch. xx, wild cats, ch. iv.

Colonel Sykes

Professor Temminck (Coenraad Jacob), 'cats' quoted in Broderip, 197–98. See *Monographies de Mammalogie.* Paris and Leiden, 1827–41, t.1 (1827) includes more than 70 references to cats.

M. Vaillant

Mr Waterton (Charles), on 'Vampires', in Hamilton Smith, Lieut. Col. Charles, *Introduction to the Mammalia, The Naturalist's Library,* vol. xiii. Edinburgh: W. H. Lizars and London: S. Highly, 1842, 126 '*Vampyrus Spectrum,* Leach' and Plate 2 Cheiroptera, the source of Fig. 7 in the *Elements.*

———, *Wanderings in South America, the North-West of the United States, and the Antilles, in the Years 1812, 1816, 1820 and 1824: With Additional Instructions for the Perfect Preservation of Birds &c. for Cabinets of Natural History.* London: J. Mawman, 1825.

Mr Wentzel (Frederick) see Franklin

Mr Williamson (Thomas, Capt.), *Oriental Field Sports, being a Complete, Detailed and Accurate Description of the Wild Sports of the East.* 2 vols. London: Edward Orme, 1807, vol. 1 and 1808, vol. 2.

Mr Wilson

Captain Woodhouse

BIBLIOGRAPHY

'360o Ciência Descoberta' Exhibition Podcast. Lisbon: Fundação Calouste Gulbenkian, 2013. https://www.youtube.com/watch?v=HZwX2VdmtPE.

Abir-Am Pnina, G. and Dorinda Outram, eds. *Uneasy Careers and Intimate Lives: Women in Science 1789–1979*. New Brunswick: Rutgers University Press, 1987.

Adams, Capt. John. *Remarks on the Country Extending from Cape Palmes to the River Congo*. London: G. & W. B. Whittaker, 1823.

Aderinto, Saheed. 'Journey to Work: Transnational Prostitution in Colonial British West Africa'. *Journal of the History of Sexuality* 24, no. 1 (Jan. 2015): 99–124.

Albaret, Jean-Jacques, Monique Simier, Famara Smbou Darboe, Jean-Marc Ecoutin, Jean Raffray and Luis Tito de Morais. 'Fish Diversity and Distribution in the Gambia Estuary, West Africa, in Relation to Environmental Variables'. *Aquatic Living Resources* 17 (2004): 35–46.

Alibrandi, Rosamaria. 'British Parliamentary Abolitionists: Sir Thomas Fowell Buxton (1786–1845) and the Political and Cultural Debate on Abolitionism in the Nineteenth Century'. *Parliaments, Estates and Representation* 40, no. 1 (2020): 21–34.

Alic, Margaret. *Hypatia's Heritage: A History of Women in Science from Antiquity to the Late Nineteenth Century*. London: The Women's Press, 1986.

Allen, David Elliston. *The Naturalist in Britain: A Social History*. Princeton: Princeton University Press, 1994.

Andrews, Stuart. *Unitarian Radicalism: Political Rhetoric, 1770–1814*. Basingstoke: Palgrave Macmillan, 2003.

Anon. 'Review of *The Excursions to Madeira and Porto Santo*'. *The New Monthly Magazine and Literary Journal* 15 (1 July 1825): 317.

Anon. 'Art. I. *Memoirs of Baron Cuvier*. By Mrs R. Lee'. *The Monthly Review* iv (Oct. 1833): 159–78.

Anon. 'Art. III. *Memoirs of Baron Cuvier*. By Mrs R. Lee'. *The Eclectic Review* x (Sept. 1833): 228–39.

Anon. 'Art. IV. Life and Labors of Cuvier'. *The Foreign Quarterly Review* xiv (Dec. 1834): 164–85.

Anon. 'The African Wanderers'. *Dublin University Magazine* 31, no. 182 (Feb. 1848): 252–57.

Ashley, Melissa. 'Elizabeth Gould, Zoological Artist 1840–1848: Unsettling Critical Depictions of John Gould's "Laborious Assistant" and "Devoted Wife"'. *Hecate* 39, nos. 1/2 (2014): 101–22.

Augstein, Hannah Franziska, ed. and intro. *Race: The Origins of an Idea, 1760–1850*. Bristol: Thoemmes Press, 1996.

Baiesi, Serena. *Letitia Elizabeth Landon and Metrical Romance: The Adventures of a Literary Genius*. Bern: Peter Lang, 2009.

Balon, Eugene K., Michael N. Bruton and David L. G. Noakes. 'Women in Ichthyology: An Anthology in Honour of ET, Ro and Genie'. *Environmental Biology of Fishes* 41, nos. 1–4 (1994): 7–125.

Barker, Thomas P. 'Spatial Dialectics: Intimations of Freedom in Antebellum Slave Song'. *Journal of Black Studies* 46, no. 4 (2015): 363–83.

Barnett, Clive. 'Impure and Worldly Geography: The Africanist Discourse of the Royal Geographical Society, 1831–73'. *Transactions of the Institute of British Geographers* NS 23 (1998): 239–51.

Barton, Ruth. '"Men of Science": Language, Identity and Professionalization in the Mid-Victorian Scientific Community'. *History of Science* xli (2003): 73–119.

Bastin, John. 'Sir Stamford Raffles and the Study of Natural History in Penang, Singapore and Indonesia'. *Journal of the Malaysian Branch of the Royal Asiatic Society* 63, no. 2 (1990): 1–25.

Bauchot, Marie-Louise, Jean Daget and Roland Bauchot. *L'Ichthyologie en France au début du xixe siècle: l'Histoire naturelle des poissons de Cuvier et Valenciennes.* Paris: Bulletin du Muséum national d'histoire naturelle, 1990.

Beaver, Donald deB. 'Writing Natural History for Survival 1820–1856: The Case of Sarah Bowdich, later Sarah Lee'. *Archives of Natural History* 26, no. 1 (1999): 19–31.

———. 2021. 'Lee [*Née* Wallis; *Other Married Name* Bowdich], Sarah'. Oxford Dictionary of National Biography. Oxford: Oxford University Press. https://doi.org/10.1093 / ref:odnb/16310.

Beer, Gillian. *Darwin's Plots: Evolutionary Narrative in Darwin, George Eliot and Nineteenth-Century Fiction.* 3rd edn. Cambridge: Cambridge University Press, 2009.

Bell, Thomas. *A History of British Quadrupeds.* London: John Van Voorst, 1837.

———. *A History of British Reptiles.* London: John Van Voorst, 1839.

Belloc, Gérard. 'Rapport général sur la cinquième croisière du navire "Président-Théodore-Tissier"'. *Revue des Travaux de l'Institut des Pêches Maritimes* 10, no. 3 (1937): 269–325. https://archimer.ifremer.fr/doc/00000/5755/.

Ben-David, Joseph. 'The Rise and Decline of France as a Scientific Centre'. *Minerva* 8, no. 2 (April 1970): 160–79. https://www.jstor.org/stable/41822018.

Biller, Sarah. *The Memoirs of the Late Hannah Kilham.* London: Darnton & Harvey, 1837.

Birkett, Dea. *Spinsters Abroad: Victorian Lady Explorers.* Oxford: Basil Blackwell, 1994.

Blankaert, Claude, Claudine Cohen, Pietro Corsi and Jean-Louise Fischer, eds. *Le Muséum au premier siècle de son histoire.* Paris: Éditions du Muséum National d'Histoire Naturelle, 1997.

Blunt, Alison. *Travel, Gender and Imperialism: Mary Kingsley and West Africa.* New York: The Guilford Press, 1989.

Bondeson, Jan. *The Feejee Mermaid and Other Essays in Natural and Unnatural History.* Ithaca: Cornell University Press, 1999.

Bourgeois, Bertrand. *Poétique de la maison-musée 1847–1898.* Paris: Harmattan, 2009.

Braithewaite, Colin J. R. 'Transactions and Neglected Data'. *Scottish Journal of Geology* 47, no. 2 (2011): 179–88.

Bredekamp, Horst, Vera Dünkel and Birgit Schneider, eds. *The Technical Image: A History of Styles in Scientific Imagery.* Chicago: University of Chicago Press, 2015.

Brisson, Ulrike. 'Fish and Fetish: Mary Kingsley's Studies of Fetish in West Africa'. *Journal of Narrative Theory* 35, no. 3 (2005): 326–40. http://www.jstor.org/stable/30225805.

Brown, Stephen W. 'William Smellie and Natural History: Dissent and Dissemination'. In *Science and Medicine in the Scottish Enlightenment,* edited by Charles W. J. Withers and Paul Wood, 191–214. East Linton: Tuckwell Press, 2002.

Bucher, Henry H. 'Mpongwe Origins: Historiographical Perspectives'. *History in Africa* 2 (1975): 59–89.

Byrne, Angela. 'The Scientific Traveller'. In *The Routledge Research Companion to Travel Writing,* edited by Alasdair Pettinger and Tim Youngs, chapter 1. Abingdon and New York: Routledge, 2019.

Cadenat, Jean. *Poissons de mer du Sénégal.* Dakar: IFAN, 1950.

Carpenter, William B. *Scripture Natural History: Containing a Descriptive Account of the Quadrupeds, Birds, Fishes, Insects, Reptiles, Serpents, Plants, Trees, Minerals, Gems and Precious Stones Mentioned in the Bible.* Lincoln: Edmands & Company, 1833.

Carpenter, William B. *Zoology: A Systematic Account of the General Structure, Habits, Instincts, and Uses of the Principal Families of the Animal Kingdom.* London: William S. Orr, 1848.

Carron, Helen. '"Rare Book Series". Continuing the Exploration of Women Illustrators' (17 Sept. 2020). https://www.emma.cam.ac.uk/members/blog/?id=472

Casson, Max, Gérôme Calvès, Mads Huuse, Ben Sayers and Jonathan Redfern. 'Cretaceous Continental Margin Evolution Revealed using Quantitative Seismic Geomorphology, Offshore Northwest Africa'. *Basin Research* 33 (2021): 66–90.

Catlow, Agnes. *Drops of Water: Their Marvellous and Beautiful Inhabitants Displayed by the Microscope*. London: Reeve and Benham, 1851.

Ceesay, Hasoom. *Gambian Women: An Introductory History*. Bundung: Fatoumatta's Print and Communication Centre, 2007.

Cook, L. M., R. A. D. Cameron and L. A. Lace. 'Land Snails of Eastern Madeira: Speciation, Persistence and Colonization'. *Proceedings of the Royal Society of London Series B* 239 (1990): 35–79.

Church, R. J. Harrison. *West Africa: A Study of the Environment and of Man's Use of it*. Seventh Edition. London: Longman, 1974.

Cleevely, R. J. 2004. 'Bell, Thomas (1792–1880)'. Oxford Dictionary of National Biography. Oxford: Oxford University Press. https://doi.org/10.1093 /ref:odnb /2029.

'CLOFFA': Check-list of the Freshwater Fishes of Africa (CLOFFA) at http://vmcloffa-dev.mpl.ird.fr/.

'Colchester Unitarians'. https://www.ukunitarians.org.uk/colchester/colhistory.html.

Coleman, Deirdre, ed. *Maiden Voyages and Infant Colonies: Two Women's Travel Narratives of the 1790s*. Leicester: Leicester University Press, 1998.

Collini, Stefan, ed. *C. P. Snow: The Two Cultures (Canto Classics)*, 1–21. Cambridge: Cambridge University Press, 2012.

Cooter, Roger and Stephen Pumfrey. 'Separate Spheres and Public Places: Reflections on the History of Science Popularization and Science in Popular Culture'. *History of Science* 32, no. 3 (Sept. 1994): 237–67.

Cosslett, Tess. *Talking Animals in British Children's Fiction, 1786–1914*. Farnham: Ashgate, 2006.

Couch, Jonathan. *A History of the Fishes of the British Islands*. 4 vols. London: Groombridge and Sons, 1860–65.

Creese, Mary R. S. 'BOWDICH LEE, Sarah Eglonton (née Wallis: 1791–1856)'. In *The Dictionary of Nineteenth-Century British Scientists*. Vol. 1 (A-C), edited by Bernard Lightman, 243–44. Bristol: Thoemmes Continuum, 2004.

Creese, Mary R. S. with Thomas M. Creese. *Ladies in the Laboratory? American and British Women in Science, 1800–1900: A Survey of their Contributions to Research*. Lanham and London: The Scarecrow Press, 1998 (revised 2004).

Crow, Britt and Judith Carney. 'Commercializing Nature: Mangrove Conservation and Female Oyster Collectors in The Gambia'. *Antipode* 45, no. 2 (2012): 275–93.

Curtin, Philip D. '"The White Man's Grave": Image and Reality, 1780–1850'. *Journal of British Studies* 1, no. 1 (Nov. 1961): 94–110.

Cuvier, Georges. *Prospectus: Histoire naturelle des poissons*. Paris: Imprimerie de Mallet-Bachelier, 1862.

Dahlstrom, Michael F. 'Using Narrative and Storytelling to Communicate Science with Nonexpert Audiences'. *PNAS* 111, suppl. 4 (16 Sept. 2014): 13614–20.

Darnell, Regna. 'History of Anthropology in Historical Perspective'. *American Review of Anthropology* 6 (1977): 399–417.

Darwin, Charles. *Narrative of the Surveying Voyages of His Majesty's Ships* Adventure *and* Beagle *between the years 1826 and 1836, Describing their Examination of the Southern Shores of South America, and the Beagle's Circumnavigation of the Globe*. London: Henry Colburn, 1839. The facsimile is available online at the Darwin Correspondence Project at http://darwin-online.org.uk/content/frameset?itemID=F10.3&viewtype=text&pageseq=545.

Daston, Lorraine. 'Science, History Of'. *International Encylopedia of the Social and Behavioural Sciences* 21, second ed. (2015): 241–47.

'Database of Scientific Illustrators, 1450–1950 (DSI)'. https://dsi.hi.uni-stuttgart.de/index.php?function=show_static_page&id_static_page=1.

Davy, Sir Humphry. *Salmonia: Or Days of Fly-Fishing*. London: John Murray, 1828.

Debarbieux, Bernard. 'The Various Figures of Mountains in Humboldt's Science and Rhetoric'. *Cybergeo*, 2012. https://doi.org/10.4000/cybergeo.25488 (last consulted 6 July 2022).

Deligeorges, Stéphanie Alexandre Gady and Françoise Labelette. *Le Jardin des Plantes et le Muséum National d'Histoire Naturelle*. Paris: Éditions du Patrimoine, 2004.

Desmarest, A. G. *Bulletin des sciences naturelles et de géologie* 6 (1825): 396–98.

Desmond, Adrian. 'The Making of Institutional Zoology in London, 1822–1836: Part 1'. *History of Science* 23, no. 2 (June 1985): 153–85.

———. 'The Making of Institutional Zoology in London, 1822–1836: Part 2'. *History of Science* 23 no. 3 (Sept. 1985): 223–50.

Dickson, Mora. *The Powerful Bond: Hannah Kilham, 1774–1832*. London: Dobson, 1980.

Dolan, Brian.'Pedagogy through Print: James Sowerby, John Mawe and the Problem of Colour in Early Nineteenth-Century Natural History Illustration'. *The British Journal for the History of Science* 31, no. 3 (Sept. 1998): 275–304.

Doubrovsky, Serge. *Autobiographiques: De Corneille à Sartre*. Paris: PUF, 1988.

Douglass, Frederick. *Narrative of the Life of Frederick Douglass. An American Slave*. Boston: Published at The Anti-Slavery Office, 1845.

Driver, Felix. *Geography Militant: Cultures of Exploration and Empire*. Oxford: Blackwell, 2001.

Dumeril, Auguste. *Histoire naturelle des poissons ou Ichthyologie générale*. 2. Vols. Paris: Librairie encyclopédique de Roret. Vol. 1, 1865; Vol. 2, 1870.

Durant, John. 'What is Scientific Literacy?' *European Review* 2, no. 1 (1994): 83–89.

Eaton, Amos. *Zoological Text-Book Comprising Cuvier's Four Grand Divisions of Animals, also Shaw's Improved Linnean General System*. Albany: Websters and Skinners, 1826.

Eddy, Matthew D. 'Natural History, Natural Philosophy and Readers'. In *The Edinburgh History of the Book in Scotland. Vol. 2 'Enlightenment and Expansion'*, edited by Bill Bell, Stephen W. Brown and Warren McDougall, 297–308. Edinburgh: Edinburgh University Press, 2012.

Edgerton, Robert B. *The Fall of the Asante Empire: The Hundred-year War for Africa's Gold Coast*. New York and London: The Free Press, 1995.

Eger, Elizabeth, Charlotte Grant, Clíona Ó Gallchoir and Penny Warburton, eds. *Women, Writing and the Public Sphere, 1700–1830*. Cambridge: Cambridge University Press, 2001.

Ehrenberg, Christian Gottfried. *Recherches sur l'organisation des animaux infusoires*. Paris: J. B. Baillière, 1839.

'Engraving of Thomas Edward Bowdich, after a painting by William Derby'. Source http://www.britannica.com/ebc/art-9018.

Everill, Bronwen. 'Bridgeheads of Empire? Liberated African Missionaries in West Africa'. *The Journal of Imperial and Commonwealth History Commonwealth History* 40, no. 5 (2012): 789–805.

Falconbridge, Anna Maria. *Two Voyages to Sierra Leone during the Years 1791–2-3*. London: Printed for the Author and sold by different booksellers throughout the Kingdom, 1794.

Fara, Patricia. *Pandora's Breeches: Women, Science and Power in the Enlightenment*. London: Pimlico, 2004.

'F'[Férussac, Baron André Étienne d'Audebert de]. 'Description et figures de plusieurs hélices, découvertes à Porto Santo par T. Edward Bowdich Esq, par G. B. Sowerby'. *Bulletin des sciences naturelles et de géologie* 3, no. 74 (1824): 92–93.

Fineman, Joel. 'The History of the Anecdote: Fiction and Fiction'. In *The New Historicism*, edited by H. Aram Veeser, 49–76. New York and London: Routledge, 1989.

'FishBase': the FishBase Database at https://www.fishbase.org.

Fleming, Paul. 'The Perfect Story: Anecdote and Exemplarity in Linneaus and Blumenberg'. *Thesis Eleven* 104, no. 11 (2011): 72–86.

Flint, John. 'Freeman, Thomas Birch (1809–1890)'. https://doi.org/10.1093/ref:odnb/47629.

Ford, Brian J. *Images of Science: A History of Science Illustration*. London: British Library Publications, 1992.

Fowler, Henry Weed. 'The Fishes of the United States Eclipse Exhibition to West Africa'. *Proceedings of the United States National Museum* 56 (1919): 195–292.

France, Peter. 'From Eulogy to Biography: The French Academic Éloge'. In *Mapping Lives: The Uses of Biography*, edited by Peter France and William St Clair, 83–101. Oxford: Oxford University Press, 2002.

Frederiks, Martha T. *'We Have Toiled all Night': Christianity in the Gambia*, 1456–2000. Zoetermeer: Uitgeverij Boekencentrum, 2003.

Fulford, Tim, Debbie Lee and Peter J. Kitson, eds. *Literature, Science and Exploration in the Romantic Era*. Cambridge: Cambridge University Press, 2004.

Fyfe, Christopher. 2021. 'Falconbridge [*née* Horwood], Anna Maria (*b.* 1769, *d.* in or after 1802?)'. Oxford Dictionary of National Biography. Oxford: Oxford University Press.

Gailey, Harry R. Jr. *A History of the Gambia*. London: Routledge & Kegan Paul, 1964.

Gates, Barbara T. *Kindred Nature: Victorian and Edwardian Women Embrace the Living World*. Chicago: The University of Chicago Press, 1998.

Gates, Barbara T. and Ann B. Shteir, eds. *Natural Eloquence: Women Reinscribe Science*. Madison: University of Wisconsin Press, 1997.

Gaugain, Mrs Jane. *The Lady's Assistant for Executing Useful and Fancy Designs on Knitting, Netting and Crochet*. Edinburgh and London: Ackermann & Co., 1841.

Geoffroy Saint-Hilaire, Isidore. *Description des mammifères nouveaux ou imparfaitement connus de la collection du Muséum d'Histoire Naturelle et remarques sur la classification et les caractères des mammifères*. Vol. 10. Paris: Archives du Muséum d'Histoire Naturelle, Gide Libraire, 1858–61.

Gomon, Martin F. 'A Revision of the Labrid Fish Genus *Bodianus* with Descriptions of Eight New Species'. *Records of the Australian Museum Supplement* 30 (2006): 1–333.

Gray, J. M. *A History of the Gambia*. Cambridge: Cambridge University Press, 1940.

Green, Toby. *A Fistful of Shells: West Africa from the Rise of the Slave Trave to the Age of Revolution*. London: Allen Lane, 2019.

Greenstein, Susan. 'Sarah Lee: The Woman Traveller and the Literature of Empire'. In *Design and Intent in African Literature*, edited by David F. Dorsey, Phanuel A. Egejuru and Stephen H. Arnold, 133–37. Washington, DC: African Literature Association & Three Continents Press, 1982.

Griffith, Edward and others, trans. *The Animal Kingdom Arranged in Conformity with its Organization. By Baron Cuvier... With Additional Descriptions of all the Species hitherto Named, and of Many not before Noticed*. London: G. B. Whittaker, 1827–35.

Gruber, Jacob W. 2006. 'Owen, Sir Richard (1804–1892)'. Oxford Dictionary of National Biography. Oxford: Oxford University Press. https://doi.org /10.1093 /ref:odnb /21026.

Guillemin, Antoine. *Bulletin des sciences naturelles et de géologie* 5 (1825): 347–50.

Hankins, Thomas L. 'In Defence of Biography: The Use of Biography in the History of Science'. *History of Science* xvii (1979): 1–16.

Harris, Katherine D. *Forget Me Not: A Hypertextual Archive of Ackermann's Nineteenth-Century Literary Annual, An Edition of the Poetess Archive* (2001), revised 2007. http://www.orgs.miamioh.edu/anthologies/FMN/Site%20Index.htm)/.

Higgitt, Rebekah and Charles W. J. Withers. 'Science and Sociability: Women as Audience at the British Association for the Advancement of Science, 1831–1901'. *Isis* 99 (2008): 1–27.

Hill, Matthew. 'Towards a Chronology of the Publication of Francis Moore's "Travels into the Inland Parts of Africa..."'. *History in Africa* 19 (1992): 353–68.

'History from Below'. *Theme link within the Online Archive, 'Making History: The Changing Face of the Profession in Britain'*. London: The Institute of Historical Research, School of Advanced Study, University of London, 2008. https://archives.history.ac.uk/makinghistory/themes/history_from_below.html.

Hogendorn, Jan and Marion Johnson. *The Shell Money of the Slave Trade*. Cambridge: Cambridge University Press, 1986.

Holmes, Richard. 'In Retrospect: On the Connexion of the Physical Sciences'. *Nature* 514 (2014): 432–33. https://doi.org/10.1038/514432a.

Hughes, Arnold and David Perfect. *A Political History of the Gambia: 1816–1994*. Rochester: University of Rochester Press, 2006.

Hureau, Jean-Claude. '"Un Exceptionnel Naturaliste éclectique". In "Théodore Monod: un homme curieux"'. *Autres Temps. Cahiers d'éthique sociale et politque* 70, no. 1 (2001): 25–38.

Hyam, Ronald. *Britain's Imperial Century, 1815–1914*. BT Batsford: University of Michigan Press, 1976.

Ingold, Tim. 'Hau Debate. Anthropology Contra Ethnography'. *Hau: Journal of Ethnographic Theory* 7, no. 1 (2017): 21–26.

Inter-religious Dialogue. https://www.newworldencyclopedia.org/entry/Inter-religious_Dialogue.

Jackson, Christine E. *Fish in Art*. London: Reaktion Books, 2012.

Jansen, Justin J. E. F. 'The Bird Collection of the Muséum National d'Histoire Naturelle, Paris, France: The First Years (1793–1825)'. *Journal of the National Museum (Prague)* 184, no. 5 (Nov. 2015): 81–111.

Jarosz, Lucy. 'Constructing the Dark Continent: Metaphor as Geographic Representation of Africa'. *Geografiska Annaler: Series B, Human Geography* 74, no. 2 (1992): 105–15.

Jefferson, Anne. *Biography and the Question of Literature in France*. Oxford: Oxford University Press, 2007.

Jenkins, Alice. 'Writing the Self and Writing Science: Mary Somerville as Autobiographer'. In *Rethinking Victorian Culture*, edited by Juliet John and Alice Jenkins, 162–78. Basingstoke: Macmillan Press Ltd, 2000.

———. 'Alexander von Humboldt's "Kosmos" and the Beginnings of Ecocriticism'. *Interdisciplinary Studies in Literature and Environment* 14, no. 2 (Summer 2007): 89–105. https://www.jstor.org/stable/44086615.

Jenkins, Bill. 'The Platypus in Edinburgh: Robert Jameson, Robert Knox and the Place of the *Ornithorhynchus* in Nature, 1821–24'. *Annals of Science* 73, no. 4 (2016): 425–41.

Jerdan, William. *The Autobiography of William Jerdan*. 4 Vols. London: Arthur Hall, Vertue Co., 1852–53. It is consultable online at http://www.lordbyron.org/contents.php?doc=WiJerda.1852.Contents.

Jones, Kathryn N., Carol Tully and Heather Williams. *Hidden Texts, Hidden Nation: (Re)Discoveries of Wales in Travel Writing in French and German (1780–2018)*. Liverpool: Liverpool University Press, 2020.

Jordan, David Starr. 'The History of Ichthyology'. *Science* 16, vol. 398 (15 Aug. 1902): 241–58.

———. *The Genera of Fishes: A Classification of Fishes*. Repr. Stanford: Stanford University Press, 1963.

Kehlmann, Daniel. *Measuring the World*. Trans. Carol Brown Janeway. London: Quercus Publishing, 2007.

Keller, Evelyn Fox. *Reflections on Gender and Science*. New Haven and London: Yale University Press, 1985.

Kete, Kathleen. *Making Way for Genius: The Aspiring Self in France from the Old Regine to the New*. New Haven and London: Yale University Press, 2012.

Kinane, Ian, ed. and intro. *Didactics and the Modern Robinsonade*. Liverpool: Liverpool University Press, 2019.

Klein, Emily S. and Ruth H. Thurstan. 'Acknowledging Long-Term Ecological Change: The Problem of Shifting Baselines'. In *Perspectives on Oceans Past: A Handbook of Marine Environment History*, edited by Kathleen Schwerdtner Máñez and Bo Poulsen, 11–29. Dortrecht: Springer, 2016. https://doi.org/10.1007/978-94-017-7496-3_2.

Kohlstedt, Sally Gregory. 'In from the Periphery: American Women in Science, 1830–1880'. *Signs* 4, no. 1 (1978): 81–96.

Kozma, Liat. 'Prostitution and Colonial Relations'. In *Selling Sex in the City: A Global History of Prostitution, 1600s–2000s*, edited by Magaly Rodríguez García, Lex Heerma van Voss and Elise van Nederveen Meerkerk, 730–47. Leiden: Brill, 2017.

Kramer, Lloyd S. 'The Declining Study of Nineteenth-Century France'. *Central European History* 51, no. 4 (Dec. 2018): 640–45.

Kret, Mariska E. and Tom S. Roth, eds. 'Anecdotes in animal behaviour'. *Behaviour*, 157, no. 5 (2020): 385–86.

Kuhn, Thomas S. *The Structure of Scientific Revolutions*. Chicago: University of Chicago Press, 1962.

Lambert, Miss Francis. *My Crochet Sampler*. London: John Murray, 1844.

Le Gal, Yves. '2009: Le laboratoire de biologie marine de Concarneau a 150 ans'. *La Lettre du Collège de France* 26 (June 2009): 49–52.

Leach, Fiona. 'Resisting Conformity: Anglican Mission Women and the Schooling of Girls in Early Nineteenth-Century West Africa'. *History of Education: Journal of the History of Education Society* 41, no. 2 (2012): 133–53.

Leask, Nigel. *Curiosity and the Aesthetics of Travel Writing, 1770–1840*. Oxford: Oxford University Press, 2002.

Leitão, Henrique, Teresa Nobre de Carvalho, Joaquim Alves Gaspar and Antonio Sánchez, eds. *360º Ciência Descoberta*. Lisbon: Fundacão Calouste Gulbenkian, 2013.

Levine, Philippa. *The Amateur and the Professional: Antiquarians, Historians and Archaeologists in Victorian England, 1838–1886*. Cambridge: Cambridge University Press, 1986.

Lightman, Bernard. *Victorian Popularizers of Science: Designing Nature for New Audiences*. Chicago and London: University of Chicago Press, 2007.

Lindstrom, Lamond. 'Sophia Elau, Ungka the Gibbon and the Pearly Nautilus'. *Journal of Pacific History* 33, no. 1 (June 1998): 5–27.

Lipscombe, Susan Bruxvoort. 'Introducing Gilbert White: An Exemplary Natural Historian and his Editors'. *Victorian Literature and Culture* 35 (2007): 551–67.

Livingstone, David N. 'The History of Science and the History of Geography: Interactions and Implications'. *History of Science* 22, no. 3 (Sept. 1984): 271–302.

———. *The Geographical Tradition: Episodes in the History of a Contested Discipline*. Oxford: Blackwell, 1992.

———. *Putting Science in its Place: Geographies of Scientific Knowledge*. Chicago: Chicago University Press, 2003.

Livingstone, David N. and Charles W. J. Withers, eds. *Geographies of Nineteenth-Century Science*. Chicago: University of Chicago Press, 2011.

Long, Arthur J. *Faith and Understanding: Critical Essays in Christian Doctrine*. London: The Lindsey Press, 1963. https://www.unitarian.org.uk/pages/faith.

Louca, Vasilis, Steve W. Lindsay, Silas Majambere and Martyn C. Lucas. 'Fish Community Characteristics of the Lower Gambia River Floodplains: A Study in the Last Major Undisturbed West African River'. *Freshwater Biology* 54 (2009): 254–71.

Lowe, R. T. 'A Synopsis of the Fishes of Madeira; with the Principal Synonyms, Portuguese Names and Characters of the new Genera and Species', Corr. Memb. of the Zool. Soc. (Paper communicated on 28 March 1837). *The Transactions of the Zoological Society of London* 2, no. 3 (June 1839): 173–200.

Lowther, David. 'Un-gentlemanly Science: Rhetoric and Rivalry in the Codification of British Zoology, 1830–1840'. In *Victorian Culture and the Origin of Disciplines*, edited by Bernard Lightman and Bennett Zon, 111–134. New York: Routledge, 2020.

Luroth, S. G. *Bulletin des sciences naturelles et de géologie* 10 (1827): 125–6.

Mabberley, David J. 'Edward and Sarah Bowdich's Names of Macaronesian and African Plants, with Notes on those of Robert Brown'. *Botanica Macaronesica* 6 (1978): 53–66.

Mackenzie, John M., ed. *Imperialism and Popular Culture*. Manchester: Manchester University Press, 1986.

Mackenzie, John M. *The Empire of Nature: Hunting, Conservation and British Imperialism*. Manchester: Manchester University Press, 1988.

———. *Imperialism and the Natural World*. Manchester: Manchester University Press, 1990.

———. *Museums and Empire: Natural History, Human Cultures and Colonial Identities*. Manchester: Manchester University Press, 2009.

Madge, Clare. 'Collected Food and Domestic Knowledge in the Gambia, West Africa'. *The Geographical Journal* 160, no. 3 (Nov. 1994): 280–94.

———. 'Therapeutic Landscapes of the Jola, The Gambia, West Africa'. *Health & Place* 4, no. 4 (1998): 293–311.

Mahoney, Florence. 'Review of *A History of the Gambia*, by Harry R. Gailey'. *The Journal of African History* 6, no. 3 (1965): 428–9.

———. *Stories of the Gambia, with Supplement.* Bathurst: Government Printer, 1967.

———. *Creole Saga: The Gambian Liberated African Community in the Nineteenth Century.* Privately printed, *c.* 2006.

———. *Gambia Studies.* Privately printed, *c.* 2007.

Mallalieu, Huon. 'The Compleat Ichthyologist'. *Country Life* (26 Feb. 2004): 78.

Martin, Alison E. *Nature Translated: Alexander von Humboldt's Works in Nineteenth-Century Britain.* Edinburgh: Edinburgh University Press, 2018.

Maynard, Douglas H. 'The World's Anti-Slavery Convention of 1840'. *The Mississippi Valley Historical Review* 47, no. 3 (Dec. 1960): 452–71.

McClellan III, James E. and François Regourd. 'The Colonial Machine: French Science and Colonization in the Ancien Regime'. *Osiris* (2001): 31–50.

McClenachan, Loren, Francesco Ferretti and Julia. K. Baum, 'From Archives to Conservation: Why Historical Data Are Needed to Set Baselines for Marine Animals and Ecosystems'. *Conservation Letters* 5, no. 5 (2012): 249–59.

McEwan, Cheryl. 'Gender, Science and Physical Geography in Nineteenth-Century Britain'. *Area* 30, no. 3 (1998): 215–23.

———. *Gender, Geography and Empire: Victorian Women Travellers in West Africa.* Aldershot and Burlington: Ashgate, 2000.

Milne-Edwards, Henri. *Elemens de zoologie ou leçons d'anatomie, de physiologie, la classification et les mœurs de animaux.* Paris: Chez Crochard, 1834.

Milne-Edwards, Henri and Achille Comte. *Cahiers d'histoire naturelle à l'usage des collèges, des institutions religieuses et des écoles normales primaires. vol. 1 cahier Zoologie.* Paris: Fortin, Masson et Cie, 1840 (Second Edition 1844).

Monod, Théodore. '*Brevorha* Gill, 1861 et *Ethmalosa* Regan 1917'. *Bull. IFAN* 23 (1961): 306–417.

Moore, Alfred and Jack Stilgoe. 'Experts and Anecdotes: The Role of "Anecdotal Evidence" in Public Scientific Controversies'. *Science, Technology & Human Values* 34, no. 5 (Sept. 2009): 654–77.

Moore, Francis. *Travels into the Inland Parts of Africa: Containing a Description of the Several Nations for the Space of Six Hundred Miles up the River Gambia…* Second Edition. London: D. Henry and R. Cave, 1740.

Mouchard, Christel. *Adventurières en crinolines.* Paris: Éditions du Seuil, 1987.

Moussa, Sarga, ed. *L'idée de "race" dans les sciences humaines et la littéerature (xviiie et xixe siècles).* Paris: L'Harmattan, 2003.

Mulligan, Shane and Peter Stoett. 'A Global Bioprospecting Regime: Partnership or Piracy?' *International Journal* 55, no. 2 (Spring 2000): 224–46.

'ncse: nineteenth-century serials edition'. Online database of 'Research Resources for Nineteenth-Century Periodicals'. https://ncse.ac.uk/reference/bibliography.html.

Njie, M. and H. Mikkola. 'A Fisheries Co-Managements Case Study from The Gambia'. *The ICLARM Quarterly* 24, nos. 3–4 (July–Dec. 2001): 40–49.

Oldroyd, David R. ed. and intro. Special Number on 'Biography'. *Earth Sciences History* 13, no. 1 (2013): ii–xiii.

Orr, Mary. 'Pursuing Proper Protocol: Sarah Bowdich's Purview of the Sciences of Exploration'. *Victorian Studies* 49, no. 2 (Winter 2007): 277–85.

———. 'Keeping it in the Family: The Extraordinary Case of Cuvier's Daughters'. In *The Role of Women in the History of Geology*, edited by Cynthia V. Burek and Bettie M. Higgs, 277–86. London: Geological Society, Special Publications 281, 2007.

———. *Flaubert's Tentation: Remapping Nineteenth-Century French Histories of Religion and Science.* Oxford: Oxford University Press, 2008.

———. 'Rethinking the Pioneering Text: Sarah (Bowdich) Lee's "Playing at Settlers, or the Faggot House" (1855)'. *International Journal of Children's Literature* 5, no. 2 (Dec. 2012): 135–50.

———. 'Fish with a Different Angle: *The Fresh-Water Fishes of Great Britain* (1828–1838) by Mrs Sarah Bowdich'. *Annals of Science* 71, no. 2 (2014): 206–40.

———. '"Adventures in Australia (1851) by Mrs R. Lee: Thinking Girl Readers at Home and Abroad". In "Rethinking the Representation of Colonial Girls", edited by Tamara Wagner'. *Women's Writing* 21, no. 2 (May 2014): 48–65.

———. 'New Observations on a Geological Hotspot Track: *Excursions in Madeira and Porto Santo* (1825) by Mrs T. Edward Bowdich'. *Centaurus* 56, no. 3 (August 2014): 135–66.

———. '"Women Peers in the Scientific Realm: Sarah Bowdich (Lee)'s Expert Collaborations with Cuvier, 1825–1833". In "Women and Science", edited by Claire Jones and Susan Hawkins'. *Notes and Records* 69, no. 1 (March 2015): 37–52.

———. '"The Stuff of Translation and Independent Female Scientific Authorship: The Case of Taxidermy (1820) by Mrs R. Lee". In "Ingenious Minds: British Women as Facilitators of Scientific Knowledge Exchange, 1750–1900", edited by Alison Martin'. *JLS* 8, no. 1 (2015): 27–48.

———. 'Amplifying Women's Intelligence through Travel: Inna's Tale in "The Booroom Slave" by Sarah Bowdich'. *Forum for Modern Language Studies* 51, no. 3 (July 2015): 269–86.

———. 'Les *Mémoires du baron Georges Cuvier* (1833) de Mistress Lee: Mémoires scientifiques, pacte biographique, ou réécriture des savoirs?' In *Littérature Française et Savoirs Biologiques au XIXe Siècle: Traduction, Transmission, Transposition*, edited by Thomas Kinkert and Gisèle Séginger, 183–200. Berlin and Boston: de Gruyter Open Access, 2020.

———. 'Collecting Women in Geology: Opening the International Case of a Scottish 'Cabinétière, Eliza Gordon Cumming (*c* 1798–1842)'. In *Celebrating 100 Years of Female Fellowship of the Geological Society: Discovering Forgotten Histories*, edited by Cynthia V. Burek and Bettie M. Higgs, 63–73. London: Geological Society, Special Publications 506, 2021.

———. 'Catalysts, Compilers and Expositors: Rethinking Women's Pivotal Contributions in Nineteenth-Century "Physical Sciences"'. In *The Palgrave Handbook of Women and Science Since 1660*, edited by Claire G. Jones, Alison E. Martin and Alexis Wolf, 505–28. Switzerland: Springer Nature/Palgrave Macmillan, 2022.

Outram, Dorinda. 'Scientific Biography and the Case of Georges Cuvier: With a Critical Bibliography'. *History of Science* 14, no. 2 (1 June 1976): 101–37.

———. 'The Language of Natural Power: The "Eloges" of Georges Cuvier and the Public Language of Nineteenth-Century Science'. *History of Science* 16, no. 3 (1 Sept. 1978): 153–78.

———. *Georges Cuvier: Vocation, Science and Authority in Post-Revolutionary France.* Manchester: Manchester University Press, 1984.

'Owen, Lady Caroline: Commonplace Book'. AIM25 Archives in London and the M25 area. https://www.aim25.com/cgi-bin/vcdf/detail?coll_id=10218&inst_id=9&nv1=search&nv2=.

Owen, Richard. 'On the Mammary Glands of the Ornithorhynchus Paradoxus'. (Paper communicated by J. H. Green FRS 21 June 1832). *Philosophical Transactions of the Royal Society* 122 (1832): 517–83.

———. 'On the Ova of the Ornithorhynchus Paradoxus'. (Paper communicated by Sir Anthony Carlisle FRS 19 June 1834). *Philosophical Transactions of the Royal Society* 124 (1834): 555–66.

———. 'On the Young of the Ornithorhynchus Paradoxus, Blum'. (Paper read May 27 1834). *The Philosophical Transactions of the Zoological Society of London* 1, no. 3 (March 1835): 221–30.

Padian, Kevin. 'Biology in History: The Rehabilitation of Sir Richard Owen'. *American Institute of Biological Sciences* 47, no. 7 (July–Aug. 1997): 445–53.

Palmer, Caroline. '"I will tell nothing that I did not see": British Women's Travel Writing, Art and the Science of Connoisseurship, 1776–1860'. *Forum for Modern Language Studies* 51, no. 3 (July 2015): 248–68.

Parenti, Paulo. 'On the Status of Some Nominal Species of Fishes Described by Sarah Lee Bowdich (*sic*) in the Account "Excursions in Madeira and Porto Santo During the Autumn of 1823"'. *Boletim Museu do História Natura do Funchal* lxix (2019): 5–12.

Parenti, Paolo and John E. Randall. 'An Annotated Checklist of the Species of the Labroid Fish Families Labridae and Scaridae'. *Ichthyological Bulletin of the J. L. B. Smith Institute of Ichthyology* 68 (2000): 1–97.

Parker, John. 'The Death of Adumissa: A Suicide at Cape Coast, Ghana'. *Africa: The Journal of the International African Institute* 91, no. 2 (Feb. 2021): 205–25.

Paul, Harry W. *From Knowledge to Power: The Rise of the Science Empire in France, 1860–1939.* Cambridge: Cambridge University Press, 1985.

Pels, Peter and Oscar Salemink, eds. *Colonial Subjects: Essays on the Practical History of Anthropology.* Ann Arbor: University of Michigan Press, 1999.

Pennell, H. Cholmondeley. *The Angler-Naturalist: A Popular History of British Fresh-Water Fish. With a Plain Explanation of the Rudiments of Ichthyology.* London: John Van Voorst, 1863.

Pérez, Salvdor. 'Evocadora imagen biográfica de Sarah Bowdich Lee' in the blog https://www.taxidermidades.com/2016/07/taxidermy-obra-de-la-naturalista-e-ilustradora-sarah-bowdich-lee.html.

Phillips, Patricia. *The Scientific Lady: A Social History of Women's Scientific Interests, 1520–1918.* London: Weidenfeld and Nicolson, 1990.

Pietsch, Theodore W., ed. *Historical Portrait of the Progress of Ichthyology: From its Origins to Our Own Time by Georges Cuvier.* Trans. Abby J. Simpson. Baltimore: Johns Hopkins University Press, 1995.

Pimm, Stuart L. 'Africa: Still the "Dark Continent"'. *Conservation Biology* 21, no. 3 (June 2007): 567–69.

Poirier, Jean-Pierre. *Histoire des Femmes de Science en France: Du Moyen Age à la Révolution.* Paris: Pygmalion Gérard Watelet, 2002.

Poovey, Mary. *The Proper Lady and the Woman Writer: Ideology as Style in the Works of Mary Wollstonecraft, Mary Shelley and Jane Austen.* Chicago: Chicago University Press, 1984.

Porter, Andrew. '"Cultural Imperialism" and Protestant Missionary Enterprise, 1780–1914'. *The Journal of Imperial and Commonwealth History* 25, no. 3 (1997): 367–91.

Pratt, Mary Louise. *Imperial Eyes: Travel Writing and Transculturation.* London and New York: Routledge, 1992.

Provost, Elizabeth. 'Assessing Women, Gender, and Empire in Britain's Nineteenth-Century Protestant Missionary Movement'. *History Compass* 7, no. 3 (2009): 765–99.

Pycior, Helena M., Nancy G. Slack and Pnina Abir-Am, eds. *Creative Couples in the Sciences.* New Brunswick: Rutgers University Press, 1996.

Rak, Julie. 'Are Memoirs Autobiography? A Consideration of Genre and Public Identity'. *Genre* 34 (Fall/Winter 2004): 305–26.

Rappoport, Jill. 'Buyer Beware: The Gift Poetics of L.E.L'. *Nineteenth-Century Literature* 58, no. 4 (2004): 441–56.

———. *Giving Women: Alliance and Exchange in Victorian Culture.* New York: Oxford University Press, 2012.

Reade, W. Winwood. 'Efforts of Missionaries among Savages'. *Journal of the Anthropological Society* 3 (1865): clxiii–clxxxiii.

'Recherches systématiques sur les poissons de la côte occidentale d'Afrique. Liste des poissons littoraux recoltés par le navire "Président-Théodore-Tissier" au cours de sa cinquième croisière'. *Revue des Travaux de l'Institut des Pêches Maritimes* 10, no. 4 (1937): 425–564. https://archimer.ifremer.fr/doc/00000/5775.

Renier, Anne. *'Friendship's Offering': An Essay on the Annuals and Gift Books of the Nineteenth Century.* London: Private Libraries Association, 1964.

Riego de la Branchardière, Mlle Eléonore. *Knitting, Crochet and Netting.* London: S. Knights, 1846.

Rowe, Natalie. 'Sweet Ordering, Arrangement and Decision: The Domestic Nature of Science Illustration by Women in the Eighteenth and Nineteenth Centuries'. *Journal of Illustration* 1, no. 2 (Oct. 2014): 211–31.

Russell, Mary. *The Blessings of a Good Thick Skirt: Women Travellers and their World.* London: Collins, 1988.

Saint-Martin, Arnaud. 'Autorité et grandeur savantes à travers les éloges funèbres de l'Académie des sciences à la Belle Époque'. *Genèses* 87, no. 2 (2012): 47–68.

Sartori, Eric. *Histoire des Femmes Scientifiques de l'Antiquité au XXe siècle: Les Filles d'Hypatie.* Paris: Plon, 2006.

Schiebinger, Londa. 'The History and Philosophy of Women in Science: A Review Essay'. *Signs* 12, no. 2 (1987): 305–32.

———. 'Gender and Natural History'. In *Cultures of Natural History*, edited by Nicholas Jardine, James A. Secord and Emma Spary, 163–77. Cambridge: Cambridge University Press, 1996.

———. 'Has Feminism Changed Science?' *Signs* 25, no. 4 (Summer 2000): 1171–75.

Schneller, Beverly E. 2006. 'Jerdan, William (1782–1869)'. Oxford Dictionary of National Biography. Oxford: Oxford University Press. http://www.oxforddnb.com/view/article/14770.

Scrope, William. *Days and Nights of Salmon Fishing in the Tweed: with a Short Account of the Natural History and Habits of Salmon, Instructions to Sportsmen, Anecdotes, Etc.* London: John Murray, 1843.

Secord, Anne. 'Science in the Pub: Artisan Botanists in Early Nineteenth-Century Lancashire'. *History of Science* 32, no. 3 (Sept. 1994): 269–315.

———. 'Corresponding Interests: Artisans and Gentlemen in Nineteenth-Century Natural History'. *The British Journal for the History of Science* 27, no. 4 (Dec. 1994): 383–408.

Serge, Marc, ed. and trans. *A Woman of Courage: The Journal of Rose de Freycinet on her Voyage around the World, 1817–1820.* Canberra: National Library of Australia, 1996.

Shapin S. and A. Thackray. 'Prosopography as a Research Tool in History of Science: The British Scientific Community 1700–1900'. *History of Science* 12, no. 1 (1974): 1–28.

Sheets-Pyenson, Susan. 'War and Peace in Natural History Publishing: *The Naturalist's Library*, 1833–1843'. *Isis* 72, no. 1 (March 1981): 50–72.

Shortland, Michael and Richard Yeo, eds. *Telling Lives in Science: Essays on Scientific Biography.* Cambridge: Cambridge University Press, 1996.

Shteir, Ann B. *Cultivating Women, Cultivating Science: Flora's Daughters and Botany in England, 1760–1860.* Baltimore: Johns Hopkins University Press, 1996.

———. 'Elegant Recreations? Reconfiguring Science Writing for Women'. In *Victorian Science in Context*, edited by Bernard Lightman, 236–55. Chicago and London: University of Chicago Press, 1997.

Shteir, Ann B. and Bernard Lightman, eds. *Figuring it Out: Science, Gender and Visual Culture.* Hanover and London: University Press of New England, 2006.

Simier, Monique, Charlene Laurent, Jean-Marc Ecoutin and Jean-Jacques Albaret. 'The Gambia River Estuary: A Reference Point for Estuarine Fish Assemblages Studies in West Africa'. *Estuarine, Coastal and Shelf Science* 69 (2006): 615–28.

Skelton, P. H. and E. R. Swartz. 'Walking the Tightrope: Trends in African Freshwater Systematic Ichthology'. *Journal of Fish Biology* 79 (2011): 1413–35.

Smith, Roger. 2021. 'Carpenter, William Benjamin (1813–1885)'. Oxford Dictionary of National Biography. Oxford: Oxford University Press. https://doi.org/10.1093/ref:odnb /4742.

Smith, Simon. *British Imperialism, 1750–1970.* Cambridge: Cambridge University Press, 1998.

Sodeman, Melissa. 'Gilbert White, Anecdote, and Natural History'. *Studies in English Literature 1500–1900* 60, no. 3 (2020): 507–28.

Söderqvist, Thomas. *The History and Politics of Scientific Biography.* Aldershot: Ashgate, 2007.

Southorn, Lady Bella. *The Gambia: The Story of the Groundnut Colony.* London: George Allen and Unwin Ltd., 1952.

Stafford, Barbara Maria. *Voyage into Substance: Art, Science, Nature and the Illustrated Travel Account.* Cambridge, MA and London: MIT Press, 1994.

Strickrodt, Silke. *'Those wild Scenes': Africa in the Travel Writings of Sarah Lee (1791–1856).* Gliennicke/Berlin and Cambridge, MA: Galda-Witch Verlag, 1998.

Svensson, G. S. O. *Fresh-Water Fishes from the Gambia River.* Stockholm: Almquist & Wiksells Boktrycheri A-B, 1934.

Swindell, Kenneth and Alieu Jeng. *Migrants, Credit and Climate: The Gambian Groundnut Trade, 1834–1934.* Leiden: Brill, 2006.

Taquet, Philippe. *Georges Cuvier.* Paris: Odile Jacob, 2006.

Thompson, Carl. *The Suffering Traveller and the Romantic Imagination.* Oxford: Clarendon Press, 2007.

———. 'Earthquakes and Petticoats: Maria Graham, Geology, and Early Nineteenth-Century "Polite" Science'. *Journal of Victorian Culture* 17, no. 3 (Sept. 2012): 329–46.

———. 'Women Travellers, Romantic-era Science and the Banksian Empire'. *Notes and Records* 73 (2019): 431–55.

———. 'Maria Graham and the Chilean Earthquake of 1822: Contextualizing the first Female-Authored Article in *Transactions of the Geological Society*'. In *Celebrating 100 Years of Female Fellowship of the Geological Society: Discovering Forgotten Histories*, edited by Cynthia V. Burek and Bettie M. Higgs, 117–24. London: Geological Society, Special Publications 506, 2021.

Topham, Jonathan R. 'Science Publishing and the Reading of Science in Nineteenth-century Britain: A Historiographical Survey and Guide to Resources'. *Studies in History and Philosophy of Science, Part A* 31, no. 4 (2000): 559–612.

Turner, Katharine, ed. and intro. 'Women Writing Travel: A Virtual Special Issue'. *Forum for Modern Language Studies* 59, no. 2 (April 2023): 315–19.

Twells, Alison. '"So distant and wild a scene": Language, Domesticity and Difference in Hannah Kilham's Writing from West Africa, 1822–1832'. *Women's History Review* 4, no. 3 (Dec. 1995): 301–18.

Vaughan, Megan. 'Healing and Curing: Issues in the Social History and Anthropology of Medicine in Africa'. *The Society for the Social History of Medicine* 7, no. 2 (1994): 283–95.

Wagner, Tamara. 'Silver Fork Novel (Fashionable Novel)'. https://www.oxfordbibliographies.com/view/document/obo-9780199799558/obo-9780199799558-0136.xml (last accessed 3 August 2022).

Warren, Mrs Eliza. *The Court Crochet Collar and Cuff Book*. London: Ackermann & Co., 1847.

Watts, Michael. *The Dissenters. Vol. II, The Expansion of Evangelical Nonconformity*. New York: Oxford University Press, 1995.

Wessels, Q. and A. M. Taylor. 'Anecdotes to the Life and Times of Sir Richard Owen (1804–1892) in Lancaster'. *Journal of Medical Biography* 25, no. 4 (2017): 226–33.

Westby-Gibson, John and Felix Driver. 2021. 'Bowdich, Thomas Edward (1791–1824)'. Oxford Dictionary of National Biography. Oxford: Oxford University Press. https://doi.org/10.1093/ref:odnb /3027.

Whewell, William. 'On the Connexion of the Physical Sciences. By Mrs Somerville'. *The Quarterly Review* 51, no. 101 (1834): 54–68.

Whitehead, Peter J. P. 'The West African Shad, Ethmalosa Fimbriata (Bowdich, 1825): Synonymy, Neotype'. *Journal of Natural History* 4 (1967): 585–93.

Willem, Jean-Pierre. *L'Ethnomédicine, une alliance entre science et tradition*. Genève-Bernex: Éditions Jouvence/Gaillac: Biocontact, 2006.

Williams, R. B. 'The Artists and Wood-Engravers for Thomas Bell's *History of British Quadrupeds*'. *Archives of Natural History* 38, no. 1 (2011): 170–72.

Withers, Charles W. J. 'Mapping the Niger, 1798–1832: Trust, Testimony and "Ocular Demonstration" in the Late Enlightenment'. *Imago Mundi* 56, no. 2 (2004): 170–93.

'Women Shellfishers and Food Security Project. Participatory Assessment of Shellfisheries in the Estuarine and Mangrove Ecosystems of the Gambia'. Report 2021. https://www.crc.uri.edu/download/WSFS2021-The-Gambia-Report-FIN508.pdf.

Wood, James R. *Anecdotes of Enlightenment: Human Nature from Locke to Wordsworth*. Charlottesville: University of Virginia Press, 2019.

Wright, Donald R. *The World and a very Small Place in Africa: A History of Globalization in Niumi, The Gambia*. Second Edition. Armonk, New York and London: M. E. Sharpe, 2004.

Zgórniak, Marek, Marta Kapera and Mark Singer. 'Fremiet's Gorillas: Why Do They Carry off Women?' *Artibus et Historiae* 27, no. 54 (2006): 219–37.

Zimmermann, Maurice. 'Afrique Occidentale. Mission Gruvel et Chudeau'. *Annales de Géographie* 18, no. 99 (1909): 278–80.

INDEX

As the principal subject of this book, name references to 'Sarah' (née Wallis, then Bowdich, then Lee) would include almost every page. Readers should therefore consult the indexed topic – e.g. 'ichthyology' – to find Sarah's connections to it. For readers new to her works, the indexed entry for 'Bowdich, Sarah' covers indicative points of interest that interconnect across the book.

Index compiled by Dr. Matthew W. Ylitalo.

Printed in the USA
CPSIA information can be obtained
at www.ICGtesting.com
JSHW020419100924
69520JS00001B/1

9 781839 986093